DIMENSIONS OF TIME
AND LIFE

The Study of Time VIII

# DIMENSIONS OF TIME AND LIFE

## The Study of Time VIII

*Edited by* J. T. Fraser and M. P. Soulsby

INTERNATIONAL UNIVERSITIES PRESS, INC.
*Madison, Connecticut*

**Library of Congress Cataloging-in-Publication Data**
Dimensions of time and life / edited by J.T. Fraser and M.P. Soulsby.
    p.   cm. — (The study of time, ISSN 0170-9704 ; 8)
    Selected papers from the 8th Conference of the International
Society for the Study of Time.
    Includes bibliographical references and indexes.
    ISBN 0-8236-1295-3
    1. Time—Congresses. 2. Life—Congresses. I. Fraser, J. T.
(Julius Thomas), 1923- . II. Soulsby, M. P. (Marlene Pilarcik)
III. International Society for the Study of Time. Conference (8th :
1994? : Cerisy-la-Salle, France) IV. Series.
QB209.S85 vol. 8
[BD638]
304.2'3—dc20                                                        95-25389
                                                                   CIP—

Manufactured in the United States of America

TO THE MEMORY OF
GEORGE H. FORD (1914–1994)
President, ISST, (1979–1983)
WHO TAUGHT DIGNITIY AND NOBILITY
IN THOUGHT, WORD, AND CONDUCT

# Table of Contents

# Contributors

*Barbara Adam*  Currently a lecturer in social theory at the University of Wales, Cardiff, she has written extensively on social time. Her book *Time and Social Theory* has won the Philip Abrams Memorial Prize, awarded by the British Sociological Association for "best first book." She is editor of the interdisciplinary journal *Time & Society*.

*Hoyt Alverson*  Professor at Dartmouth College where he teaches anthropology and linguistics. Author of numerous publications in these two fields, his book, *Mind in the Heart of Darkness* (1978) describing three years of anthropological research in Southern Africa, won both the Herskovits Prize of the African Studies Association and First Prize in the University of Chicago Folklore Competition. His most recent book is *Semantics and Experience* (1994).

*M. H. Oliva Augusto, Ph.D.*  Professor and Researcher in the Department of Sociology, School of Philosophy, Literature and Human Sciences, University of Sao Paulo. Her research has been directed at understanding the relationship between the individual and society. She has published various articles in Brazilian journals, as well as a book, *Intervencionismo estatal e ideologia desenvolvimentista* (State Intervention and Developmentalist Ideology, 1978).

*Mark H. Aultman L.L.D.*  A lawyer who practices in the area of legal ethics and professional responsibility in Columbus, Ohio. In addition to writing articles and teaching courses in legal ethics, he has also written a work of fiction, *Nightfire*, a philosophical fantasy story. He has been an ISST member for over a decade, he was Treasurer for six years, and serves on the ISST Council.

*Richard A. Block*  Professor of Psychology at Montana State University and treasurer of the International Society for the Study of Time. His research focuses on human memory and cognition, especially concerning issues related to the psychology of time.

*Michel Bonnet*  Professor of Neurosciences at the University of Provence, Marseilles. His research field concerns the processing of information in motor control mechanisms. He uses chronometric, reflexological, and electrophysiological methods in human subjects.

*Jens Brockmeier*  Faculty member of Linacre College, University of Oxford, England; and Institute for Psychology, University of Innsbruck, Austria.

*T. L. Christie*    Faculty member of Wilfrid Laurier University, Waterloo, Ontario, Canada.

*Sylvie Droit, Ph.D.*    Lecturer at the University of Clermonts-Ferrard. Her dissertation entitled "From Enacted to Represented Time" focused on acquired temporal regulations in the three- to six-year age range.

*J. T. Fraser*    Founder of the International Society for the Study of Time; author of *Time, the Familiar Stranger; Of Time, Passion, and Knowledge*; and other time-related books, and editor of the nine volumes of *The Study of Time* series.

*Sabine Gross, Ph.D.*    Assistant Professor of German at the University of Wisconsin in Madison. Her publications include articles on literary canon formation, time and media, textual representations of body and gender, and Bertolt Brecht. Her book *Zeichenstruktur, Kognition, Medium: Zur Materialität des Lesens* [Sign Structure, Cognition, Medium: Towards a Materiality of Reading] was published in 1994. She has also taught dance, movement, and improvisation.

*J. M. Halpern*    Faculty member of the University of Massachusetts, Amherst.

*Paul A. Harris*    Faculty member in the English Department at Loyola Marymount University in Los Angeles, specializing in 20th-century literature, and studies in literature and science. He has published interdisciplinary articles on writers including Faulkner, DeLillo, Morrison, and Calvino. His dissertation *Time Spaced Out in Words: From Physics to Faulkner* is under contract with University of Michigan Press. He is also the editor of *The Textures of Time: Chaos and Poiesis*.

*Rémy Lestienne, Ph.D.*    Senior researcher at CNRS (the French fundamental research agency), has performed work in elementary particle physics, epistemology, and neuroscience. The notion of time and its relations to the concepts derived from it in physics and in biology are central to his research program. He has published about 70 scientific articles and two books about time and chance, respectively. He is presently Science Attaché at the French Embassy, Washington and CNRS representative to the United States.

*Judy Lochhead*    Associate Professor of Music, State University of New York at Stony Brook. Research areas: phenomenological approaches to music analysis; recent music in the Western, classical tradition; the music of Joan Tower and Roger Sessions; musical expressivity.

*Lennart Lundmark*    Docent in History at Umẽ University, Sweden. He has published a number of books and articles in Swedish on conceptions of time, and will shortly publish a book about how technological change has influenced time-conceptions from the Middle Ages to the present.

*Françoise Macar*    Director of Research at the Centre National de la Recherche scientifique, Laboratory of Cognitive Neurosciences, Marseille. She is author of *Le Temps: Perspectives Psychophysiologiques* and coeditor of *Time, Action and Cognition*.

*Samuel L. Macey*    The author of over 50 publications, mainly in Restoration and 18th-century literature and in time studies. His more recent books include *Henry Carey's Dramatic Works, Studies in Robertson Davies' Deptford Trilogy* (with R. G. Lawrence), and *Clocks and the Cosmos: Time in Western*

*Life and Thought*, as well as *Money and the Novel; Patriarchs of Time: Dualism in Saturn-Cronus, Father Time, The Watchmaker God, and Father Christmas; The Dynamics of Progress: Time, Method, and Measure*; and *Time: A Bibliographic Guide*. He is General Editor of the *Encyclopedia of Time* (1994). Dr. Macey is a Professor Emeritus of English at the University of Victoria, Fellow of the Institute of Management Services, Founding Editor of the English Literary Studies Monograph Series, and was President of the International Society for the Study of Time (1989–1992).

*Masaki Miyake, Ph.D.*    Professor of International History at the School of Political Science and Economics, Meiji University, Tokyo. He is a member of the International Committee of Historical Sciences (Lausanne-Paris). He has a special interest in contemporary Japanese-German history. He also has a parallel and complementary interest in the methodology and theory of history, especially in the issue of the concept of time.

*David A. Park*    Professor Emeritus of Physics at Williams College. He is the author of three books on technical aspects of theoretical physics and three more on cultural and historical aspects of science. He is a Fellow of the American Physical Society.

*Viviane Pouthas*    Student of Fraisse, first studied timing in animals. She redirected her attention to human subjects and is now head of a group studying the developmental psychobiology of time in infants, young children and, more recently, in elderly people. In addition to publishing her experimental results, she has coedited *Guyau and the Idea of Time* (North-Holland, 1988) and the proceedings of the Nato workshop, *Time, Action and Cognition* (Kluwer Academic, 1992).

*Joëlle Provasi*    Lecturer at L'Ecole Pratique des Hautes Etudes à Paris. In her doctoral dissertation she studied the acquisition of temporal regulations of nonnutritive sucking activity in young infants.

*Lewis Rowell*    Associated with the International Society for the Study of Time since 1976 and has served as Treasurer, a member of the ISST council, President (1986–1989), and Executive Secretary (1992–present). He is the author of five books and numerous articles on time and rhythm in music, ancient and oriental music, and the philosophy of music. His 1992 book on *Music and Musical Thought in Early India* has received the Otto Kinkeldey Award from the American Musicological Society. Rowell is currently a Professor of Music at Indiana University, Bloomington, where he has taught since 1979.

*Marlene Pilarcik Soulsby, Ph.D.*    Associate Professor of German and Comparative Literature, Penn State University, Worthington Scranton Campus; editor *Time's News* coeditor *Dimensions of Time and Life* (1996) and member of the Council of the International Society for the Study of Time; author of articles on time and literature, East-West literary comparisons, and foreign language pedagogy.

*John J. Stuhr*    Professor and Head of the Department of Philosophy at the Pennsylvania State University. Educated at Carleton College and Vanderbilt University, he has held fellowships in Germany, Australia, and Russia. He writes on pragmatism, European philosophy, ethics, and politics—most recently, *Philosophy and the Reconstruction of Culture.*

*Robert Thornton*    Faculty member, Department of Anthropology, University of Witwaters and, Johannesburg, South Africa.

*Frank Vidal*    Medical Officer (Lieutenant-Commander) in the Institute of Naval Medicine, Toulon, France (Service de Santée des Armées). His research field concerns the mechanisms of attention and duration programming in motor timing tasks. He studies behavioral and electrophysiological indices on human subjects.

# Foreword and Acknowledgments

The practitioners of modern research and technology would seem to thrive through achieving more and more precision on subjects of less and less magnitude. The concomitant demise of the polymath and the master craftsman further demonstrates our increasing inability to view the world and its artifacts from a wider perspective. Both as a teacher and as a dean I have often advised students who despaired of ever finding a thesis, that if they looked closely enough at the wooden table between us—that is to say at its atoms, molecules, organic matter, structure, and so forth—it could afford them at least as many doctorates as Faust gained in Wittenberg. But this is surely not all that the consideration of a table can provide, and some of us at least might profit from a wider perspective. A table is also an imperfect copy of one of Plato's ideal patterns, an excellent example of Kant's *Ding an sich,* and an object which has borne the full weight of the immense hecatomb that human beings are constructing from all other forms of life.

The International Society for the Study of Time (ISST) is largely composed of the very academics whose early training and research have encouraged them to specialize in subjects of less and less scope. But the very interdisciplinary nature of a society of heterogeneous scholars from over 30 countries in more than 20 disciplines committed to the study of time is bound to broaden the perspective of those who are thus engaged. Furthermore, the insistence on delivering papers in plenary session—and in a milieu which has always proved to be a delight to the eyes, the mind, and the body—invariably encourages highly productive, jargon free, and very profitable discussions throughout the day and in smaller groups often late into the night.

In every respect, "Dimensions of Time and Life" as the primary focus of the eighth international conference of ISST, proved to be an inspired choice. With the exception of biologists, participants would find no self-evident entry into the subject from their own disciplines. How does time and life apply to music, technology, philosophy, physics, film, and so much more? The pleasant pain and the resulting creative energy that derive from being obliged to view one's subject in a newer and broader perspective produced the excellent papers chosen for presentation at the conference. And that same creative energy is here presented in the more Apollonian form of the papers selected by their external readers and the editors.

Despite and perhaps even because of the need to overcome unforeseen emergencies, the eighth conference of ISST proved to be a resounding success. Of course the setting in the château of Cerisy-la-Salle in Normandy helped greatly. And we were also indebted to Hervé Barreau and his Local Organizing Committee including Marianne Barreau, Vivian Pouthas, François Macar, and Joel Pouthas, as well as to Rémy Lestienne who gave us considerable help and CNRS whose generous funding supported the endeavors of the French Committee. In particular, we will all remember with special pleasure the wonderful excursion organized by Joel Pouthas which included the Bayeux tapestry, L'Abbaye-aux-Dames, L'Abbaye-aux-Hommes, the Landing Beaches, and a *vin d'honneur* and lunch that were *sans pareil.*

The generous and unstinted help that I received from members, sometimes at very short notice, bodes well for the collegial spirit of ISST as it enters its second quarter-century. It does full credit to the viability of the institution founded by J. T. Fraser, to whom so many of us continue to look as an incomparable mentor and guide. Let me thank first those who undertook special responsibilities at our eighth triennial conference: Paul Harris (poster exhibit), Robert Grudin with Nicholas Tresilian (book exhibit), David Burrows (Evaluation Committee), and Rick Block's magnificent job throughout the whole process of registration. Let me also thank (in no particular order) the many members who ungrudgingly chaired sessions, introduced speakers, and gave and read papers for absent colleagues: Michael Young, Paul Harris, Marlene Soulsby, Fred Turner, Lew Rowell, George Ford, Rémy Lestienne, Charles Sherover, Donald Miller, David Park, Gert Müller, Gerald Whitrow, Jack Arlow, and Barbara Adam. Last but by no means least my heartfelt thanks go to all the members present at the conference who through their attendance at Cerisy-la-Salle, their papers, their posters, their questions, their lively conversation, and above all their excellent spirit made that meeting such a credit both to the study and to the studiers of time.

My very special thanks go to those who served ISST for the period 1989 to 1992, while I was President, but who will not be continuing on the Council. My equally warm thanks to those who served well and will now be continuing on the Council in one capacity or another, including, in particular, Lew Rowell who is now taking on the responsibility of Executive Secretary. My thanks also to the Nominating Committee, and I was delighted that our deliberations made possible the subsequent election of an excellent new Council, the average age of whose members is some 12 years fewer than that of the one which it supersedes.

Finally, with a mixture of deep relief and even deeper confidence, I was happy to hand over my responsibilities and those of the past Council at the Business Meeting to the new Officers and Members of the Council for 1992 to 1995: Helga Nowotny (President), Lew Rowell (Executive Secretary), Richard A. Block (Treasurer), Marlene Soulsby (Editor of *Time's News*), and Robert Thornton (Membership), plus the following Members of the Council: Barbara Adam, Hans Ågren, Mark Aultman, Olga Hasty, Rémy Lestienne, and Albert Rabin. Our job has been to provide a strong close to the first quarter-century of ISST; their job is to ensure an even stronger continuation in the years that follow.

*Samuel L. Macey*
*President, 1989–1992*

# Introduction

For the reflective mind the theme time and life is likely to appear challengingly ill-defined. For the analytic mind, tying the idea of time and life together is likely to appear disturbingly broad. But the rich confusion of notions associated with time and life should not prevent a search for an understanding of their relations; the multiplicity of stars only encouraged people in their pursuit of astronomy.

The Eighth Conference of the International Society for the Study of Time offered a platform for its members to discuss issues about time and life as seen from their particular disciplines. Selected papers from that conference, published in this volume, fall into three broad categories: (1) life and time as they are understood in their biological, cognitive, and psychological dimensions; (2) the experience of time and life in words, sounds, and images; and (3) time and life as ordered according to sociological, historical, and anthropological perspectives.

It seemed to the editors that this spontaneous grouping bore a resemblance to the spirit, though not to the number, of T. S. Eliot's *Four Quartets*, a poetic and intellectual masterpiece dealing with time in its relation to the human experience of life (1943). We decided, therefore, to use brief, representative segments of Eliot's work to serve as a lyrical and philosophical backdrop against which the trains of specialized thinking may be placed. As will be seen, at times the words of the poet capture intents and visions similar to those of the papers. At other times Eliot's lines draw a counterpoint to the academic comments, creating yet other, new perceptions of time and life.

These brief interludes convey something of the mood of the conference. It was a meeting of about a hundred scholars, set in the idyllic surroundings of the 17th-century château at Cerisy-la-Salle, Normandy, France. There, time and life have been etched into the facades; memories of the French past are kept alive; and worn stone stairs recall those who climbed them in centuries past. At the conference, the presence of the past folded into the spontaneous energy of the present and inspired the search for new understanding. There was, in the words of Eliot in "Little Gidding," an "easy commerce of the old and the new."

> And every phrase
> And sentence that is right (where every word is at home,
> Taking its place to support the others,
> The word neither diffident nor ostentatious,
> An easy commerce of the old and the new,
> The common word exact without vulgarity,
> The formal word precise but not pedantic,
> The complete consort dancing together)
> Every phrase and every sentence is an end and a beginning. . . .
>
> ["Little Gidding," 1943, p. 58]

## Reference

Eliot, T. S. (1943), *Four Quartets*. New York: Harcourt, Brace, & World, 1971.

# I.
# Time and the Life Process

1

# Time and the Origin of Life

*J. T. Fraser**

*Abstract*   This paper interprets the coming about of life entirely in temporal terms. It is an attempt to formulate a principle—that of biogenesis—in a manner that is uncommitted to any of its specific manifestations. Seeking such a general model of a process is a practice traditional in natural science.

The argument begins with a survey of the spectrum of biological cycles in species alive today. The findings suggest that oscillatory processes in living organisms are much more than adaptive measures. Instead, biological oscillations, observed across a frequency range of 24 orders of magnitude and synchronized from instant to instant, constitute the life process. In the phenomenal world, the inner synchronization is manifest as the organic present. It is thus that life creates a "now" in the presentless world of nonliving matter. The organic present so born and maintained allows for a distinction to be made between present and nonpresent conditions. In their turn, the nonpresent categories of time may be separated into futurity and pastness in terms of the present needs and available means of an organism.

When this model of the life process is reduced to its simplest schematic form, a set of minimal requirements obtain to which even the earliest molecular aggregates had to conform if they were to be called alive. According to this understanding, the perpetuation of life is seen as the passing along of the biological skills that are necessary for maintaining the organic present against external and internal perturbations.

The burden of the paper is the testing of the proposed model of biogenesis in terms of three criteria. Does a system so envisaged have the potentiality for complexifying through evolution by natural selection? Can it define self-directed purpose in terms of its needs, and with it, distinguish between future and past? Is the functioning of the model consistent with the principles of thermodynamics? The paper concludes by giving affirmative answers to these questions.

This paper outlines an understanding of life and its origin with the assistance of the hierarchical theory of time.[1] In so doing, it follows a type of reasoning which

---

*The Founder's Lecture, July 4, 1992.

[1]The hierarchical theory of time, also known as the theory of time as conflict, has been elaborated in the author's books and published papers. A convenient entry may be had through Fraser (1987, 1990).

3

is traditional to natural science. It formulates an interpretation of a process in a manner that is independent of, and is uncommitted to, any of its specific manifestations.

## The Notion of Biogenesis

Opinions about the origin of life have varied from epoch to epoch, and differ from culture to culture.[2] Creation legends suggest intellectual readiness to distinguish between living and nonliving objects and to identify the position of life with respect to the world at large. Early narratives about the origin of living things matured into reasoned arguments in the thoughts of the Presocratics such as when Anaximander in the 6th century B.C. reasoned, rather than declared, that animals came into being from slime (Kirk and Raven, 1975, p. 141). Aristotle in *The History of Animals* took the idea of spontaneous generation of primitive organisms for granted, as did almost everyone else until Pasteur succeeded in demonstrating that examples of presumed, spontaneous generation were instants of reproducing microorganisms.[3] Darwin imagined ''some warm, little pond with all sorts of ammonia, phosphoric salts, light, heat [and] electricity in it'' as a likely setting of the origin of life (from an editorial note in Darwin (1887) v. III, p. 18, from an otherwise unidentified letter by Charles Darwin written in 1871). In the 1930s Haldane and Oparin suggested that the appearance of life was preceded by a long period of chemical synthesis, during which polymerization proceeded at a slow pace, creating, eventually, the first prototypic cells (Oparin, 1966). Contemporary biologists, interested in the origin of life, study cosmochemistry and the geological and physical conditions of the biogenetic earth (for current work see the journal *Origin of Life* and the publications of the International Society for the Study of the Origin of Life).

The ancestry of their approach to biogenesis may be found in the thought of Democritus of Abdera, a Greek atomist who lived around the turn of the 5th century B.C. and maintained that atoms were very small and simple, alike in qualities but different in their relationships. He also believed that some groups of atoms are subtler than others. For instance, those of the soul were like those of fire: very small, smooth, and spherical, so as to secure for them the mobility that is necessary

---

[2]Though often spoken of as new life, the birth of a child does not amount to the creation of life; it only illustrates a particular method of reproduction. The same holds for test-tube babies: they attest to technical virtuosity in stealing human eggs and sperms. One would be closer to demonstrating biogenesis if one could construct a phoenix from chemicals on the shelf. But the genetic alchemist would still have to convince others that his bird can reproduce by some known means and not by self-immolation and resurrection, and that it is a member of a species that could have been, but was not, created by organic evolution.

[3]Pasteur did not say anything about the origins of life. What he asserted, with reference to experiment and general scientific knowledge, was that there were no known circumstances in which it could be affirmed that microscopic beings came into the world without germs, without parents similar to themselves (Pasteur, 1864).

for penetration.[4] The molecules of life, as understood today, are small by ordinary measures, though they are gigantic in atomic terms, having molecular weights a billion times that of hydrogen. But they are neither round nor smooth. They resemble, instead, very thin and long solids.

In approaching the task of this paper, I take my cue from the dynamics of these molecules. I note that a DNA molecule is a system of hundreds of millions of atoms which continuously wiggle, vibrate, and oscillate at vibration rates which span the electromagnetic spectrum from radio waves to the infrared. Groups of oscillating patterns, quantized vibrations called phonons, wander around the molecule as if looking for a place to settle, but as long as the DNA remains an integral, functioning unit, the phonons never stop moving.[5] If, for whatever reason, the DNA falls apart, its collective oscillations vanish as water waves do when the water vanishes. These molecules can keep on carrying their coded messages only as long as their intricate dances are kept coherent according to the laws of chemistry and physics.

As a first step in trying to trace the origin of life in the temporal behavior of certain molecules, it will be necessary to understand the relationship between biological oscillations and the life process.

## Biological Clocks and the Life Process

The terms *biological oscillations, biological rhythms,* and *biological clocks* are not used uniformly across the literature.[6] I am going to use them interchangeably and mean by them all cyclic phenomena that are involved in maintaining life. The spectrum of biological clocks in species alive today is spread across 24 orders of magnitude.[7] The human body possesses clocks across 22 orders of magnitude in

---

[4]"Democritus . . . says that of all the shapes the spherical is the most mobile, and that this is the shape of the particles of both fire and mind," Aristotle, *De Anima*, 405a–11. He regarded soul as something distributed throughout the body, resembling what today is loosely called human life.

[5]For guidance to recent work on long-range correlations that guarantee coherence, see the editorial "Long-range correlations within DNA" (Maddox, 1992, p. 103). The DNA molecule is structurally stable in spite of its wild dance because potentially reactive subgroups of the nucleotide bases are tucked inside and immobilized by its geometrical tightness.

[6]Here are some samples. "Biological clock: an innate mechanism by which living organisms are able to perceive the lapse of time" (Wallace, King, and Sanders, 1986, p. 1138). "The internal mechanism of an organism that regulates circadian rhythms and various other periodic cycles" (Toothill, 1981, s.v. "Biological Clock"). Biological rhythms are "periodic biological fluctuations in an organism that correspond to, and are in response to periodic environmental change. . . . The internal mechanism by which such a rhythmic phenomenon occurs and is maintained even in the absence of the apparent environmental stimulus is termed a biological clock" (*Encyclopedia Britannica Micropaedia*, 1983a), s.v. "Biological rhythm"). "Living things are extraordinarily well adapted to their rhythmic environment and have become periodic in diverse aspects of their physiology" (*Encyclopedia Britannica Macropaedia*, 1983b, s.v. "Periodicity, Biological").

[7]The fastest ticking clocks are the molecules of the body, such as those of the skin. They respond to ultraviolet light at $10^{16}$ Hertz. Retinal cells respond to light between $10^{15}$ and $10^{14}$ Hertz. Photosynthetic

frequency. The morphologies of the different clocks vary substantially, but they all share a common dynamics: they all oscillate at their particular frequencies.

With few exceptions (Bonner, 1974; Goodwin, 1976; Winfree, 1980) all definitions of biological oscillations imply that living organisms, having come about through some yet unidentified steps, have acquired through natural selection a store of cyclic variations as parts of their adaptive strategies. In other words, it is almost universally maintained that life is historically prior to the cyclic processes of life. I believe that this assumption is erroneous. It is analogous to claiming that the sounds of the instruments of an orchestra assist the orchestra in making music. But those sounds do not *assist* an orchestra in its performance, they constitute the music, provided that they are kept correlated from instant to instant according to selected principles of harmony.

Likewise, biological clocks do not merely assist life in its adaptive endeavor but, more fundamentally, they comprise—they make up—living organisms. In this view, life is seen as a process that consists of the instant to instant coordination of chemical and physical oscillations, according to principles that will help maintain them in mutually supporting rather than destructive relationships. I call this schematic representation *the coordinated clockshop model of life* and find it useful as a conceptual tool for understanding biogenesis. In pursuit of that understanding, we leave behind the 6 million named species alive today, and direct our attention backward along the history of life. During the journey we watch the contemporary forms of life devolve toward their common origin.

First the hominids vanish, then the primates, the mammals, the vertebrates, the invertebrates, and the protists (bacteria and algae) disappear. Then, far below the almost naked DNA known as prokaryotes, we arrive at the roots of the phylogenetic tree of life. At that point of the journey, as an exercise in purposeful curiosity, we take the imagined ancestor of all life forms, and place it under the microscope of our analytical capacities.

## The Primeval Clockshop of Life

The clockshop beneath the microscope is a self-organizing system, able to maintain its dynamic balance against internal and external perturbations. It shares this capacity for homeostasis with all stable inorganic compounds, but with an important

---

processes that capture light energy and change it into forms of energy useful in the synthesis of organic compounds involve cyclic reactions with periods of $10^{-12}$ seconds. Periods of insect wing beats range between $10^{-4}$ and $10^{-3}$ seconds. Human vocalization occupies a frequency range between about 20,000 and 100 Hertz; that is, they have periods between $10^{-5}$ and tenths of a second. Periods of neural signals are between 3 and 10 seconds. The fastest growing bacteria reproduce every 600 seconds, cells divide at rates from $10^3$ to $10^5$ seconds. All living organisms, probably down to the genes, display circadian periods just below $10^5$ seconds, hundreds of thousands of species show lunar periods of about $10^6$ seconds, circannual rhythms of $10^7$ seconds, and there are plants which flower every 13 or 17 years; that is, with a period around $10^8$ seconds; for a graphical summary see Fraser (1987, p. 127).

difference. The living clockshop has crossed a threshold of complexity beyond which the short-range atomic and molecular forces ceased to be adequate to insure ordering and integrity. For that reason, in addition to short-range forces, it had to have appropriate methods for long-range dynamic coordination, perhaps of the type I mentioned when discussing the DNA.

The need for coordination among molecules that are far apart in terms of atomic distances is a demand of life for the creation and maintenance of *simultaneities of necessity*. Here "necessity" refers to the securing of sufficient coherence to help hold the microscopic world of the giant molecule together. The referent of necessity is the integrity of the living molecular group, that is, of the life process itself. In the much simpler world of inorganic matter one can find only *simultaneities of chance*. By this term I mean any chance combination of elements and molecules which, by the laws of chemistry and physics, can maintain their stability until some other chance event disrupts it, but which do not demand long-range coherence. In the phenomenal world the condition I called simultaneities of necessity corresponds to the *organic* or *living present* of life, that is, to the instant by instant, nondestructive synchrony of the life process.

The organic present is the simplest form of the now. Just as life may be identified with the creation and maintenance of the organic present, so the mind may be identified with the creation and maintenance of the *mental present*. And again, a group of people may be said to constitute a society only if, and only as long as, they are able to create and maintain a *social present*. These three presents are hierarchically nested in that each higher one subsumes the one or ones beneath it and, for that reason, they are necessarily simultaneous. The reasoning given earlier for the introduction of organic present into the presentless world of physical time may therefore be extended to the evolution of a nested hierarchy of presents. This claim needs a bit of elaboration.

There is nothing in the nonliving world, as revealed through the equations of physics, that could be used to define a now.[8] But, future, past, and the metaphor of time's flow—which is the changing of future into past—make sense only with respect to a now. Consistently, there is nothing among physical processes that could be held responsible for the experience of the passage of time. As I shall show later, the secular increase of entropy in thermodynamically closed systems, often called the thermodynamic arrow of time, is also inadequate for the definition of a direction of passage. As one would expect from the logic of the situation, the absence of a definable now and that of directed time are complementary aspects of time in the physical world. This does not mean that the physical world is timeless, but only

[8]Einstein avoided dealing with the "now" by maintaining that although there is something essential about it, identifying what that essential feature is, is not within the tasks of science (Schilpp, 1963, p. 37). He took time to be absolute as far as its flow is concerned, though relative in its measure. Building on the intuitive obviousness of time's passage and excusing himself from dealing with the "now," he constructed Special Relativity Theory by defining what one is to mean by now at a distance, provided one already knows what to mean by the "now" here. This was a way of smuggling biotemporality (and the higher temporalities) into the description of physical change.

that its temporalities are undirected and that the metaphor of the flow of time cannot be applied to physical processes when they are considered in themselves.

It was the coming about of the organic present that broke the symmetry of physical time by making it possible for the information content of a complex system to acquire meaning. Specifically, the definition of the now allowed a distinction to be made between present and nonpresent conditions. It is with respect to nonpresent conditions that a particular type of meaning, called purpose, could arise. With the ideational category of purpose came the possibility of distinguishing among different kinds of nonpresent conditions. Those that related to the unfulfilled needs of an organism could be separated from those that pertained to fulfilled ones. Events and conditions which were not present but which could satisfy unmet needs were assigned to a category of time called the future and became targets of intentions. Nonpresent conditions responsible for the means that were useless for reaching future satisfaction were assigned to the past. The ceaseless rearrangement of intentions with respect to the future, narratives about the past, and sense impressions about the present is what we call the flow of time.

What is to be meant by the organic present being inserted into a presentless world? Let me offer a visual metaphor as an aid to its understanding. The San Francisco cable cars move by attaching themselves to an underground cable. The cable, which is hidden from sight, is moved by some distant and invisible machinery. The present of life does not attach itself to a cosmic flow of time whose motion is maintained by unknown powers. Instead, life creates those conditions and operational properties of matter which give rise to our experiences and ideas of the present and of the nonpresents.

Keeping these comments in mind, let us assume that miniature synchronized clockshops, as here envisaged, did come about. Then, inquire into the manner in which they were likely to have evolved under the selective forces of the temporal organization of their environments. If such a path permits organic evolution to have taken place the way we know it did, then the coordinated clockshop model of life has some merit, and in search for the origin of life, we may begin asking questions about those primeval clockshops.

## The Evolution of the Primeval Clockshops

I submit that an assembly of molecular oscillators, complex enough to define an organic present, will necessarily become subject to natural selection under the pressure of the cyclic complement of environmental changes.

To begin in the beginning: there are good reasons to believe that our nearest nonliving ancestors resembled crystals.[9] Perhaps they occupied protective microscopic niches in geological formations, as blue-green algae occupy air spaces today

---

[9]Joseph Needham, writing about mesoforms that occur between successive stable levels of organization, noted that between the living and nonliving realms the crystalline represented the highest degree of organization (1944, p. 255). See also his *Order and Life* (1968, p. 158 and passim). Bernal called for a generalized crystallography as the key to the biology involved in the origins of life (1951, p. 34;

in the rocks of the Dry Valleys of Antarctica. That environment was unfriendly to later forms of life: the days were 18 hours or shorter, the atmosphere was mostly hydrogen, the surface of the earth was bathed in ultraviolet radiation. Natural radioactivity was strong, there were continuous electric discharges, and hosts of obscure phenomena, such as sonoluminescence and possibly its inverse, sound produced by light. The ground continuously shook, boulders rolled, and the earth vibrated at many frequencies.

In that environment, improved adaptation meant the acquisition of such collective oscillatory behavior as would enhance the chances of survival of the system as a unit. I can imagine, for instance, a crystalline structure lodged in a crevice whose width oscillated at a stable rate. Crystals that could match that rate of oscillatory change would have a better chance for survival than other units which did not.

Since the regularities of these primeval clocks are unlikely to have exactly fitted the temporal niches of the oscillatory spectrum of the environment, there existed an error signal through which the environment could exert selection pressure on the clocks. The selective forces would favor those configurations which were able to provide better fits to the cyclic variations of the surroundings. By favoring certain frequencies over others, natural selection supplanted accident. During this selection process life was born when the molecules crossed a threshold of complexity such that, above it their integrity came to depend on long-range ordering, as discussed before. With that ordering came the capacity for defining self-interest with reference to an organic present.

These clockshops were bound to complexify for reasons similar to those which make the history of timekeepers show increasing complexity in the construction of clocks. Namely, each step taken to improve the precision with which an external cycle could be modeled—whether by a human clockmaker or by organic evolution—served as a means for recognizing more refined cyclic structuring in the environment. Each improvement opened up the system to the possibility of, and need for, finer tuning.

Imagine, for instance, that the proto-organisms evolved sufficiently to take advantage of the tides which carried the nutrients they needed. After the tidal rhythm was internalized, then expressed through external behavior, the diurnal and semidiurnal inequalities and the lunisolar variations of the tides began to serve as new agents of selection. The difference between what an organism was capable of doing and what the environment demanded it ought to be doing, was an early example of the perennial existential tension of the life process.

As the spectral analysis of environmental periodicities continued, the complement of internalized rhythms had to expand. Also, as the clocks of the clockshop widened their frequency spectrum, the inner coordination of the assembly had to become increasingly more precise. The increased precision is recognized in the

---

1967, p. 192). A. G. Cairns-Smith carried these arguments further by maintaining that the ancestors of life were, in fact, crystalline structures that stood in for the later DNA-RNA-protein system of modern biochemistry (Maddox, 1985, p. 197).

evolutionary narrowing of the width of the organic present.[10] Also, the internal cycles created beat frequencies to which nothing in the external world needed to correspond. Some of these had to be control functions that assured the continued viability (autonomy) of the organism. Henceforth, natural selection while acting upon the externally manifest (phenotypic) rhythms came to effect the internal (geno-typic) rhythms.

The early Precambrian earth, upon which life was born, was a busy but stable place; its climate and geological conditions took millions of years to change. During these eons, some regions of the earth became populated with blue-green algae and prokaryotic bacteria. Their rate of evolution, though slow, was certainly faster than that of the nonliving earth because inanimate matter could change only through simultaneities of chance rather than simultaneities of need, defining a purpose. Somewhere along this history, the environment to which species had to adapt be-came biotic; the target of adaptation itself began to evolve. This novel condition made life subject to a Malthusian principle in evolutionary rates: in the intensifying evolutionary race, rapidly adapting species began to gain advantages over those that changed slowly. The same type of conflict between different evolutionary rates is being replayed in our epoch upon a higher integrative level. Namely, the living species on earth today, including our own bodies, are unable to change biologically at the rate required by the cultural changes that are being created by the collective power of human minds.

Let me turn to an early example of a conflict between different evolutionary rates and illustrate how such a conflict gave rise to something unpredictably new, namely, sexual reproduction.

To do so, I have to appeal to a principle in the theory of automata, formulated by von Neumann (1969, pp. 79–80). According to that principle, a system must reach a certain threshold of complexity before it can produce others of its own kind that are of equal or increased complexity and of higher potentialities. If this principle is taken to be valid for living systems, then only after a species has reached a certain level of complexity did it become advantageous for it to reproduce through heirs that developed instead of replicating through identical heirs, as do crystal-like structures which split or cells that duplicate by mitosis.[11]

That structural and functional complexity seems to have been reached during the early Paleozoic, perhaps 500 million years ago, resulting in the appearance of sexual reproduction. There is no general agreement on the immediate causes of the emergence of sexual reproduction. But a representative view, consistent with von Neumann's speculation, maintains that sexually reproducing species could adapt to a changing environment much more rapidly than could asexually reproducing ones.[12]

---

[10]This evolutionary change has surviving examples. For instance, a tree needs a few years to notify its roots that its head has been cut off whereas, in advanced species, the notification proceeds rapidly. The tree's organic present is much broader than that of a man.

[11]This is analogous to the cost-of-reproduction argument of J. T. Bonner (1974) introduced on p. 68 and recurring throughout the book.

[12]Sexual union keeps reshuffling the genes of all successful individuals so that a virtually infinite number of combinations is produced, making the species prepared for a large variety of contingencies.

In the sexual division of biological labor, the task of securing the continuity of the organic present became a joint enterprise of two types of specialized cells. They are the male and female germ cells, known collectively as gametes. The husbanding of the gametes became the task of large assemblies of somatic cells. Gametes and somatic cells bear different relations to time. Germ cells are potentially immortal, as are single-celled organisms that reproduce asexually by mitosis, and multicellular organisms that reproduce by splitting (after reproduction there are only daughter cells: there is no body left to be buried). In contrast, organisms that reproduce sexually by interchanging genetic material do age and do die by aging.[13]

Thus, while all living organisms may be killed, members of many species do not and cannot die by aging. Aging and death are, therefore, not necessary correlates of time and life, notwithstanding all the philosophical and religious traditions which contemplate time exclusively in relation to the dreams of everlasting life. But all species live by temporal ordering referred to the organic present. It follows that time relates to life through the maintenance of the organic present, and *not* through the inevitability of aging and death.

In the course of organic evolution, with the appearance of the lower invertebrates, came the development of the nervous system, including the brain, the seat of its central command. It had, as one of its tasks, the coordination of the cyclic processes of organisms. Also, as the species became more complex and their behavior less predictable, being able to anticipate the conduct of friend, foe, and food became advantageous and hence, one must assume, it was favored by natural selection. In other words, it is reasonable to assume that the ability of the mind to prepare for future contingencies has evolved in response to the need for anticipation. The mind made it possible for humans to create a world of imagined conditions and work out strategies through the manipulation of symbolic representations of reality, based on past experience.

The emergence of the mind reminds us of von Neumann's idea of complexification thresholds. Only above a certain complexity of the central nervous system was it possible for individuals of a species to assist other individuals in becoming as or more advanced in their skills than were their teachers. Crossing that theshold made it possible to create cumulative knowledge in the form of symbols and through them, pass on to later generations the fruits of acquired characteristics. This feat cannot be accomplished by biological means alone.

Let me sum up this section on the evolution of a primeval molecular assembly, schematically represented as a coordinated clockshop. The section traced the presumptive fate of such a clockshop, along a trajectory of necessary complexification. It found that the history of life, interpreted in terms of natural selection working upon that clockshop, coincided with what we know about the history of organic

---

Asexual reproduction remains useful in stable environments. A reliable introduction to the issues involved may be found in the works of John Maynard Smith.

[13]The essence of sexual reproduction is the exchange of genetic material. The development of two different forms of germ cells, a large, stationary one with food material and a small, mobile sperm is a later development. With the coming about of sexual reproduction the cyclic order ceased to be the only form of life; to it was added the linear or aging order of life.

evolution, from biogenesis all the way to the mental functions of the human brain. In search for new perspectives upon the origin of life, we are justified, therefore, in seeking an understanding of the way in which the earliest miniature clockshops are likely to have come about.

## The Origins of the Primeval Clockshops: A Perspective

Whenever a natural process is unavailable to direct human sense experience—as is the case with the creation of the universe or the structure of the atom or the origin of life—then the scientific method suggests the formulation of suitable abstract concepts. To make defensible guesses about the origins of coordinated, living clockshops, it will be necessary to employ one of these concepts. Conclusions reached through its use may later be translated back into statements about structures and functions of the phenomenal world.

In this case, that conceptual tool is the notion of entropy. For a definition of entropy we can do no better than to turn to one given by Professor Whitrow: "The concept of entropy is a mathematical measure of the disorganization of a system. The idea first arose as a part of the theory of heat, but a similar notion can be associated with probability distributions of any kind" (Edwards, 1972; s.v. "Entropy").

I would like to amplify this definition as well as sketch the relation of entropy change to time and to the life process. Then, I would like to use whatever shall have been learned to arrive at a particular view of biogenesis in terms of the nature of time.

The measure of anything is a comparison expressed in numbers. The entropy of a system is a numerical comparison, usually between two different states of the same system. A measure of entropy is not intrinsic to a system, just as the number of length units is not a property of the length of a road: the numbers depend on the units used. Also, since there are many different processes to which the idea of an organization may be applied, there are many different ways of defining entropy.[14] Furthermore, the reference state of a system to which zero entropy is assigned, remains arbitrary.

Consider, for instance, a container with two compartments separated by a wall, where each compartment contains a different gas. Remove the wall and watch the two noninteracting gases mix. If the initial condition is said to have had zero entropy then, at any later time, it should be possible to assign a value of nonzero entropy to the gas, measured in suitable units.

Consider next a freshly slaughtered sheep. Zero entropy may be assigned to its carcass as to its most highly organized state, because we know from experience

---

[14]Specific formulations of entropy principles exist for such diverse uses as steam tables, transfer of messages along radio links, learning behavior of rats, population pressure, and the economics of commodity production. There is no single unit or even physical dimension of entropy that would be common to these different uses.

that, beginning with that state, the body will decay. Again, it should be possible to measure the increase in the disorganization of the carcass, at least in principle.

Instead of a carcass, let us think of a living sheep. To give any meaning to its present degree of disorganization, one must have as a reference, a final and complete sheep. But there is no final sheep because evolution is open-ended. The task must, therefore, be reversed. One has to measure the degree of organization of the sheep with respect to one of its own, less organized states, such as when it was an embryo. Or, we may go further back and compare its present organization with a sampling of the primeval chaos. The new variable, necessary for the measurement, is called negentropy or information. Obviously, the gases with their increasing entropy, and the development of the sheep in terms of its increasing information content are oppositely directed changes.[15] Now, let us relate the concept of entropy time and organic evolution.

The long-term entropic behavior of systems that are thermodynamically isolated from the rest of the world are governed by a principle known as the Second Law of Thermodynamics. There are many ways of stating that principle, depending on how entropy is being measured. But common to all of them is the rule that the total entropy of an isolated, or closed system can, in the long run, only increase.[16] There are yet other systems, said to be open, such as living sheep. These are not defined by a box within which they are enclosed, but by their geometrical boundaries. Being open means that energy and information from the rest of the world may freely cross those boundaries.

Thus, both closed and open systems assume an outside world from which they are—or are not—closed off.

The Second Law of Thermodynamics caught the attention of Eddington, who perceived in its statement about the universality of decay, the physical basis of the human sense of time. He associated the decay process with the human experience and idea of passage, and named it the arrow of time (Eddington, 1958, pp. 69, 79, 101). But, for the Second Law to be applicable, it is necessary to begin with systems that may become disorganized. A principle that governs decay must necessarily assume some other principles that govern growth. This simple reason is a powerful argument against received teachings on time and thermodynamics. Namely, Eddington's arrow may be attached, with equal justification, to the entropy increasing or entropy decreasing processes of the world, to the decaying carcass, no less than to the developing embryo, to the mixing of gases no less than to the emergence of

---

[15]Information theory saves the idea of using entropy as a measure of directed change by using a sleight of hand. It is a particular representation of the future. It postulates an ensemble of possible future events and determines current entropy in terms of the uncertainties associated with them. This approach has proved itself useful in handling messages in languages whose statistical properties are known, but has little usefulness for calculating the entropy of a living sheep, for the statistical properties of sheep yet to evolve are not known.

[16]The actual changes inside the closed container are governed by the laws of the physical, biological, mental, and social worlds existing within, including an increase in negentropy. Since the Second Law is a statistical statement, it allows for transient trends of decreasing entropy for the processes taking place in the container.

macroscopic order from cosmic chaos. Since the association is arbitrary, neither growth nor decay, neither entropy decrease nor increase can serve, in itself, as the ultimate agency responsible for the direction of time and with it, the ultimate root of the passing of life.

We should have been able to reach the same conclusion, without knowing anything about entropy or thermodynamics, because no arrow in itself can define a direction, be it one in space or in time. Up needs down, right needs left, decay needs growth. The directedness of change we describe as the passage of time cannot come, therefore, either from decay or from growth considered in itself, because they both point in the same direction—and that direction may be arbitrarily selected. The cells of my body manufacture enzymes and create decay products in the same sense of time's passage.

The sources of time's flow must be sought, instead, in the purposeful behavior of living systems. It is the goal-directedness of life, referred to the organic present, that creates the distinction between what is judged as present and nonpresent, and divides the nonpresents into future and past, as discussed earlier. But then, how does one explain all of life having a common beat as it were, in respect to a shared organic present? The answer, which I proposed elsewhere, is that families of living organisms share a common present only, and only to the extent that they communally establish that present through intra- and interspecies communication (see Fraser [1987, pp. 192–201; 1978, p. 90 and passim]).

The view that time acquires its directedness with the life process is an unpopular one for an age that is more impressed by the mechanical and physical than with the organic. Many clever proposals have been made to rescue the belief that the ultimate roots of time's passage must necessarily be found among the governing principles of physics.[17] Some proposals maintain that, at the time of its creation, the universe was given a great degree of ordering, and since then, history has but lighted fools to dusty death. Others argue for a continuous creation of macroscopic information from the chaos of the atemporal substratum of the world serving as the source of our sense of time, driven, ultimately, by the expansion of the universe. There are statements in the literature of time in physics that maintain that time is entirely a mental phenomenon, or else a mystery, and will forever remain beyond the reach of science.

But none of these ploys, erected on the Platonic foundations of physical thought, can account for the phenomenon of the present. For that reason, none may be used to define future, past, and the passage of time. As I asserted earlier, directed time was born with life itself as a phenomenon of inner coordination.

To integrate this view of life and directed time with the undirected time of the physical world, whence life arose, it will be necessary to consider the creation of the world as it is interpreted in the theory of time as conflict (Fraser, 1982, p. 104).

---

[17]Why is it that the literature of time in physics, with very few exceptions, insists that our sense of time must be attached to increasing disorder? One can think of at least three reasons: (1) the provincialism and metaphysics of physical science that holds that anything as important as time must necessarily derive from physical phenomena; (2) a cultural setting which is much more impressed by the inorganic than by the organic; and (3) the association of time with the basically tragic view of life implicit in Christianity.

In all sciences, religions, and philosophies which concern themselves with the origin of the world, that origin is usually identified with the emergence of order out of chaos. At variance with this view, I submit that the aspect of Creation upon which we ought to concentrate is not that of order, but that of conflict between ordering and disordering. Ever since Creation, conflict has remained the fundamental dynamics of the universe (this is the basic tenet of *Time as Conflict* [Fraser, 1978]). The ordering and disordering processes of the inorganic world are uncorrelated. There are only few and only transient physical processes whose integrity depends on the coordinated decrease *and* increase of entropy. The undirected character of the three physical temporalities demonstrates a statistical averaging, or random mixing of growth and decay, conditions which are among the hallmarks of the inorganic world.

It is out of this matrix of directionless time that life first arose. The crucial prebiotic step of inorganic evolution was the appearance of nucleotides or, in any case, of large molecules whose structures and chemistry allowed them to be strung together into longer and longer polymers. In this process of expansion the significant variable was, as already stressed, that of complexity. Somewhere along the history of complexification, the ordering and disordering processes within the giant molecules have become so intricately interwoven that they could not be thought of anymore as separate—either theoretically or experimentally—without destroying the system. The mutual definition of the two opposing thermodynamic arrows is, therefore, a corollary of the creation of the organic present.[18] Life was born when an organic present came to be defined (as sketched) and thus introduced into the presentless and undirected temporalities of the physical world.

In the unity of opposing trends we may recognize a late and sophisticated example of orderability and disorderability, identified above with cosmic creation. This immanent dialectical contradiction is manifest in the existential tension of all life forms. In an easily recognizable manner, they are the conflicts between the needs of organisms in terms of their self-directed purpose, pitted against the possibilities determined by the environment.

It is often said that life is fleeting, ephemeral, easily cut off by sword or snuffed out by ill wind. But life is neither ephemeral nor is it fleeting. It is robust and lasting. The organic skill of combining growth and decay in the living present is now 3.5 billion years old, which is three-fifths the age of the universe.

---

[18]We learn from Shakespeare's *As You Like It* (1600 ca.) that

> And so from hour to hour, we ripe and ripe,
> And then from hour to hour, we rot, and rot,
> And thereby hangs a tale.

(Act II, scene 7, line 26)

Although youth is taken to be the age of ripening and old age that of rotting, in the life process ripening and the rotting take place simultaneously, kept coherent by the synchronization of its clockshops.

## Summary

What did this exercise in the natural philosophy of time reveal?

It demonstrated that it is possible to interpret life and its evolution in terms of time. To do so, it was necessary to abandon the idea of a qualitative homogeneity of time, and replace it by the recognition that time is comprised of a nested hierarchy of qualitatively different temporalities. It was also necessary to adopt the coordinated clockshop model, as the most general representation of life. In the course of reasoning it was learned that the coming about of life was—or seems to have been—a necessity, once molecular evolution crossed a certain threshold of complexity. Self-directed behavior that emerged and remained under the selective forces of the cyclic spectrum of the environment henceforth guaranteed a process of complexification and through it, the stability of life itself.

The arguments offered remain independent of any particular chemistry or spatial structuring of biological matter, and hence are sufficiently general not to limit the phenomenon of life to the particular path that it has taken on earth. Life was positioned, within the broadest conceivable horizons, which are those of the universe.

Anyone aware of the immense variety of organic forms may wonder how such a schematic understanding could assert anything about the community of God's vast zoo of life. The answer is that it is the task of science to identify the universal in the uncountably many of its particulars, and that the relationship between time and life constitutes one of those universals. Specifically, the reasoning suggests that from among the chemically and physically possible structures and functions, life has selected those that could serve in the creation and maintenance of the organic present.

I would like to close with a criticism of the ideas presented. It comes from Christopher Fry's "*The Lady Is Not for Burning*" (1950, p. 53).

> We have given you a world as contradictory
> As a female, as cabalistic as the male,
> A conscienceless hermaphrodite who plays
> Heaven off against hell, hell off against heaven,
> Revolving in the ballroom of the skies,
> Glittering with conflict as with diamonds;
> When all you ask us for, is cause and effect.

## References

Aristotle *De Anima*.
————— *Historia Animalium*, 539b, Book V, Ch. 1.
Bernal, J. D. (1951), *The Physical Basis of Life*. London: Routledge & Kegan Paul.
————— (1967), *The Origins of Life*. Cleveland, IL: World.

Bonner, J. T. (1974), *On Development*. Cambridge, MA: Harvard University Press.

Cairns-Smith, A. G. (1981), Beginnings of organic evolution. In: *The Study of Time*, Vol. 4, ed. J. T. Fraser, N. Lawrence, & D. Park. New York: Springer.

Darwin, F., ed. (1887), *The Life and Letters of Charles Darwin*. London: John Murray.

Eddington, A. S. (1958), *The Nature of the Physical World*. Ann Arbor: University of Michigan Press.

Edwards, P., ed. (1972), *The Encyclopedia of Philosophy*. New York: Macmillan.

Encyclopedia Britannica (1983a), Biological rhythm. In: *Encyclopedia Britannica, Macropaedia*, 15th ed.

—— (1983b), Periodicity, biological. In: *Encyclopedia Brittanica, Macropaedia*, 15th ed.

Fraser, J. T. (1978), *Time as Conflict*. Boston: Birkhäuser Verlag.

—— Lawrence, N., & Park, D., eds. (1981), *The Study of Time*, Vol. 4. New York: Springer.

—— (1982), *The Genesis and Evolution of Time*. Amherst: University of Massachusetts Press.

—— (1987), *Time, the Familiar Stranger*. Amherst: University of Massachusetts Press.

—— (1990), *Of Time, Passion, and Knowledge*. Princeton, NJ: Princeton University Press.

Fry, C. (1950), *The Lady Is Not for Burning*. New York: Oxford University Press.

Goodwin, B. C. (1976), *Analytical Physiology of Cells and Developing Organisms*. New York: Academic Press.

Kirk, G. C., & Raven, J. E. (1975), *The Presocratic Philosophers*. Cambridge, U.K.: Cambridge University Press.

Maddox, J. (1985), Editorial. The DNA as a kind of solid. *Nature*, 317:197.

—— (1992), *Nature*, 358:103.

Needham, J. (1944), Integrative levels: A reevaluation of the idea of progress. In: *Time the Refreshing River*. London: Allen & Unwin.

—— (1968), *Order and Life*. Cambridge, MA: MIT Press.

Oparin, I. A. (1966), *Life: Its Nature, Origins, and Development*, tr. A. Synge. New York: Academic Press.

Pasteur, L. (1864), Des générations spontanées. In: *Oeuvres de Pasteur*, Vol. 2. Fermentations et Générations dites Spontanées, ed. V.-R. Pasteur. Paris: Masson, 1922.

Schilpp, P. A., ed. (1963), *The Philosophy of Rudolf Carnap*. La Salle, Quebec: Open Court.

Smith, J. M. (1978), *The Evolution of Sex*. New York: Cambridge University Press.

Toothill, E., ed. (1981), *Dictionary of Biology*. New York: Facts on File.

von Neumann, J. (1969), *Theory of Self-Reproducing Automata*, ed. & completed A. W. Burns. Urbana: University of Illinois Press.

Wallace, R. A., King, J. L., & Sanders, G. P. (1986), *Biology: The Science of Life*. Glenview, IL: Scott, Foresman.

Winfree, A. T. (1980), *The Geometry of Biological Time*. New York: Springer.

2

# Biobehavioral Rhythms: Development and Role in Early Human Ontogenesis

*V. Pouthas, J. Provasi, S. Droit*

*Abstract*   Rhythmicity is so fundamental a feature of life that we frequently take it for granted. Biobehavioral rhythms, however, emerge during the fetal period and continue to develop during the entire lifespan. Furthermore, they are thought to play an important role in the neuromotor, cognitive, and affective development of the human infant. Some illustrative examples will be drawn from current research to describe when and how various behavioral periodicities, whose frequencies range across different orders of magnitude, evolve during the first months of life. This brief overview shows that early biobehavioral rhythms serve a regulatory function before the emergence of more goal-directed behaviors, and provide young infants with the bases for the acquisition of more complex temporal regulations.

Common sense tells us that rhythmicity is a fundamental feature of life. One doesn't need to be a scientist to know that rhythms are present at every level of life:

1. The biological level. Our heart beats at a certain rhythm, we breathe at a certain rhythm, we are awake during the day and asleep at night.

2. The psychological level. We know that we are more efficient in athletics or cognitive activities at certain hours of the day. We are also aware that our level of arousal can fluctuate even in a 1- to 2-hour period. The reader may be less attentive to what he or she reads in the next few pages, not because of lack of interest, but simply because attentional capacities may not be as efficient as they are at this point in the text. On the other hand, some of our motor activities have a spontaneous tempo. If someone asks you to tap with your finger on the table, you will first tap irregularly but will then rapidly establish a regular tapping rate, with

*Acknowledgments.* We are grateful to A.-Y. Jacquet for her indispensable aid in preparing the manuscript.

*19*

each tap produced about 500 to 700 milliseconds after the preceding one. This is the outcome of what has been termed the preferred tempo. If you listen to a piece of music, for example, you will tend to tap your feet or nod to the rhythm of the auditory stimulus.

3. Rhythms are also present at the social level. In Western societies work is organized around the 8-hour day, 5-day week, with a 2- to 4-week holiday per year, depending on the country. In France there has been a recent debate on the need to adjust the school day to fit biological rhythms. The rationale is that biological rhythms probably form the basis for both psychological and social rhythms.

In order for everything in an individual or social system to run smoothly there needs to be true synchronization of all these rhythms. In other words, biological, psychological, and social rhythms are an integral part of life and to some extent can be viewed as constituents of life itself. We only realize how basic these rhythms are and appreciate their adaptive value when they are disrupted. For example, recent studies have shown that biological rhythms are impaired with advancing age: their amplitude and periodicity decrease, and significant alterations in the synchronization with environmental cues are often observed. This is manifest in nocturnal sleep disruption such as shifts in sleeping and waking times to earlier hours, and hence a decline in daytime alertness. Because of this misalignment of the circadian phase, the elderly find themselves frustrated: they are awake at a time when most of their family and friends are still sleeping and in addition they are unable to remain alert in the evening hours. Thus it is difficult for them to participate in many social activities. This change in biological rhythms with aging shows that although rhythmicity is a fundamental feature of life it cannot be taken for granted. Rather, rhythmicity emerges during the fetal period and develops during the entire lifespan.

The examples below illustrate how biobehavioral rhythms, whose frequencies range enormously in magnitude, develop during the first months of life. The emergence and the stabilization of periodicities of motility during the second part of fetal life, which are signs of normal gestational progress and good fetal health, are examined first, followed by the development of sleep–wake circadian rhythmicity and the role of social synchronizers in this development. We then turn to spontaneous cyclic motor activity and rhythmical stereotypies in young infants, and explore hypotheses regarding their role in neuromotor and cognitive development. This framework should help clarify the presentation of studies conducted in our lab which provide evidence that young infants are able to learn to modify the pace of motor rhythms in order to control their environment. Lastly, we will discuss examples of the development of early social interaction rhythms.

## Fetal Motility Rhythms

Analysis of biological rhythms in the fetus were originally motivated by the need to better understand the behavioral states of the fetus and to diagnose fetal distress. Electroencephalograms or electromyograms are used in the human adult, newborn, or child to measure sleep or wake states. This is obviously not possible in the fetal period. By taking a composite measure of overall movement, changes in heart beat, and respiratory activity, however, it has been shown that by the third trimester:

1. There is a cyclical pattern of "active" and "calm" phases. A fetus is defined as active if within a 3-minute period his or her movements last for more than 1 s and the fetal heartbeat varies by more than 10 beats per minute. The fetus is said to be calm if these criteria are not met. The length of these cycles evolves. One of the most instructive examples from the body of literature on ultradian fetal rhythms can be found in a study by Dierker, Pillay, Sorokin, and Rosen (1982). They clearly showed that cycles of activity and rest develop during the final weeks of pregnancy. They compared two groups of fetuses, one with a gestational age of 28 to 30 weeks and the other with a gestational age of 38 to 40 weeks.

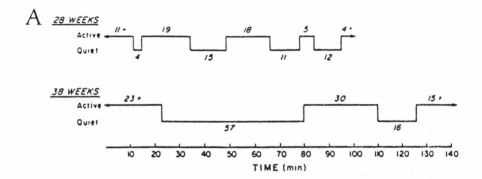

B

| Gestational age (wks) | N | Activity periods/hr (median) | Duration active period in min (median) | Duration quiet period in min (median) |
| --- | --- | --- | --- | --- |
| 28-30 | 7 | 6.6 (6.2) | 10.3 (7.0) | 9.9 (9.5) |
| 38-40 | 14 | 2.2 (2.0) | 35.0 (34.0) | 18.3 (20.5) |
|  |  | $P < .001$ | $P < .001$ | NS |

NS = not significant.

*Figure 2.1. A* Sequential epochs plotted as either active or quiet for a 28-week and a 38-week fetus. *B* Comparison of fetal activity periods (redrawn from Dierker et al., 1982).

The periods of activity of the 28- to 30-week fetuses were briefer, less well defined, and were associated with more cycle changes per hour than in the group of older fetuses. Furthermore, the duration of cycles increased significantly from 20 minutes on average at 28 weeks to 55 minutes on average at 38 weeks.

2. It has also been shown that the fetus is more active in the evening and the start of the night than at any other time of the day. It moves more, its heart beats are more variable, and the fetus also has more frequent respiratory bouts (Minors and Waterhouse, 1979; Patrick, Campbell, Carmichael, Natale, and Richardson, 1982; Visser, Goodman, Levine, and Dawes, 1982).

There is clear-cut evidence for the presence of biological rhythms during the fetal period. Naturally they are less apparent than the cycle we would like to describe now and which new parents all hope for: the sleep–wake cycle.

## The Development of Sleep–Wake Circadian Rhythmicity in the Infant

The distribution of sleep throughout the day and night changes dramatically over the first weeks of life. As we all know, the newborn spends almost an identical amount of time asleep during the day and the night, and then gradually stays awake slightly longer during the day than at night. The issue is which factors prompt the development of this circadian rhythm: a spontaneous maturation process, or the experience of external sychronizers. Several studies have provided partial responses to this question, in particular works by Martin du Pan (1970) who studied an infant maintained in continuous uniform illumination and fed on demand; in other words, with no alternation of light and darkness or a regular mealtime rhythm (every 3 or 4 hours is typical).

Figure 2.2 shows that in this infant, sleep (in white) and wakefulness (in black) were random at first, but by day 65 there was a long sleep between 6:00 and 15:00. This is a clear demonstration of the maturation of a circadian rhythm of roughly 24 hours, which, however, was not synchronized to day and night. This synchronization occurred at the age of 80 days, when he was exposed to a normal light and dark cycle. These observations, in addition to accounts by Kleitman and Engelman (1953) of an infant who was also fed on demand and who showed a "free running" rhythm during weeks 9 to 15, suggest that the circadian clock first matures independently of external synchronizers. The infant can only respond to external synchronizers when this maturation has occurred.

Nevertheless, it may be possible to help infants set up a circadian cycle of sleep and wakefulness by making babycare activities regular and by paying much more attention (physical contact, face-to-face interaction) during the day than at night. The Sander, Stechler, Burns, and Julia (1970) study provides a whole host of evidence for this. In newborns who were raised in nurseries by several caretakers, and who were cared for at fixed times, their sleep–wake cycle exhibited certain features of the adult pattern as early as the second week of life (with longer sleep times at night). This was in contrast to newborns placed in their mother's rooms and cared for by their mothers from birth. Overly early establishment of circadian rhythms may, however, be damaging to the cognitive development of the infant.

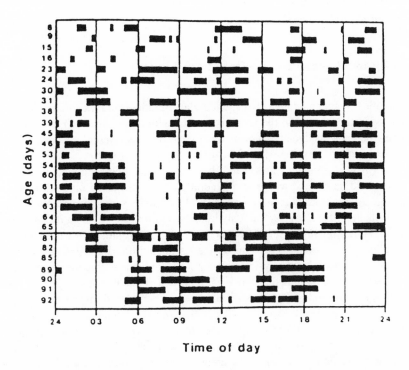

Time of day

*Figure 2.2.* Distribution of sleep (white) and wakefulness (black) in an infant who, from the eighth day of life, was kept in isolation in uniform illumination and fed on demand. After day 80 of life, the infant was exposed to normal alternation of light and dark. Note that the ordinate does not represent successive days. Data from Martin du Pan (from Pouthas, 1990).

When an infant is kept awake artificially, over a period of time which does not correspond to his or her natural rhythm, he or she cannot process the enormous variety of new sensory experiences of the surrounding world. This is not the case for less "precocious" infants since their wakeful periods are distributed over the entire day. This shows how important it is to let the circadian rhythm set in gradually. The role of the mother and the family is critical: she builds up a temporal frame which matches the infant's needs, and anticipates the periods when he or she is awake and active, determining the "right" time to initiate communication with the child. Infants can then process information available to them at their own (endogenous) rhythmic rate, and gradually increase their ability to meet the regular and consequently predictable changes in their physical and above all social environments. The result is a gradual synchronization, or in Davis's terms, a "tuning" of endogenous rhythms with external rhythms (Davis, 1981).

As stressed by Reinberg, Hallek, and Hellbrügge (1987), the development of sleep–wake circadian rhythmicity, although mostly endogenous, is also highly affected by the environment. When the caretaker (and particularly the mother) experiences difficulty in tuning to the infant, for example during extended hospital stays, the development of the ability for temporal adaptation may be slowed down and growth in general impaired.

## Cyclicity and Rhythmicity in Infant Motor Activity

Periodicities at much higher frequencies than ultradian and circadian biological rhythms, in other words, in the few second or few minute range such as cyclic and rhythmic movements, are often considered to play an important role in development, in particular motor development.

Recent work by Robertson (1987) has shown that spontaneous motor activity in fetuses and newborns, which was generally considered to be random, is in fact characterized by cyclic fluctuations on the scale of minutes, now commonly referred to as cyclic motility (CM). We can only hypothesize at this time about the functional role of CM. It probably differs as a function of the environment in which it occurs (the uterine environment for the fetus or the aerial environment of the newborn that "buzzes" with social and physical stimulation). Regular alternation of periods of intense activity and periods of relative calm may facilitate the neuromotor development of the fetus, and afford an adequate level of neural and muscular activity by distributing the gains of supraliminal activity and the benefits of rest along the temporal continuum. After birth, when the child is plunged into a new external and social environment, CM may regulate interactions with the surroundings. Social stimulations that activate the infant's motor system may fortuitously match intrinsic fluctuations in spontaneous motility and thereby provide the opportunity for interactional synchrony. The infant's motor activity also, however, calls for a social response from his or her caretaker, which occurs during the pause following activity (relative quiescence).

Thus, for example, analysis of the coupling of spontaneous motor activity and visual attention shows how CM may regulate the infant's interactions with the environment, by enabling sustained visual attention, during periods of decreasing motor activity, to be directed to the attentional targets identified during the period of increased activity (Robertson and Bacher, 1992). This does not imply that CM is the best source of regulation; however, it is operational before more intentional temporal regulations have emerged. It has even been shown that the coupling of spontaneous motor activity and visual attention persists for at least 3 months after birth, in spite of presumably increasing intentional control of each system.

On the other hand, during the entire first year of life in normal infants motor rhythmical stereotypes, such as hand waving or kicking, can be observed. As Thelen (1981) points out, "the frequency and diversity of rhythmically repetitive movements are so great that the infant appears to be following the dictum: 'if you can move at all, move it rhythmically' " (p. 239). These rhythmic movements are believed to play a major role in motor development. The clocks which time them are viewed as possible precursors of timing mechanisms that control the temporal regulation of more complex motor acts later on in development. According to

Ashton (1976), these stereotypies may play an important role in the temporal regulation of different subroutines of the mature motor act. Young children must not only select the most appropriate subroutine from their repertory and decide upon the order in which these should be performed, but also need to learn *when* to shift from one routine to another, *how long* to pause between each routine, and *when* to stop. One typical example of regular practice in regulating different subroutines of a motor act is kicking, frequently observed in 3- to 6- to 7-month olds. Thelen argues that kicking is the expression of a central motor program used later in development for locomotion.

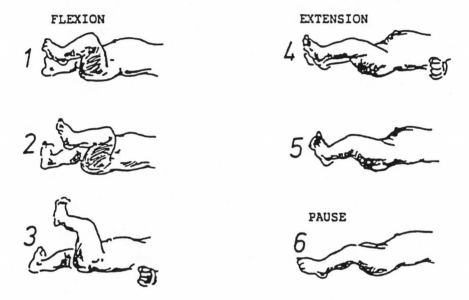

*Figure 2.3.* Successive phases of an infant single-leg kick. (1, 2) flexion phase; (3, 4, 5) extension phase; (6) interkick interval (redrawn from Thelen, 1981).

As shown in Figure 2.3, a kick is composed of rapid flexion (lasting 200–300 milliseconds), a short pause, and then an extension which lasts a little longer than the flexion (300–600 milliseconds) followed by the next flexion whose duration can vary considerably (observational data range from 140 milliseconds to 5 seconds). A parallel can be drawn between kicking and locomotion. Flexion and extension correspond to steps, whereas the interkick interval can be seen as the analog of the stance phase. It has been shown that the speed of kicking decreases when the duration of the pause between two kicks increases. Similarly, we walk more slowly when the stance phase is prolonged. Although an endogenous microrhythm underlies the timing of kicking, it may be assumed that in spontaneous kicking infants may "control" the rhythm to a certain extent, by progressively adjusting the interval duration between kicks. Our unit has shown that from birth, infants can learn to regulate a rhythmic activity underpinned by endogenous temporal mechanisms.

## Acquired Temporal Regulations of Nonnutritive Sucking Activity in Young Infants

A temporal response differentiation paradigm, an ingenious procedure developed originally by DeCasper and Sigafoos (1983), was used. It was devised to test whether an infant can modify the rhythm of his or her nonnutritive sucking, a response that can be measured at birth.

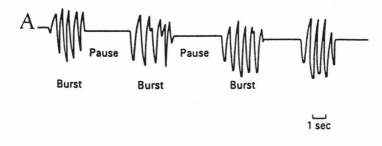

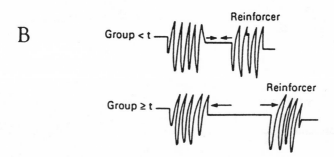

*Figure 2.4. A* An example of a 4-day-old infant's nonnutritive sucking behavior (redrawn from a polygraph recording). *B* Schema of the experimental procedure, *t* represents the reinforcement criterion imposed during the reinforcement phase. In group < *t*, a sucking burst produced the music only if it followed an interburst interval (IBI) that was less than the 30th percentile of the baseline intervals. In group > *t*, a sucking burst produced the music only if the preceding IBI was greater than the 70th percentile of the baseline intervals.

Nonnutritive sucking consists of an alternation of bursts of about 4 to 10 sucks (at a rate of 2 to 3 sucks per second) and pauses of roughly 3 to 15 seconds (see Figure 2.4A). Infants are required to shorten or lengthen the duration of their pauses. This is done by first obtaining a baseline measure of the "spontaneous" sucking activity of the infant when given a pacifer-type no-hole nipple. After this baseline value has been obtained (after 22 pauses in our case) the pause distribution is calculated and reinforcement criterion set up. In other words, the target duration value which will receive reinforcement is defined. Reinforcement was the agreeable onset of a short interlude of music. This reinforcement was delivered during the sucking phase following the pause of the "right" length. One group of babies was reinforced for making their pauses shorter than the duration of their spontaneous ones, and the other group was reinforced for making their pauses longer (see Figure 2.4B).

**Table 2.1**

**Median IBIs During the Baseline, the Beginning and the End of the Testing Session for Newborns and 2-Month-Old Infants**

| Subjects | Condition | Baseline | 22 First Pauses | 22 Last Pauses |
|---|---|---|---|---|
| | < t | 5.0 | 3.9 | 4.0 |
| Newborns | > t | 5.1 | 5.3 | 5.4 |
| | Control | 5.2 | 5.2 | 5.4 |
| | < t | 5.1 | 4.3 | 3.7 |
| 2-Month-Olds | > t | 5.1 | 4.9 | 6.1 |
| | Control | 5.0 | 4.8 | 5.2 |

Table 2.1 summarizes the findings. Newborns and 2-month-olds are able to learn to make their pauses shorter than their baselines, as of the start of the test phase. In contrast, only 2-month-olds are able to learn to slow down their sucking rhythm. In addition, even at this age, learning to make pauses longer is more difficult than to make them shorter. Babies are only able to produce longer pauses at the end of the test phase. This asymmetry—where acceleration is possible as of birth, and a slowdown is only possible at 2 months and more difficult to produce—may be due to the inherent alternation in this rhythmic activity between periods of activity and rest periods. A lengthening of pauses beyond a certain threshold may result in a loss of the rhythmic contours of the activity and finally cessation. The overall data suggest, however, that very young infants are able to learn to modify a temporal organization, and hence the pace of a hardwired rhythmic activity with an endogenous clock. They can by extension trigger an event in their environment and to a certain extent control it.

These data tend to confirm the assumption that human behavioral rhythmicities which are controlled by endogenous timing during early development come progressively under extrinsic control as development proceeds (Lewkowicz, 1989). This takes place gradually, as the infant accumulates experience of changes in the nature of control, but above all during interactional exchanges. This is because interactional and dialogue situations contain a key feature for early regulation of rhythms.

## Regulation of Pace of Adult–Infant Interaction Rhythms

The first example is a naturalistic observation of newborns and 2-week-old infants conducted by Kaye and Wells (1980) in the context of feeding. Sucking in infants consists of the alternation of bursts of sucks and pauses. During the pauses mothers tend to jiggle the baby or the bottle expressing the belief that it would wake up the infant and elicit sucking again. The results of the Kaye and Wells study show that it is not jiggling which elicits a new burst of sucking but rather its cessation. Thus, when the baby is about 2 weeks old mothers change from jiggling until the infant starts sucking again to jiggling and stopping. The smooth alternation of turns (jiggling–sucking), when it takes place, is due to the mother's social intention to accommodate her turn to the temporal organization imposed by the infant. These authors, however, suggest that, as a result of the experience of turn taking in feeding situations the infant may progressively learn to anticipate this turn taking and take on his or her own role as an active partner in the sequence.

The development of the infant's active role in exchanges with adults can also be seen in studies on interactional rhythms between mothers and 3- to 5-month-olds, observed in free play situations. In these situations, analysis focuses on the temporal structure of positive indices (vocalizations, mimics directed toward the adult) and negative indices (avoidance, moving away) that occur during the exchange.

Lester, Hoffman, and Brazelton (1985) compared the social interaction rhythms of full-term and preterm infants, aged 3 to 5 months, with their mothers (see Table 2.2).

**Table 2.2**
**Summary of Results**
**(Lester, Hoffman, and Brazelton, 1985 study)**

|  | Age (months) | Synchronization | Control Over the Exchange |
|---|---|---|---|
| Full-Term | 3 | + | + |
| Preterm | 3 | − | − |
| Full-Term | 5 | ↗ | ↗ |
| Preterm | 5 | = | = |

Their findings show that (1) there was greater synchrony in exchanges between full terms and their mothers than between preterms and their mothers; and (2) that full terms had better control over the exchange. In addition, they report that synchrony and dominance increased between the ages of 3 and 5 months in full terms. These cycles of attending/nonattending observed in infants' interactions with their mothers are thought to reflect the presence of intrinsic neural oscillators or timing mechanisms in the central nervous system. The differences between preterms and full terms could thus be due to the fact that the regulatory functions of the nervous system of preterms are still immature. Hence preterms are less able to receive any information from the environment and are not as able to modify their behaviors in a relevant way to adjust to their mothers' behaviors. In other words, preterms have more difficulty developing efficient communicative skills.

Communicative skills start to emerge as of the first months of life, as shown in studies by Arco (1983) and by Arco and McCluskey (1981). They asked mothers or fathers to play with their 3- to 5-month-olds at their own pace, as they would do spontaneously, and then at a slower and a faster tempo. The findings show that babies recognize temporal changes and respond differentially to natural, faster, and slower rhythms. Infant positive facial expressiveness, positive facial coaction, and mother–infant synchrony were significantly higher (and the quality of the exchange more easily sustained) when pace was natural or fast than when it was slow.

All these examples thus show that very early on, infants exhibit impressive capacities for regulating their behavior in social interchanges with their caregivers and particularly with their mothers. The presence of dynamic processes in the development of social interaction has been assumed to have implications for the development of communication and language. Jasnow and Feldstein's study (1986) provides evidence supporting this assumption. They examined the temporal patterning of vocal exchanges between 9-month-olds and their mothers. These vocal exchanges consisted primarily of alternating rather than simultaneous vocalizations. Each partner waited approximately .8 sec to allow the other member of the dyad to speak. The temporal properties of these dialogues correspond to the parameters of adults' conversational exchange.

## Conclusion

This brief overview of the development of biobehavioral rhythms shows that they are omnipresent, and regardless of level (biological, cognitive, emotional, or social) are systems of positive (activity, wakefulness, movement, attention) and negative (rest, sleep, quiet, nonattentiveness) cycles which resemble homeostatic mechanisms.

It is likely that these early rhythms serve multiple adaptive functions. Because of his or her temporal structure, the infant is more involved (during the positive period) and less so (during the negative period), or in Green's terms (1975) "because of its rhythmic structure . . . the human organism . . . changes and moves because it has life and it does this with a momentum which alternates thrust and recovery" (p. 1). Several authors (Robertson, 1987; Sander et al., 1970, among others) have hypothesized that nonactivity, nonmovement, and nonattentiveness periods give the infant time to assimilate the information inputs of the activity, movement, and attention phases. Furthermore, studies as a whole, whether they analyze the development of ultradian and circadian rhythms (in the low-frequency range) or motor rhythms or even social interaction rhythms (in the high-frequency range), all stress that the infant's rhythms provide a clear temporal structure which caretakers use, in most cases unconsciously, to provide care and interact with the child at the "right time." In turn, this quality of synchronization, which is devolved to the adult in very early exchanges, may affect the evolution of the child's own rhythms. The infant may gradually learn to control them and become an active partner. By doing so, the child is no longer locked into his or her own rhythms and can acquire more complex temporal regulations, which do not correspond to biological oscillators and thus require the development of the ability to perceive and conceive of the passage of time.

# References

Arco, C. M. (1983), Infant reactions to natural and manipulated temporal patterns of paternal communication. *Infant Behav. & Develop.*, 6:391–399.

—— McCluskey, K. A. (1981), "A change of pace": An investigation of the salience of maternal temporal style in mother–infant play. *Child Develop.*, 52:941–949.

Ashton, R. (1976), Aspects of timing in child development. *Child Develop.*, 47:622–626.

Davis, F. C. (1981), Ontogeny of circadian rhythms. In: *Handbook of Behavioral Neurobiology*, Vol. 4, ed. J. Aschoff. New York: Plenum Press, pp. 257–274.

DeCasper, A. J., & Sigafoos, A. D. (1983), Intrauterine heartbeat: A potent reinforcer for newborns. *Infant Behav. & Develop.*, 6:12–25.

Dierker, L. J., Pillay, S. K., Sorokin, Y., & Rosen, M. G. (1982), Active and quiet periods in the preterm and term fetus. *Obstet. & Gynecol.*, 60:65–70.

Green, H. B. (1975), Aging. Temporal stages in the development of the self. In: *The Study of Time*, Vol. 2, ed. J. T. Fraser & N. Lawrence. New York: Springer, pp. 1–19.

Jasnow, M., & Feldstein, S. (1986), Adult-like temporal characteristics of mother–infant vocal interactions. *Child Develop.*, 57:754–761.

Kaye, K., & Wells, A. J. (1980), Mothers' jiggling and the burst–pause pattern in neonatal feeding. *Infant Behav. & Develop.*, 3:29–46.

Kleitman, N., & Engelman, T. G. (1953), Sleep characteristics of infants. *J. Appl. Physiol.*, 6:269–282.

Lester, B. M., Hoffman, J., & Brazelton, T. B. (1985), The rhythmic structure of mother–infant interaction in term and preterm infants. *Child Develop.*, 56:15–27.

Lewkowicz, D. J. (1989), The role of temporal factors in infant behavior and development. In: *Time and Human Cognition. A Life-Span Perspective*, ed. I. Levin & D. Zakay. Amsterdam: North-Holland, pp. 9–62.

Martin du Pan, R. (1970), Le rôle du rythme circadien dans l'alimentation du nourrisson. *La Femme, l'Enfant*, 4:23–30.

Minors, D. S., & Waterhouse, J. M. (1979), The effects of maternal posture, meals and time of day on fetal movements. *Brit. J. Obstet. & Gynecol.*, 86:717–723.

Patrick, J., Campbell, K., Carmichael, L., Natale, R., & Richardson, B. (1982), Patterns of fetal body gross movements over 24-hour observation intervals during the last 10 weeks of pregnancy. *Amer. J. Obstet. & Gynecol.*, 142:363–371.

Pouthas, V. (1990), Temporal regulation of behaviour in humans: A developmental approach. In: *Behaviour Analysis in Theory and Practice: Contributions and Controversies*, ed. D. E. Blackman & H. Lejeune. Hillsdale, NJ: Lawrence Erlbaum Associates, pp. 33–52.

Reinberg, A., Hallek, M., & Hellbrügge, T. (1987), Rhythmes biologiques. In: *L'enfant et sa santé*, ed. Manciaux, M., Lebovici, S., Jeanneret, O., Sand, E. A. & Tomkiewicz, S. Paris: Doin, pp. 337–360.

Robertson, S. S. (1987), Human cyclic motility: Fetal-newborn continuities and newborn state differences. *Develop. Psychobiol.*, 20:425–442.

—— Bacher, L. F. (1992), Coupling of spontaneous movement and visual attention in infants. Poster presented at the International Conference on Infant Studies, Miami.

Sander, L. W., Stechler, G., Burns, P., & Julia, H. L. (1970), Early mother–infant interaction and 24-hour patterns of activity and sleep. *J. Amer. Acad. Child Psychiatry*, 9:103–123.

Thelen, E. (1981), Rhythmical behavior in infancy: An ethological perspective. *Development. Psychol.*, 17:237–257.

Visser, G. H. A., Goodman, J. D. S., Levine, D. H., & Dawes, G. S. (1982), Diurnal and other cyclic variations in human fetal heart-rate at term. *Amer. J. Obstet. & Gynecol.*, 142:535–544.

3

# Attention and Brain Activation in the Processing of Brief Durations

*F. Macar, F. Vidal, and M. Bonnet*

*Abstract*   The processing of brief durations is best explained by the attentional model of time perception, centered on the existence of specific chronometric mechanisms. Two pieces of evidence are reported. Behavioral experiments reveal that human subjects are capable of controlling the attentional resources they devote to time processing, and that subjective duration increasingly shortens as attention to time diminishes. Electrophysiological studies suggest that distinct brain sites are activated, first, while a motor response whose duration must be accurately timed is being programmed, and, second, while this duration is produced.

One of the means by which living organisms adapt their behavior to the environment is time processing. In the present paper, time processing refers to the physiological mechanisms that enable one to estimate the duration of external or internal events in the second or minute range without using a watch. On many occasions, the duration of significant signals or that of one's own acts must be accurately evaluated: Accidents may occur if proper responses to certain stimuli do not take place at the right time. The main question discussed in the present paper is whether the processing of temporal information, when it is essential in everyday life, involves the activation of specific brain structures.

A simple illustration of the conditions under which this question should be asked may be drawn from animals' struggle for life. While a predator is watching its prey and attempting to evaluate how much time will elapse before the prey reaches a particular place suitable for attack, are certain brain areas activated in

*Acknowledgments*. The authors are grateful to Monique Chiambretto for having developed the computer programs used in the experiments described. Part of this research was supported by a Grant (91200A) from the French DRET.

relation to time processing? Before examining such a possibility, it is admittedly necessary to demonstrate that temporal parameters may actually be processed in themselves, rather than being derived from the computation of other features.

There no doubt exist a number of situations where time is deduced from the processing of space and speed. Piaget (1946), among others, provided evidence on this topic. In his well-known developmental studies where cars or trains were involved, however, he typically studied situations where speed was quite salient. Other cases do exist, and Fraisse (1957) argued that even newborns have intuitions about duration while they are crying in order to get their needs satisfied. The duration of a stimulus may constitute the only feature which differentiates two otherwise equivalent signals. In melodies, the duration of each note is precisely timed, and nice expressive effects can be obtained, for instance, by slightly lengthening one particular note. All sorts of animals are able to perceive and to produce small time differences. As Michon (1972) pointed out, time is information and its processing may require controlled attention in certain cases. The term *temporal information* is currently used to account for this idea, without implying any assumption as to the nature of this information.

## Attentional Effects on Subjective Duration

Over the last 20 years, a number of experiments have been designed to manipulate the subject's attention by varying task parameters in paradigms in which a temporal and a nontemporal task must be performed simultaneously. For instance, subjects were required to estimate the duration of a period during which a certain number of stimuli were delivered. The stimuli, whose number and/or contents varied in different trials, were to be counted or classified (for a review, see Block [1989, 1990, 1992]). This has proved an efficient technique with which to demonstrate that subjective duration is influenced by the amount of attention allocated to the nontemporal task. When the stimuli increased in number or in complexity, that is, when they admittedly required much attention and left little attention to the temporal task, subjective duration diminished.

Those data indicated that subjective duration relates directly to the amount of attention devoted to time processing when this type of processing is essential to one's performance. Since the attentional effects observed were only an indirect consequence of the manipulation of task parameters, however, the question of whether temporal processing can involve *controlled* attention was still unanswered. This question refers to the distinction between automatic and controlled processes in information processing theory (see for instance Kahneman [1973]). Among other features, controlled attention is deliberate whereas automatic processing is obligatory. Michon and Jackson (1984) established that temporal coding may necessitate controlled attention in tasks involving judgments on the order and positions of items in a list. It remained necessary to demonstrate that time judgments can also be modulated by controlled processes when the duration of an event is concerned.

A simple technique was recently used with this intention (Casini, Macar, and Grondin, 1992; Grondin and Macar, 1992; Macar, Grondin, and Casini, in press). Under double-task conditions, subjects were required to allocate various amounts of attention to a temporal and a nontemporal performance. Task parameters remained unchanged from one trial to another. An attention-sharing technique such as this has been occasionally used in visual detection tasks, and the indexes of performance for detection were found to vary with the proportion of attention required (Bonnel, Possamaï, and Schmidt, 1987). Interference effects appeared between the levels of performance obtained in the two concurrent tasks; that is, one was impaired when the performance in the other improved. This finding suggests that attention is a limited-capacity system. When applying this technique to time, we expected interference effects between the temporal and the nontemporal tasks. In addition, we assumed that any reduction in the amount of attention devoted to the temporal task would shorten subjective duration.

One of the experiments achieved in this frame consisted in delivering words on a video screen during a period of 12 or 18 seconds which was delimited by two auditory clicks and corresponded to a trial. The words appeared in succession and were visible for 0.3 second each. They belonged to various semantic categories. The nontemporal task was to count animal names at each trial. The temporal task was to reproduce the duration of a trial by pressing a button for the same amount of time as soon as the click marking the termination of the trial had occurred. The subjects were instructed to share their attention between the "word task" and the "duration task" in proportions which were specified on the screen before each trial (for instance, 75% to duration and 25% to words). The main results are illustrated in Figure 3.1. As expected, underestimation increased when less attention was given to time, and interference effects between the "duration" and the "word" tasks were obtained, although the "word" task was less sensitive to the attentional instructions than the "duration" one. We confirmed these effects of controlled attention with other tasks and shorter duration ranges, in visual and auditory modalities.

These data support the attentional model of time perception, which was formulated by Thomas and Weaver in 1975. This model implies the existence of two types of central processors: a timer which processes temporal information, and a stimulus processor which deals with the nontemporal features of the stimulus, such as its color or its intensity. This is admittedly a schematic description, and several distinct processing devices should probably be classified under each of these two headings. Nevertheless, one important aspect is that attentional resources are supposed to be shared between the timer and the stimulus processor. Since these resources are limited, when much attention is given to nontemporal stimulus features, little attention is thought to be devoted to the temporal features. This provokes a shortening in subjective duration, because the timer is viewed as a cumulative processor which encodes organic pulses. When attention shifts from the accumulator, it is switched off and the pulses stop being recorded, until attention is focused on duration again.

The cumulative properties attributed to the timer are consistent with the positive relationship that has been found to occur between attention and subjective

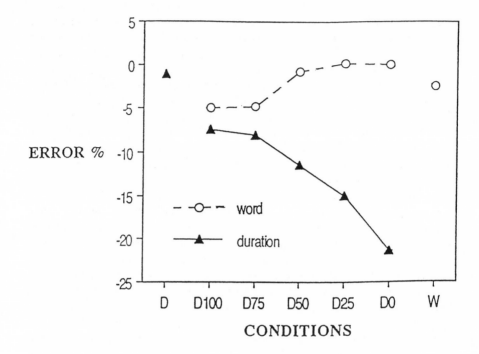

*Figure 3.1.* Attention-sharing procedure between a temporal and a nontemporal task. Percentages of time errors (solid line) and of word errors (dotted line) are given under five conditions of attention sharing (numbers refer to the percentage of attention that the subject was asked to allocate to duration; both tasks had to be done, even under the D100 and D0 conditions) and two control conditions (D and W, indicating that solely duration or solely words had to be processed). Along the abscissa, from left to right, less and less attention was allocated to the duration task. As a result, time error percentage increased whereas word error percentage decreased. The signs on the ordinate indicate the direction of the errors. For instance, a time error of −10 percent corresponds to underestimates of 1.2 and 1.8 s, respectively, in the 12 and 18 s durations. The word error percentage is the difference between the animal names counted and those to be counted, as a function of the latter. The data are averaged over 17 subjects.

duration. This is of particular interest since, when attention decreases, it seems a priori less likely to obtain this positive relationship than to observe an increase in variability, with many too short and too long subjective durations being produced (see Thomas and Weaver [1975] for a discussion of the variability issue). The nature of the organic pulses that are supposed to accumulate as a function of time is still mysterious, however. Rhythmic pulses provided by cellular oscillators as well as biochemical or bioelectrical changes involved in the transmission of information along neural paths are among the possible candidates.

## Timers and Networks

Arguments favorable to the timer concept have also been reported in studies with animals. For instance, Meck and Church (1984) compared the discrimination of duration and of number by rats. They obtained similar effects in both tasks, and concluded that duration processing is best explained on the basis of pulse counting processes. Gallistel (1990) also concluded that the estimations of duration, number, and rate of occurrence are closely related. Church and Broadbent (1990), however, recently proposed a connectionist model which does not involve cumulative processes. These authors remark that there is no known biological correlate for the accumulation of time. They also emphasize that the storing of temporal information on successive trials may be difficult to realize if it necessitates the storing of an increasingly large number of values and if hardware is supposed to be used to store each value. In Church and Broadbent's model (Figure 3.2), a set of oscillators with different and reasonably stable periods are the first component of the timer. The accumulator is replaced by status indicators, one for each oscillator. These status indicators do not record pulses, but, rather, information about the phase of the oscillators. At the start of each signal to be timed, the oscillators whose periods are in the relevant range are reset to zero. The working memory codes each duration value throughout matrices of connection weights, instead of holding it as a one-dimensional number. The same matrix can be used to store multiple values: Each duration corresponds to a particular sample of connection weights. Similar sets of mechanisms are involved to retrieve the duration stored in memory. The response is decided on the basis of a comparison between this duration and the one that just occurred. This system is designed for short as well as long periods, as the ones involved in circadian rhythms.

These hypothetical matrices might correspond to neuronal networks in the central nervous system. Several connectionist models based on the principle of learned relationships within cell populations have recently been proposed in the field of time. For instance, in Miall's (1992) network, the duration of a signal is coded by the particular oscillatory units which produce synchronous pulses at the beginning and at the end of the signal. For Dehaene, Changeux, and Nadal (1987), the network generates spontaneous temporal patterns of pulses; those patterns which coincide with the temporal structure of stimuli in the environment are selected and become stable, whereas the others are progressively eliminated. For Torras (1986), the internal oscillators may be driven by external rhythms of neighboring frequency. This accounts for such facts as the driving of circadian rhythms by external synchronizors (light or temperature among others) and the driving of electroencephalographic rhythms by visual rhythmic cues (this "photic driving," evidenced in 1944 by Popov in rabbits, may occasionally appear in man). All these networks are supposed to include oscillators, but this is not a prerequisite. In Desmond and Moore's (1991) model, duration is coded by sequences of activity propagating along neuronal chains, and each point of connection between two units—in other words, each synapse—adds a new delay. The duration coded is a function of the number of synapses and the propagation speed within the network.

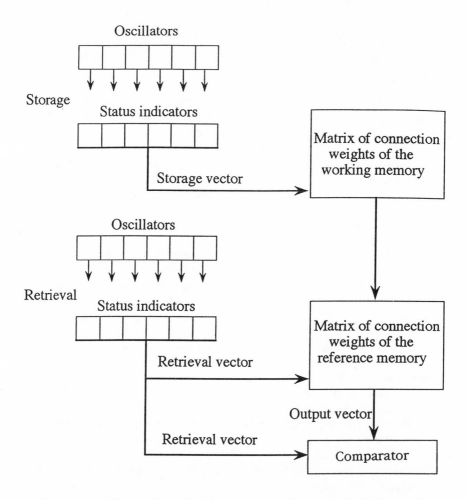

*Figure 3.2*. Schematic representation of the connectionist model proposed by Church and Broadbent (1990). At storage and retrieval levels, each status indicator yields information on the phase of its paired oscillator. Duration values are coded by matrices of connection weights. Final decision depends on comparison between the current duration and the duration stored in the reference memory matrix.

## Brain Mechanisms

Studies on brain mechanisms suggest that information may be coded by single units as well as by groups of interconnected neurons which form a functional module (see Lestienne, chapter 4, this volume). It has also been found that groups of neurons scattered on remote cerebral sites may pulse at the same frequency for a few tens of milliseconds when a signal is delivered. This synchronization may reflect the coding of relationships between various aspects of the signal, for instance its color, its intensity, and perhaps its duration.

The images of brain functioning obtained by electroencephalography, magnetoencephalography, or positron emission tomography generally rule out a strictly localized conception of the sites which appear to be involved in specific information processing. A particular process is likely subserved by interconnected neuronal networks distributed in various cerebral areas. On the other hand, certain modules must be attributed prominent functions in the networks, as suggested by the results of brain lesions in human subjects. A very localized and restricted lesion can provoke quite specific deficits, such as the loss of the capacity to speak one language and not another, or the occurrence of an agnosis limited to words belonging to only one semantic category. In the field of time mechanisms, frontal as well as temporal lesions have been associated to deficits in human subjects; precise localization has not been reported, however (see Macar [1980] for a review). The left cerebral hemisphere seems to be better than the right at discriminating the temporal order of brief sounds presented sequentially (Mills and Rollman, 1980; Lackner, 1982). In animals, experimental lesions localized in the septal and hippocampal areas induce clear deficits in temporal conditioning schedules. An important issue regarding the nature of time mechanisms is that motivational and memory effects are also produced by such lesions.

Studies on the electrophysiological activity of the human cortex during temporal tasks offer promising data. With electrodes fixed over the scalp, potential changes of positive or negative polarity and of various durations (from hundreds of milliseconds to several seconds) can be observed. "Event-related potentials" have proved to be in relation with various types of information processing. Electrophysiological experiments have revealed that different cortical sites are activated when information concerning either the intensity, the frequency, or the duration of a visual stimulus is processed, the activation of frontal sites being prominent with duration processing (Giard, Lavikainen, Reinikainen, Perrin, and Näätänen, 1992).

A series of studies where such localization was investigated in paradigms of duration production was undertaken by our team in Marseilles. These studies were grounded on behavioral experiments using the "advance information" reaction time paradigm (see Rosenbaum [1983], for a review on this method and its implications). One of the basic tasks consisted in producing intervals of accurate duration delimited by two short button presses. The intervals to be produced were either 0.7 or 2.5 seconds. The first press occurred as fast as possible after a visual response signal (S2). Two seconds before S2, another visual cue gave a preparatory signal (S1): It either indicated which duration to produce (the short or the long one) or did not yield any information on this duration. These conditions and the intervals required were randomized over trials. S2 conveyed the necessary information (short or long) in all cases. In this type of procedure, various behavioral data have indicated that the information is processed as soon as it is available. When S1 specifies which feature of the response is to be produced in the current trial, such information is processed during the preparatory period separating S1 and S2. Among other effects, the reaction time after S2 then decreases, since the time needed to process the relevant information takes place during the preparatory period and not after S2. This result has been obtained with various response parameters such as force, amplitude, direction, and effector (see Lépine, Glencross, and Requin [1989] for discussion). It also holds with response duration, whether an interval or a sustained motor press is produced, in a range of 0.2 to 5.5 seconds (Klap, Wyatt, and Maclingo, 1974; Vidal, Bonnet, and Macar, 1991; Vidal, Macar, and Bonnet, submitted).

Thus duration, like other response parameters, can be viewed as part of the ''motor program'' which specifies several motor features before response execution (Keele, 1986). In line with the timer or the connectionist approach, one may propose that the programming of duration consists in activating specific sources of pulses or specific networks, so that the proper units be ready for use before the response must actually be performed. There are innumerable oscillatory neurons in the brain. Pulse frequency seems to range mainly between 5 and 60 Hertz (e.g., Llinas, 1988; Gray, König, Engel, and Singer, 1989). If some pulse frequencies are involved in the processing of a particular duration, the first step of this processing is perhaps to select the oscillators which fit the duration and have been associated to it during training, and to ensure that other oscillators are not activated.

This line of reasoning led us to assume that in the ''advance information'' reaction time paradigm, specific brain areas are activated during the preparatory period when S1 indicates which duration must be produced. Brain activation as shown by an increase in the amplitude of event-related potential changes might reveal that in underlying sites, neural cells or networks increase their firing rate or undergo other functional modifications. To study this question, we used the ''source derivation method'' (Hjorth, 1975) meant for providing a reasonable approximation of the brain sources which emit the potential changes recorded over the scalp. The slow potentials that systematically occurred after S1 were analyzed (Macar, Vidal, and Bonnet, 1990; Vidal, Bonnet, and Macar, 1992). One question to be answered was where these potentials were prominent and revealed systematic changes in amplitude when the information concerning duration was processed. Three main sites were investigated: the supplementary motor area, which is thought to be involved in the planning of action; the primary motor area, the first area concerned with response execution; and a parietal area, which we thought likely to be involved in timing performance on the basis of preliminary data.

Figure 3.3 presents data collected on trials when S1 specified which duration to produce. The sustained negative potential shift visible between S1 and S2 was most sensitive to this factor. Over the supplementary motor area, prominent activity was observed beginning about 300 milliseconds after S1. Over the primary motor cortex, the activity level was high only at the end of the preparatory period. A more detailed analysis considering the level of activity related to each precued duration (Vidal, Bonnet, and Macar, 1992) revealed that the supplementary motor area and the primary motor cortex were sensitive to duration programming in the first and the second part of the preparatory period, respectively. Thus, a shift from the first to the second area seemed to occur, as if components of the motor program were elaborated in the supplementary motor area and next transferred to the primary motor cortex just before response execution. In contrast, no significant changes specific to the programming of response duration were observed over the parietal site.

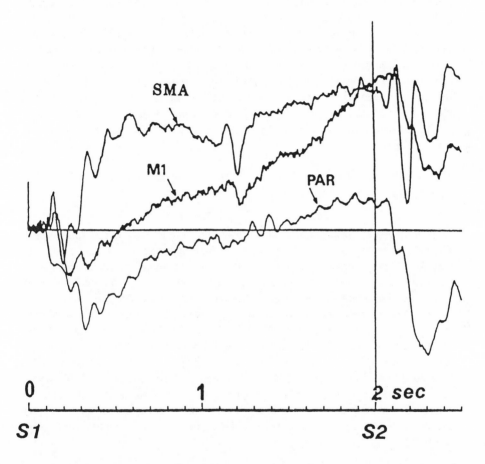

*Figure 3.3.* Potential changes recorded in an ''advance information'' reaction-time paradigm when the preparatory signal (S1) indicates the duration of the inter-press interval to be produced after the response signal (S2). The amplitude and time course of the electrophysiological data, recorded with the ''source derivation method,'' are quite different over three cortical regions: the supplementary motor area (SMA), the primary motor area (M1) and a parietal site (PAR). On the abscissa is the time (in seconds) elapsed during the ''preparatory period'' comprised between S1 and S2. On the ordinate is the amplitude of the local electric field (in microvolts/cm², negativity upward). The data are averaged over 12 subjects.

The supplementary motor area is close to the primary motor cortex, but situated more frontally. Our data, compared to some others, suggest that the function attributed to this area in motor preparatory processes may require further specification. The supplementary motor area seems to be selectively activated by situations where motor timing is involved. No differential activation appears on this area when the subject processes either the force of a motor response, or its direction, or the effector involved (Bonnet and MacKay, 1989). In contrast, the supplementary motor area seems to be involved in the processing of complex musical rhythms (Deecke and Lang, 1992). When lesioned, it yields deficits in the motor tasks which require subtle temporal coordinations between the two hands.

Finally, data analyzed *during* rather than *before* response execution revealed that when subjects produced a particular duration as accurately as possible, maximal activation occurred over frontal sites anterior to the supplementary motor area. The execution and the programming of response duration thus appear to elicit activation in distinct brain areas. Another study indicated that frontal sites are activated during the two phases of a temporal reproduction task, that is, during the presentation of the standard stimulus and during its reproduction by the subject (Casini and Macar, 1994).

## Conclusion

Behavioral and electrophysiological analyses converge toward similar conclusions. They indicate that, under conditions which make it necessary to accurately estimate durations, time processing may require attentional resources of controlled nature, and may activate specific brain sites. To test the generality of the mechanisms at work, it is important to demonstrate that they may be of identical nature when either motor timing or estimation of stimulus durations is concerned. Several data suggest that this may be the case. Keele, Pokorny, Corcos, and Ivry (1985) found a correlation between the performances of subjects involved in rhythmic motor tapping and in estimation of brief durations on different trials. In addition, under both conditions, high-level pianists yielded more accurate and less variable levels of performance than nonmusicians. On these lines, it is worth stressing that we found evidence for the existence of time processing mechanisms in quite distinct procedures involving either motor temporal regulations or perceptive time judgments.

## References

Block, R. A. (1989), Experiencing and remembering time: Affordances, context and cognition. In: *Time and Human Cognition*, ed. I. Levin & D. Zakay. Amsterdam: Elsevier Science.

———— (1990), Models of psychological time. In: *Cognitive Models of Psychological Time*, ed. R. Block. Hillsdale, NJ: Lawrence Erlbaum Associates, pp. 1–35.

———— (1992), Prospective and retrospective duration judgment: The role of information processing and memory. In: *Time, Action and Cognition: Towards Bridging the Gap*, ed. F. Macar, V. Pouthas, & W. J. Friedman. Dordrecht: Kluwer Academic, pp. 141–152.

Bonnel, A. M., Possamaï, C. A., & Schmidt, M. (1987), Early modulation of visual input: A study of attentional strategies. *Quart. J. Experiment. Psychol.*, 39A:757–776.

Bonnet, M., & Mackay, W. A. (1989), Changes in Contingent Negative Variation and reaction time related to precueing of direction and force of a forearm movement. *Brain, Behav. & Evolution*, 33:147–152.

Casini, L., & Macar, F. (1994), Behavioural and electrophysiological evidence for a specific processing of temporal information. *Psychologica Belgica*, 33:285–296.

———— ———— Grondin, S. (1992), Time estimation and attentional sharing. In: *Time, Action and Cognition: Towards Bridging the Gap*, ed. F. Macar, V. Pouthas, & W. J. Friedman. Dordrecht: Kluwer Academic, pp. 177–180.

Church, R. M., & Broadbent, H. A. (1990), Alternative representations of time, number and rate. *Cognition*, 37:55–81.

Deecke, L., & Lang, W. (1992), Supplementary motor functions: The SMA. *Proceedings of the Xth International Conference on Event-Related Potentials of the Brain*, Eger.

Dehaene, S., Changeux, J.-P., & Nadal, J.-P. (1987), Neural networks that learn temporal sequences by selection. *Proc. Nat. Acad. Sci. USA*, 84:2727–2731.

Desmond, J. E., & Moore, J. W. (1991), Single-unit activity during the classically conditioned rabbit nictitating membrane response. *Neurosci. Res.*, 10:260–279.

Fraisse, P. (1957), *Psychologie du Temps*. Paris: Presses Universitaires de France.

Gallistel, C. R. (1990), *The Organization of Learning*. Cambridge, MA: MIT Press.

Giard, M. H., Lavikainen, J., Reinikainen, K., Perrin, F., & Näätänen, R. (1992), Auditory sensory memory codes the physical attributes of acoustic stimuli in separate cortical areas. *Proceedings of the Xth International Conference on Event-Related Potentials of the Brain*, Eger.

Gray, C. M., König, P., Engel, A. K., & Singer, W. (1989), Oscillatory responses in cat visual cortex exhibit inter-columnar synchronization which reflects global stimulus properties. *Nature*, 338:334–337.

Grondin, S., & Macar, F. (1992), Dividing attention between temporal and nontemporal tasks: A Performance Operating Characteristics—POC—analysis. In: *Time, Action and Cognition: Towards Bridging the Gap*, ed. F. Macar, V. Pouthas, & W. J. Friedman. Dordrecht: Kluwer Academic, pp. 119–128.

Hjorth, B. (1975), An on-line transformation of EEG scalp potentials into orthogonal source derivations. *Electroencephalogr. & Clin. Neurophysiol.*, 39:526–530.

Kahneman, D. (1973), *Attention and Effort*. Englewood Cliffs, NJ: Prentice-Hall.

Keele, S. W. (1986), Motor control. In: *Handbook of Perception and Human Performance*, Vol. 2, ed. K. R. Boff, L. Kaufman, & J. P. Thomas. New York: John Wiley, pp. 1–60.

———— Pokorny, R. A., Corcos, D. M., & Ivry, R. (1985), Do perception and motor production share common timing mechanisms? *Acta Psycholog.*, 60:173–191.

Klapp, S. T., Wyatt, P. E., & Maclingo, W. (1974), Response programming in simple and choice reactions. *J. Motor Behav.*, 6:263–271.

Lackner, J. R. (1982), Alterations in resolution of temporal order after cerebral injury in man. *Experiment. Neurol.*, 75:501–509.

Lépine, D., Glencross, D., & Requin, J. (1989), Some experimental evidence for and against a parametric conception of movement programming. *J. Experiment. Psychol.*, 15:347–362.

Llinas, R. R. (1988), The intrinsic electrophysiological properties of mammalian neurons: Insights into central nervous system function. *Science*, 242:1654–1664.

Macar, F. (1980), *Le Temps: Perspectives psychophysiologiques*. Brussels: Mardaga.

———— Grondin, S., & Casini, L. (in press), Time estimation and attention-sharing. *Memory & Cognit.*

———— Vidal, F., & Bonnet, M. (1990), Laplacian derivations of CNV in time programming. In: *Psychophysiological Brain Research*, ed. C. H. M. Brunia, A. W. K. Gaillard, & A. Kok. Netherlands: Tilburg University Press, pp. 69–76.

Meck, W. H., & Church, R. M. (1984), Simultaneous temporal processing. *J. Experiment. Psychol.: Animal Behav. Processes*, 10:1–29.

Miall, R. C. (1992), Oscillators, predictions and time. In: *Time, Action and Cognition: Towards Bridging the Gap*, ed. F. Macar, V. Pouthas, & W. J. Friedman. Dordrecht: Kluwer Academic, pp. 215–227.

Michon, J. A. (1972), Processing of temporal information and the cognitive theory of time experience. In: *The Study of Time*, Vol. 1, ed. J. T. Fraser, F. C. Haber, & G. H. Müller. Heidelberg: Springer, pp. 242–258.

———— Jackson, J. L. (1984), Attentional effort and cognitive strategies in the processing of temporal information. In: *Timing and Time Perception*, ed. J. Gibbon & L. G. Allan. New York: New York Academy of Sciences, pp. 298–321.

Mills, L., & Rollman, G. B. (1980), Hemispheric asymmetry for auditory perception of temporal order. *Neuropsychol.*, 18:41–47.

Piaget, J. (1946), *Le Développement de la Notion de Temps chez l'Enfant*. Paris: Presses Universitaires de France.

Popov, N. A. (1944), Zur Frage der Bedeutung der Zeitfactors für die Auslegung der höchsten Nerventätigkeit. Prinzip de Zyclochronic. *Psycholgica Slovaca*, Vol. 1. Brateslava: Academia Scientarum et Artium Slovaca.

Rosenbaum, D. A. (1983), The movement precuing technique: Assumptions, applications and extensions. In: *Memory and Control in Motor Behavior*, ed. R. A. Magill. Amsterdam: North-Holland.

Thomas, E. A., & Weaver, W. B. (1975), Cognitive processing and time perception. *Percep. & Psychophysics*, 17:363–367.

Torras, C. (1986), Neural network model with rhythm-assimilation capacity. *Trans. Systems, Man, & Cybernetics*, SMC-16:680–693.

Vidal, F., Bonnet, M., & Macar, F. (1991), Programming response duration in a precueing reaction time paradigm. *J. Motor Behav.*, 4:226–234.

———— ———— ———— (1992), Can duration be a relevant dimension of motor programs? In: *Time, Action and Cognition: Towards Bridging the Gap*, ed. F. Macar, V. Pouthas, & W. J. Friedman. Dordrecht: Kluwer Academic, pp. 263–273.

———— Macar, F., & Bonnet, M. (submitted), Duration of a motor response or duration within a motor sequence: Are programming mechanisms equivalent?

4

# Chance and Time: From the Developing to the Functioning Brain

*Rémy Lestienne*

*Abstract*   Chance and timing seem to be in opposition, but a closer look at these notions unveils their kinship. This paper investigates how the biological development of the brain and its very functioning in tasks of computation make time coordination progressively emerge from initially chaotic elements. In a way, these examples extend to the domain of time the "order from noise" paradigm developed with regard to spatial organization.

## Chance and Causal Time: Contrasts and Kinships

At first glance, time stands in opposition to chance. Time in physics is identified as *the* parameter of causality. Chance, which is defined as that which escapes necessity, is therefore opposed to causal time. The search for causality in the world would be impossible without necessity.

But time should not be reduced simply to its causal interpretation, which has limits. First, the causal interpretation of this notion does not account for the direction of time, for its arrow. Second, nonseparability[1] as a general property of quantum systems limits the application of the paradigm of relativistic causality.

---

[1]Once they have interacted in the past, quantum systems, even separated by a large distance, in general cannot be described by separate quantum algorithms. This feature has observational consequences that show that the relativistic time order deals with the transmission of *information*—which cannot take place at a velocity greater than the speed of light—but does not deal with intrinsic properties of systems, that might be nonlocal; see, for instance Shimony (1988, p. 36).

Do the manifestations of chance in nature display an arrow? Mathematical chance, exemplified by tossing dice or playing cards, has no arrow in and of itself. Fortune can rebuild what it has once destroyed. But in statistical thermodynamics, entropy change, which is directed in time, has been given a probabilistic interpretation. The law of increase of entropy appears to be the concrete manifestation of randomizing processes in nature. This contradiction between natural manifestations of chance and its mathematical modeling is usually surmounted by giving entropy a subjective status: increase of entropy is accounted for by mechanisms of coarse graining, that in turn are linked to the limited precision of our instruments and to our limited ability to deal with numerical results (so that in practice *we* always define the granular structure of our investigations).

In my view, two stages should be distinguished in the usual examples of the increase of entropy. The first stage is linked to the limited precision of our measuring processes and is subjective. The second is linked to a universal sink of correlations, or a universal "ogre" for negentropy, probably cosmologically founded and perhaps linked to the expansion of the universe. This second stage is the only one to be objective, but it justifies giving an objective status to entropy. This is the thesis I proposed and defended in *Les Fils du Temps* (1990, 1994). It grants the existence of a kindred relationship between chance and time.

In my recent research I investigated processes in living organisms, looking for examples where stochasticity and timing might play an interwoven role. The building and functioning of the central nervous system (CNS) seemed to be a particularly convincing domain.

## The Role of Chance in the Central Nervous System

It has been estimated that the human brain contains about 100 billion nerve cells. Many of the cells receive synaptic contacts from 1000 cells or more. There are therefore about $10^{14}$ synaptic contacts. The precise location of these contacts, or even the specification of the neurons among which such contacts should be built, cannot be prescribed in the genome, since the DNA in its entirety, with its 50,000 to 100,000 genes (even if they were devoted entirely to specifying the construction of the CNS, which they are not) would not suffice by a large margin to serve this purpose. As a matter of fact, only the general structure of the CNS in man seems to be genetically determined. The precise details of the connections are formed partially by chance and partially through interaction with the environment. At several stages in its construction and in its functioning, the CNS of humans and of other advanced organisms uses the principles of Darwinian competition.

### *Detailed Blueprints of the Brain Are Not Encoded in the Genome*

A striking proof that the details of the nervous system are not encoded in the genome is provided by a microscopic inspection of the CNS of the small shrimp of the genus *daphnia*. These animals reproduce by way of parthenogenesis, that is, embryos develop from female eggs without a mixing of their genetic heritage with the male genome. Thus, members of the same family have exactly the same genome, like twins. Figure 4.1 presents examples of homologous nerve cells in four daphnies. In such simple animals, the number of nerve cells and the general architecture of the nervous system is strictly preserved from one individual to the next. Although the four individuals shown here bear exactly the same genome, homologous nerve cells clearly differ in the precise details of the processes of each neuron. In higher animals and especially in man, structural differences between twins are much more pronounced. The architecture of every neuron, the number of neurons, and even the details of the cerebral gyri differ.

## Swarming and Degeneration of Nerve Processes

In the early stages of the development of the CNS of embryos, the nerve cells multiply and tend to make contacts with a much greater number of cells than would be necessary for its correct functioning at an adult stage. The current theory is that this swarming is controlled by chemical signals directing the growth cones of nerve processes toward their possible targets. At later stages both the number of cells and the number of synaptic contacts between cells decline and stabilize at their normal levels.

Figure 4.2 shows the variation of the number of CNS cells during the development of chicken embryos, as a function of age. The graph shows that in the spine, 40 percent of the newly differentiated neurons die within a few days. Figure 4.3 draws our attention to the fate of synaptic contacts, in the case of the frontal cortex of monkeys. It displays the density of synapses in this cortical area as a function of age. It is not known whether, during the plateau between birth and puberty, there is a physical rapid turnover of synapses, or only a progressive fine-tuning of their efficiency (also called synaptic weights).

## Chance in Elementary Neural Events

Swarming and degeneration are not the only times when chance plays a role in the CNS. In fact, it plays a role throughout the hierarchy of neural events, beginning with the release of chemical vesicles of neuromediators in the synaptic cleft between neurons that are going to pass information from one to the other. Figure 4.4 displays inhibitory postsynaptic potential achieved by excitation of a given motor interneuron contacting the giant Mauthner cell of *teleost*, an infraclass of bony fishes (the Mauthner cell triggers an escape reaction). The histogram obtained in a series of identical electrical excitations of the interneuron fiber clearly shows that variable potentials are achieved, depending upon the number of terminals that have released a vesicle of neurotransmitter. In this case the best fit for the probability $P_j$ of releasing $j$ vesicles of neuromediator is obtained for a binomial distribution of the

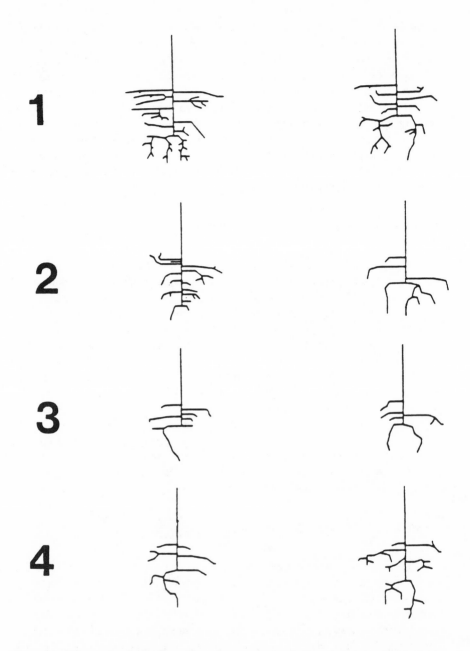

*Figure 4.1.* The axonic arborization of homologous neurons in four daphnies, with identical genomes. Left and right drawings show the configurations of similar neurons located on left side and right side of the animal's body. The variations of the axonic trees are more important from an individual to the other than from one side to the other of the same individual (after Macagno, Lopresti, and Leventhal, 1973).

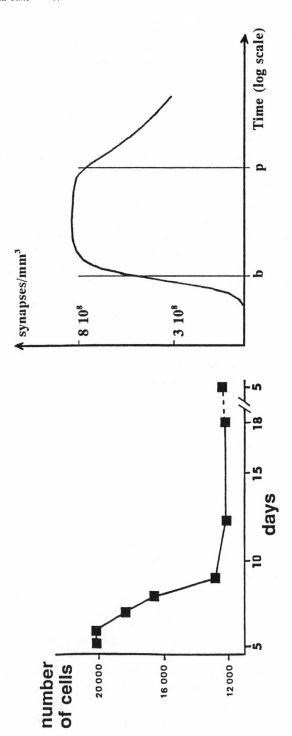

*Figure 4.2.* Total number of motoneurons in a lateral motor column of the spine of chicken embryos as a function of embryonic age (after Hamburger, 1975).

*Figure 4.3.* Density of synapses (functional contacts between neurons) in the frontal cortex of monkeys, as a function of age (logarithmic scale). *b* denotes birth age, and *p* puberty (J. P. Bourgeois, Institut Pasteur, Paris, private communication).

type $P_j = \binom{n}{j} p^j (1 - p)^{n-j}$, where $n = 6$ and $p = .46$. This result suggests that six contacts exist, each of which has the probability $p$ of expelling a neuromediator vesicle under excitation. Histology confirms the $n = 6$ value.

## *Neural Darwinism*

Neural Darwinism, a theory suggested among others by Edelman (1987), proposes that a population of neurons in a CNS is affected by the classical selective forces of organic evolution: competition and selection. Such a competition plays a role not only in the hard wiring at the development stage, but also in the enlistment of neural groups in short-lived tasks and the constituency of short-lived networks. In such a competition, usage or nonusage play a major role. It was suggested a long time ago by Donald Hebb (1949) that synaptic efficiencies between two neurons may be modified by a process of synaptic weight correction: if the firing of the second cell is successfully achieved when the first cell is fired, the synaptic weight of the synapses joining them is increased and the two cells tend more and more to fire synchronously. If, on the contrary, the firing of the first cell does not succeed in triggering the firing of the second, the synaptic weight is decreased and eventually the contact becomes totally ineffective. In a group of nearby neurons of similar neurophysiological characteristics, the Hebb's mechanism acts as a selection mechanism, leading to the constituency of highly transient but coherent assemblies, whose boundaries are rapidly changing with time and with the computational exigencies of the brain.

## Early Uses of Timing in the Developing Brain

### *Spontaneous Activity in Embryonic Visual Systems and the Wiring of Binocular Inputs*

In a series of beautiful experiments, Carla Shatz and her colleagues at Stanford (1992) were able to show that the spontaneous activity of the visual system in embryos of cats (and newborn ferrets) is necessary for a correct wiring and development of the visual system. In very early embryonic stages, the ganglion cells of the retina begin to fire spontaneously, in bursts of a few discharges separated by silences that can be as long as a minute or so. Such discharges come not at random, but are more or less periodic, and are moreover correlated in space: the entire retina is swept by waves of discharges traveling in apparently random directions, changing from one wave to the other (Figure 4.5). When the histogram of the time differences (or the correlogram) between discharges of two ganglionic cells at a given distance $d$ of each other is plotted, one finds that close ganglionic cells tend to fire in synchronism (with a dispersion of about 1 second), whereas distant cells have a

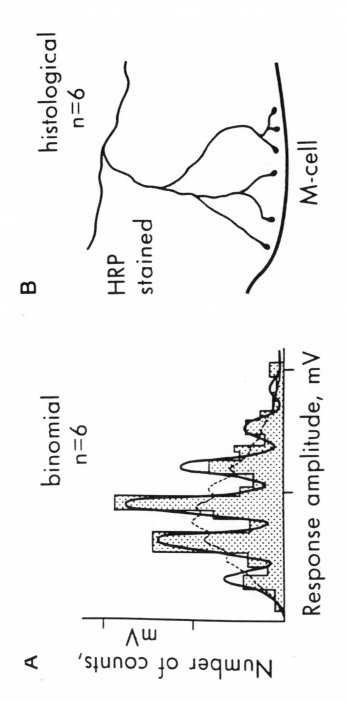

*Figure 4.4.* Following successive stimulations of a physiologically identified interneuron (rate: 1 per s), the resultant amplitude histogram of fluctuating unitary inhibitory postsynaptic potentials (shaded) was analyzed with a computer program that, taking noise into consideration, gave the best fits based upon the theoretical Poisson (dashed line) and binomial (continuous curve) distributions. Obviously, the latter provided a better approximation, indicating that variations in this case should be attributed to fluctuations in the number of terminal knob releasing vesicles of neuromediators in an all or none mode, rather than to fluctuations in the contribution from each knob. The six peaks correspond to the binomial term $n$, which defines the number of quanta. Histological studies (*B*) confirmed that there were indeed six terminal boutons between this interneuron and the Mauthner cell of a *teleost* here studied (after D. S. Faber and H. Korn, 1982).

much more dispersed correlogram (Figure 4.6). The mechanism of this loose temporal correlation of the firing of neighboring cells is unknown; it might imply $Ca^{++}$ or a neurotransmitter diffusion in the space between ganglionic cells. One should note that the correlogram of spontaneous activity in the retina (Figure 4.6A) does not shrink with time in the course of maturation, implying that the mechanism involved cannot be tuned up.

This spontaneous, correlated activity is necessary for the correct development of the visual system, in the sense that if the spontaneous discharges of the retina are blocked by injecting an appropriate drug into the eye (tetrodotoxin), then the normal segregation of cells in the lateral geniculate nucleus (LGN) into right and left eye domains does not take place.[2] The next step of this research will be to show that not only spontaneous activity but correlated spontaneous activity is necessary for the segregation of binocular inputs in the LGN, by artificially exciting cells throughout the retina, in an erratic way, and checking whether ocular segregation of cells is thus impeded.

## *Further Use of Temporally Correlated Activity in the Development of the Higher Centers of Vision*

Visual activity is also known to be necessary for the formation of binocular stripes in the visual cortex. In cats whose eyelids are sewed together at birth, this segregation is gravely impaired. Young animals surgically deprived of a visual cortex usually develop aberrant visual systems at the expense of other specialized sensory systems.

In the elaboration of thalamocortical circuits, an intermediary stage is passed through, where a neuronal "scaffold" is built, in the form of connections between the thalamus and the cortical subplate. Subplate neurons mature to function in a complex microcircuit in the fetal brain. It is not yet proven that a *timely correlated firing* of these neurons is necessary for the subsequent correct wiring of thalamocortical and corticothalamic pathways, but it has been proven that their firing is necessary for the adult pattern of connections to take place.

## Synchronism Is Enlisted in the Processing of Sensory Data

### *Coherent Oscillations in the Visual System*

Different features of a perceived object, such as color, form, or brightness, are analyzed in different and separate cortical areas. About 10 years ago, Christoph

---

[2]The two lateral geniculate nuclei, located in the thalamus, are relay centers in the pathway from the eye to the visual cortex. Each of them receives fibers coming from the two eyes, but corresponding only to the contralateral visual field. Moreover, they are normally organized in laminae: input coming

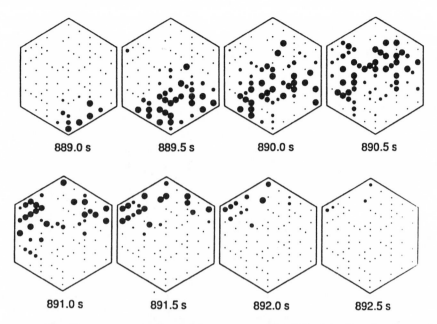

889.0 s          889.5 s          890.0 s          890.5 s

891.0 s          891.5 s          892.0 s          892.5 s

*Figure 4.5.* Time course of spike activity over an array of 61 electrodes spread over the retina of a neonatal (P5) ferret, during a burst covering the time interval from 889 to 893 s. Successive frames show the averaged firing rate over successive 0.5 s intervals. Each neuron recorded is represented with a dot at its approximate spatial location, and with a dot area proportional to the average firing rate of the neuron during the repetitive 0.5 s interval (after Meister, Wong, Baylor, and Shatz, 1991).

Von der Malsburg (1981) conjectured that synchronism and precise coherent firing of cell assemblies might be the way in which the brain resolves ambiguities as to which feature should be associated with which object in the visual field. In 1989, examples of such coherent firing began to be observed in multiunit recordings in the visual cortical area of anesthetized cats. Figure 4.7 gives an example of such observations. Multiunit activities were recorded from two sites in the cortical visual area of a cat, separated by 7 millimeters. The left column of figures displays the receptive fields, the preferred orientation and (circle) center of the visual field. The right column shows the cross-correlograms between the two groups of cells. In *D*, using a single continuous bar as a stimulus, clear coherent oscillations were obtained. When the continuity of the bar was suppressed (*E*), the synchronization between the two groups became weaker, and totally disappeared when the bars moved in opposite directions (*F*). This change of the stimulus configuration affected neither the strength nor the oscillatory nature of the responses.

Coherent oscillations are characterized by a latency time of 30 to 50 milliseconds, a precision in phase locking better than 3 milliseconds, and a duration of 50

---

from the contralateral eye contacts neurons in laminae A, C, and C2. Fibers coming from the ipsilateral eye make contacts with neurons in laminae A1 and C1.

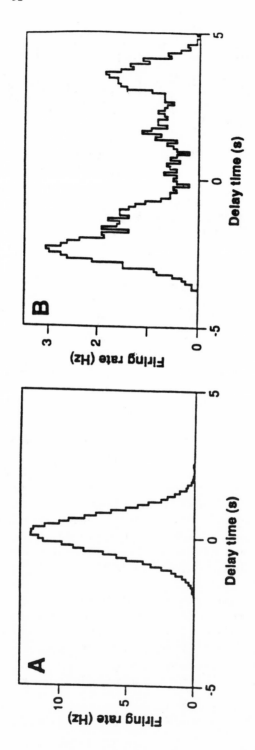

*Figure 4.6.* A Correlation between the spike trains of two neurons of the retina of the same ferret, located about 40 μm apart. *B* Similar correlogram obtained for two neurons in the same retina as *A*, but located 370 μm apart (after Meister et al., 1991).

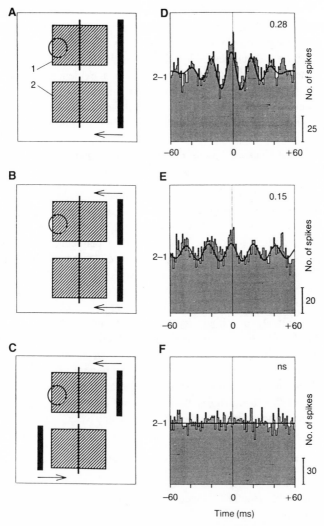

*Figure 4.7.* Evidence for coherence of oscillations in neuronal populations stimulated by feature related visual stimulations in an anesthetized cat. Multiunit activities were recorded from two sites in a particular area of the cat's visual cortex, separated by 7 mm. The two cells groups preferred vertical orientations of the bars used as visual stimulation. *A, B, C*: plots of the receptive fields and sketch of the used stimulation: a long continuous light bar moving across both fields in *A*, two independent light bars moving in the same direction in *B*, and the same bars moving in opposite directions in *C, D, E,* and *F* display the cross-correlogram between the two recording electrodes obtained with each stimulus paradigm. Using the long light bar, the two oscillatory responses were clearly synchronized. When the continuity of the stimulus was interrupted, the synchronization became weaker (*E*), and totally disappeared when the motions of the stimuli were incoherent (*F*). The graph superimposed on each stimulus configuration represents a Gabor function that was fitted to the data to assess the strength of the modulation, and the number in the upper corner indicates the relative modulation amplitude (ns: nonsignificant) (after Engel, Köning, Kreiter, Schillen, and Singer, 1992).

to 75 milliseconds. Theoretically, it is possible to account for such coherent oscillations if the two groups of cells have both feedforward and feedback connections,
with highly nonlinear dynamics of synaptic weights and conduction delay times
less than T/4, the period of the oscillations. Anatomical evidence for such reciprocal
connections abounds in the superficial layers of the cortex. Further studies seem
necessary, however, to ensure that coherent oscillations are a general phenomenon
in perceptual computation, and that their aim is indeed to bind features.

### Synchronism in the Hearing System at the Cochlear Level

Frederic Berthommier (1990) has developed a plausible model for auditory processing in the nerve fibers and in the cochlear nucleus. Information carried by the
auditory nerve has a temporal structure driven by the auditory stimulation. A pure
tone produces a maximum of activity in a frequency dependent place in the cochlea.
Each auditory fiber is similar to a bandpass filter endowed with a characteristic
frequency (CF). Below 4 kilohertz, spike trains are synchronized and phase locked
on the stimulus frequency. However, a given fiber does not convey spikes that are
absolutely phase locked. Filtering is probably obtained through correlating, in time,
a population of afferent nerve fibers, by performing computations of the type $C_{i,j}(t)$
$= 1/T \int P_i(\theta)P_j(\theta) \, d\theta$ where $C_{i,j}$ is the common output between two fibers $i$ and $j$
and $P_i(\theta)$ or $P_j(\theta)$ is the probability of an input spike from fiber $i$ (or $j$) at time $\theta$.
As a consequence, synchronism of discharges in a bundle of nerve fibers would
allow a very large increase in the selectivity of the auditory system. The system is
further refined in order to maximize the transported information, coding, and con-
veying information about features other than frequency.

### Synchronism and Delay Lines in the Hearing System of the Owl

The owl localizes the azimuthal direction of its prey by measuring the interaural
delay time of the sound stimuli.

The general scheme of the decoding system is shown on the left in Figure 4.8.
Neurons at the appropriate neural nucleus (the so-called nucleus laminaris) act as
coincidence detectors. Delays in the arrival of sounds at the right and left ears are
compensated by the conduction delays from one side of the nucleus laminaris to
the other. By means of slow conduction velocities in these small fibers, delays up
to 180 μ seconds, corresponding to the time it takes sound to travel over 6 centime-
ters in the air (more than the interaural distance), are easily achieved.

Further filtering occurs at higher neuronal levels, to filter out noise and remove
phase ambiguities (in the case of pure tones there is no possibility of deciding
whether the detected delay should be shifted by an integer number of periods of
the stimulus). Finally, a very high selectivity is obtained, endowing the owl with a
very precise means of locating its prey's azimuthal direction in the darkness of
the night.

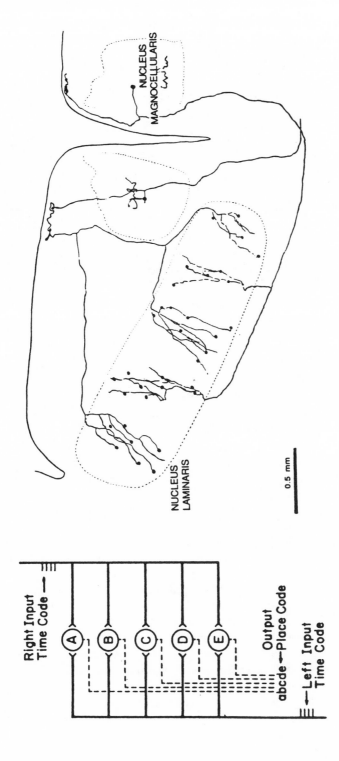

*Figure 4.8.* Left figure: model of neuronal circuit for measuring and encoding the interaural time delays in the owl. Excitations coming from the right and left ears impinge on a column of neurons acting as coincidence detectors, and fires when synchronous signals come from both inputs. Due to the geometrical arrangement of input axons, the firing of cells *A*, *B*, etc., correspond to increasing algebraic delays between the time of arrival of excitations coming from each ear. Right figure shows the actual realization of the model in the owl's nucleus laminaris. Innervation from the ipsilateral ear come from the dorsal boundary of the nucleus, while those coming from the contralateral ear come from its ventral boundary. Actual innervation paths of two neurons located in the ipsi and contra magnocelularis nuclei are shown, as revealed by HRP injections (after Konishi, Takahashi, Wagner, Sullivan, and Carr, 1988).

## Sychronism and Delay Lines in the Sonar of the Bat

The bat emits complex sounds characterized by a constant frequency pulse (CF) over a few tens of ms, followed by a brief frequency modulated signal (FM) (see Figure 4.9). Both sounds have four detectable harmonics (the second harmonic is the most intense), and the same is true for the echoes. Bats use mostly $CF_1/CF_n$ doppler shifts to compute velocities of their targets, and $FM_1/FM_n$ delays, that is, the delay between the emitted pulse of the first harmonic of the frequency modulated phase of the squeal and the returned echo of the nth harmonic, in order to compute the ranges. Here, delays may be as long as hundreds of ms, limited only by the range of the bat's sonar. Conduction delay lines are created in the inferior colliculus, up to 7 milliseconds. Further delay lines are built in the medial geniculate body, where the $FM_1/FM_n$ neurons are located, by use of intermediary inhibitory neurons. The $FM_1/FM_n$ neurons act as coincidence detectors, in a manner similar to that which has been observed in the owl. The precision with which the system, after several additional steps of filtering and comparing results from various sets of $FM_1/FM_n$ neurons, is able to discriminate differences in delays is amazing: it is on the order or less than a *tenth of a millisecond*, corresponding to a precision in the location of the target that is better than 2 cm.

## Replicating Patterns in the LGN and the Visual Cortex of the Cat

As early as 1969 Bernard Strehler suggested that fast plasticity of synaptic weights might allow the existence of quite sophisticated and very precise mechanisms of time coding in the brain, using precisely replicating temporal patterns of intervals in the spike trains. Thus, neurons could be trained and specialized in the production of specific patterns of pulses, triplets, for instance, of precisely determined constituent intervals. A similar anatomical configuration of another neuron downstream in the information flow (identified by a Darwinian type of selection mechanism as suggested above) might conversely be trained and become specialized for a while in the detection of such precisely timed triplets.

Precisely replicating triplets, with a precision on the order of 0.1 ms, have been detected in spike trains of the visual cortical areas of monkeys (Strehler and Lestienne, 1986) and cats, as well as in the LGN of cats. Experimental tests tend to show that the ability of an LGN neuron to produce such characteristic repeated patterns is gravely impaired if the return connection from cortex to thalamus is pharmacologically silenced (Beaux, Lestienne, Imbert, and Grandjean, 1992). The possibility that corticothalamic loops are intended to "carve" temporal motives in the LGN discharges is supported by consideration of the anatomical disposition of synaptic corticothalamic contacts. These are grouped in a series of distal clusters on the LGN neuron dendrites.

Although the coding function of these replicating patterns has not yet been demonstrated, it is worth noting that these fine temporal structures were found in

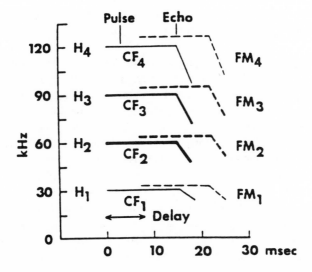

*Figure 4.9.* Sketch of the biosonar pulse of the mustached bat. Each pulse contains four harmonics (indicated by subscripts). Echoes are delayed by a period needed by the sound to travel from the bat to the target and back; the frequency of the echo is Doppler shifted according to the relative velocity between the bat and the target (after Suga, 1990b).

adult animals, but not in kittens. This is consistent with the idea of a progressive mobilization during the maturation of the CNS, step by step and level by level, of the various mechanisms of temporal coding here evoked, with a precision increasing from 1 s up to a tenth of a ms or less. It might be that this parallels the psychological trait showing that infants gradually acquire a more precise perception of time over the course of their development.

## Conclusion

Chance is present and plays a role both in the development of the brain according to the principles of neural Darwinism and in its functioning as a web of local neural networks. The number and position of neural processes (dendritic and axonal arborizations), of synaptic contacts, of released neurotransmitter vesicles, the temporal patterns of discharges at single neurons, and the contour of local networks all reveal a great deal of variation.

But nonstochastic temporal mechanisms are also present. A correct wiring of the brain is impossible without a precisely temporally coordinated endogenous activity of the neurons, as has been clearly demonstrated in the case of the visual system. Synchronization mechanisms are proposed as necessary ingredients of the functioning of the hearing system. Coherent oscillations in the firing of populations of neurons have been documented in recent years, and their role in the integration of multimodal perceptions advocated. Precisely replicating patterns of discharges have been detected in the infragranular layers of the visual cortex and thalamic nuclei.

Of outstanding interest is the progressive turning on and fine tuning of temporal mechanisms at work in the brain from embryonic to adult stages, at least in mammals. While cross-correlation analyses of spontaneous discharges in the embryonic visual system have a typical width of 1 second, other temporal mechanisms of ever increasing precision seem to be progressively activated, and reach, in some examples (the binaural system of orientation of the owl, the echolocation system of the bat, the precisely replicating patterns in the visual cortex) precisions that are quite extraordinary for a machinery built up with living neurons (about 0.1 ms). Such experimental facts seem to parallel the long known psychological traits of human beings, where infants have a very loose sense of time and poor capacities for temporal coordination. In inanimate nature time order (causality) and time arrow (entropy increase) build themselves from a chaotic substrate of molecular agitation. In living beings also, time emerges from chance and randomness.

## References

Beaux, J. C., Lestienne, R., Imbert, M., & Grandjean, B. (1992), Modulations corticales de la structure temporelle fine des trains d'impulsions dans le corps genouillé latéral dorsal du chat. *CR Acad. Sci. Paris*, 3:31–36.

Berthommier, F. (1990), Neurons sensitive to time structured inputs. Paper presented at an International Neural Network Conference, Paris.

Edelman, G. (1987), *Neural Darwinism*. New York: Basic Books.

Engel, A. K., Köning, P., Kreiter, A. K., Schillen, T. B., & Singer, W. (1992), Temporal coding in the visual cortex: New vistas on integration in the nervous system. *Trends Neurosci.*, 15:218–225.

Faber, D. S., & Korn, H. (1982), Binary mode of transmitter release at central synapses. *Trends Neurosci.*, 5:157–159.

Hamburger, V. (1975), Cell death in the development of the lateral motor column of the chick embryo. *J. Comp. Neurol.*, 160:535–546.

Hebb, D. (1949), *The Organization of Behavior*. New York: Wiley.

Konishi, M., Takahashi, T. T., Wagner, H., Sullivan, W. E., & Carr, C. E. (1988), Neurophysiological and anatomical substrates of sound localization in the owl. In: *Auditory Function—Neurobiological Bases of Hearing*, ed. G. M. Edelman, W. E. Gall, & W. M. Cowan. New York: Wiley, pp. 721–745.

Lestienne, R. (1990), *Les Fils du Temps*. Paris: Les Presses du CNRS.

——— (1994), *The Children of Time*. Urbana, IL: University of Illinois Press.

Macagno, E., Lopresti, U., & Leventhal, C. (1973), Structural development of neuronal connections in isogenic organisms: Variations and similarities in the optic system of *Daphnia magna*. *Proc. Nat. Acad. Sci. USA*, 70:57–61.

Meister, M., Wong, R. O. L., Baylor, D. A., & Shatz, C. J. (1991), Synchronous burst of action potentials in ganglion cells of the developing mammalian retina. *Science*, 252:939–943.

Shatz, C. J. (1992), The developing brain. *Sci. Amer.* (special issue, "Mind and Brain"), Sept., pp. 61–67.

Shimony, A. (1988), The reality of the quantum world. *Sci. Amer.*, 258:36–43.

Strehler, B. L. (1969), Information handling in the nervous system: An analogy to molecular-genetic coder-decoder mechanisms. *Perspect. Biol. Med.*, 12:584–612.

———— Lestienne, R. (1986), Evidence on precise time-coded symbols and memory of patterns in monkey cortical neuronal spike trains. *PNAS,* 83:9812–9816.

Suga, N. (1990a), Biosonar and neural computation in bats. *Sci. Amer.,* 262:60–66.

———— (1990b), Cortical computational maps for auditory imaging. *Neural Networks,* 3:3–21.

Von der Malsburg, C. (1981), *The Correlation Theory of Brain Function.* Internal Report 81-2. Göttingen, Germany: Max Planck Institute for Biophysical Chemistry, Department of Neurobiology.

5

# Psychological Time and Memory Systems of the Brain

*Richard A. Block, Ph.D.*

*Abstract* Psychological time involves four partially dissociable memory systems of the brain. Procedural memory, subserved mainly by the cerebellum, controls timing of learned movements. Semantic memory, subserved by several areas of the cerebral cortex, concerns linguistic and conceptual information, presumably including general knowledge concerning time. Working memory, which involves the functioning of the dorsolateral prefrontal cortex, maintains a representation of current temporal contextual information. Episodic memory, which requires the functioning of the hippocampus and other medial temporal lobe structures, is necessary to encode long-term memory for personal experiences, including temporal information about them.

People continually engage in one or more of several different kinds of temporal activities, including controlling movement timing, expressing general temporal knowledge, representing present events, and remembering past durations (Block, 1979, 1990a). Tulving (1972, 1985, 1991) distinguished several memory systems, each differing in "its brain mechanisms, [the] type of information it handles, and the principles of its operations" (Tulving, 1991, p. 10). At least four partially dissociable memory systems are involved to greater or lesser extents in different temporal experiences, judgments, and behaviors.[1] These memory systems are: (1)

---

*Acknowledgments.* Fusako Matsui and Lynn Maulding provided able assistance. Several people made helpful comments on earlier versions of this article. I am especially indebted to John Moore, Franck Vidal, and Françoise Macar. I also thank Michael Babcock, Hannes Eisler, Wesley Lynch, Robert Patterson, Viviane Pouthas, and Albert Rabin.

[1] I will not discuss the sensory processes involved in experiencing and judging simultaneity and successiveness of stimuli presented for brief durations and at very short intervals (Block and Patterson, in press; Carr, 1993). These processes may involve a fifth memory system, which Tulving (1991) called the *perceptual representation system.*

procedural memory, which contains information vital to the performance of learned movements and relatively automatic procedures; (2) semantic memory, which processes information about concepts, facts, linguistic expressions, and so on; (3) working, or short-term, memory, which contains highly accessible information about present (or very recently past) events; and (4) episodic memory, which contains information about past personal experiences, including recency, order, and duration information.

Evidence from neuropsychological, psychopharmacological, and cognitive neuroscience research suggests that anatomically and functionally separate, yet interconnected, brain areas subserve the functioning of these different memory systems that are involved in time-related tasks. Figure 5.1 shows the location of some brain areas or structures that are critically involved in psychological time. Some of the factors that influence different temporal functions (Block, 1989) may reflect the workings of neural networks, or information-processing modules, in these and other anatomically related areas. Each brain structure normally interacts directly with several other structures and indirectly with the rest of the brain. Although the present review does not describe these multiple interconnections, the interested reader may find neuroanatomical details in several sources (Kolb and Wishaw, 1990; Kandel, Schwartz, and Jessell, 1991).

The present article reviews evidence and theories on these memory systems and the brain structures that are critically involved in them. (Other recent reviews that focus on many of the issues described herein include those by Weiskrantz [1987], Melges [1989], and Kosslyn [1992].) I present a particular view on several unresolved issues, but I also discuss lingering theoretical disagreements. The review focuses mainly on evidence from studies of brain damage, neurotransmitters and drug effects, and physiological recording. In interpreting the findings from any source of evidence concerning brain function, caution must be exercised. Localized damage to a particular brain region or structure may not influence performance on a particular task, or it may produce impaired performance. This kind of evidence suggests that the damaged region or structure either does or does not participate in the interacting network of brain activity underlying task performance. Because the brain functions as a system containing highly interconnected structures, or modules, damage to any particular region or structure can also affect functioning of other regions or structures. If damage to a particular structure leads to impaired performance on a task, it may only be concluded that the structure is a critical component of the information processing required for task performance. If damage to a structure does not lead to impaired performance, then either the structure is not normally involved in performance of that particular task or some other brain structures can be deployed to perform the task. Similar caution must be exercised in interpreting studies of neurotransmitters and drug effects. A particular brain region normally contains several neurotransmitters, each of which is also distributed throughout various interconnecting brain regions. Most physiological recording techniques have a more basic source of limitation. For technical reasons that need not concern us here, these methods all suffer from limitations in ability to localize events, either in brain space (e.g., event-related potentials) or in time (e.g., positron emission tomography [PET]).

Although one may legitimately question any particular source of evidence, using any particular methodology, converging evidence from studies using different methodologies suggests a reasonably coherent account. The present account discusses certain brain regions and structures that are most critically involved in psychological time. Future studies will need to specify more precisely the exact information processing that occurs in each region or structure in order to subserve performance of psychological functions related to time or timing.

## Procedural Memory

In some temporal tasks, such as those in which a person must control the sequencing and production of movements, the supplementary motor area, premotor cortex, primary motor cortex, and related cortical areas subserve the planning and control of movement timing (Ghez, 1991b; Kosslyn, 1992; Vidal, Bonnet, and Macar, 1992). It appears that as movements become more well learned, control of movement sequencing and timing shifts to the cerebellum (see Figure 5.1). In particular, the lateral portion of the cerebellum (the so-called *cerebrocerebellum)* subserves procedural memory for timing relative intervals in highly automatized movement sequences. At present, it is unclear whether the cerebellum stores these procedural memories or whether other structures that are intimately connected with the cerebellum store the information (Leiner, Leiner, and Dow, 1991). The lateral cerebellum may then transmit the information on movement sequencing and timing to the premotor and motor cortical areas for execution. In addition, the spinocerebellum (i.e., the intermediate cerebellar hemisphere) controls some movements, and it also apparently corrects or adjusts movement sequences in progress (for a review, see Ghez, [1991a]). The extent to which the cerebral system or the cerebellar system primarily controls movement sequence timing may depend on the timing requirements of the task and the degree of automaticity of the motor program involved.

Evidence that the lateral cerebellum plays a critical role in movement timing which is dissociable from the role of the semantic, working, and episodic memory systems comes mostly from studies of brain damage. Lateral cerebellar damage does not affect other memory functions, but it produces deficits in timing movement execution and perhaps also in performing temporally predictive computations in other behavioral, perceptual, and cognitive situations (Ivry, Keele, and Diener, 1988; Ivry and Keele, 1989; Leiner et al., 1991). In contrast, damage to specialized cerebral structures that produce various types of amnesia does not impair procedural memory functions such as movement timing (Mayes, 1988).

## Semantic Memory

Several areas of both cerebral cortices, but especially of the left hemisphere, subserve language. These include two classical areas: Broca's area, which is located

Dorsolateral
prefrontal cortex
(working memory)

Hippocampus
(episodic memory)

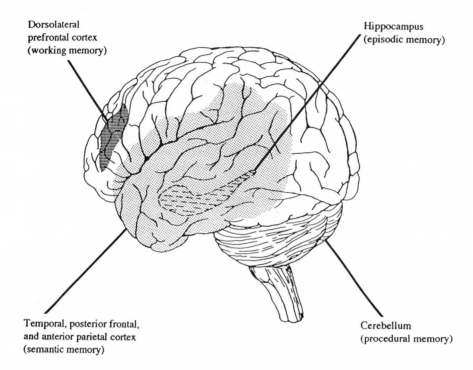

Temporal, posterior frontal,
and anterior parietal cortex
(semantic memory)

Cerebellum
(procedural memory)

*Figure 5.1.* Lateral view of the human brain (left hemisphere) showing the approximate location of several areas that subserve aspects of psychological time. See text for further explanation, and see other treatments (e.g., Kandel, Schwartz, and Jessell, 1991; Kolb and Wishaw, 1990) for additional neuroanatomical details. Clockwise from lower right: (1) The cerebellum, which is involved in procedural memory, is a complex subcortical structure. (2) Widespread areas of the temporal cortex, posterior inferior frontal cortex, and anterior inferior partietal cortex, especially of the left hemisphere, subserve semantic memory; corresponding areas of the right hemisphere are apparently not as heavily involved in semantic memory. (3) The dorsolateral prefrontal cortex of both cerebral hemispheres subserves working memory. (4) The hippocampus, which along with nearby structures is critically involved in episodic memory, is a subcortical structure underlying the temporal cortex of the left and right hemispheres in the approximate location depicted here. Only the left-hemisphere half of the hippocampus is depicted.

in the left inferior frontal lobe and is involved in linguistic production; and Wernicke's area, which is located in the left superior posterior temporal lobe and is involved in linguistic comprehension. It is now clear that many nearby cortical areas subserve various aspects of semantic memory, including widespread areas of the temporal, parietal, and frontal cortices (see Figure 5.1).

Studies of people with brain damage suggest that different memory systems may be selectively impaired, producing different kinds of amnesia (Mayes, 1988). I will call two main kinds *semantic amnesia* and *episodic amnesia* (Nielsen, 1958). Damage to any of several regions in the left cerebral hemisphere typically produces semantic amnesia, which may be quite specific (Weiskrantz, 1987). For example, a man known as M.D. suffered cerebral damage as a result of stroke (Hart, Berndt, and Caramazza, 1985). Afterwards, he had considerable difficulty naming fruits and vegetables, such as a visually presented apple. He could easily name other kinds of objects, however, and he could categorize pictures of fruits and vegetables that he could not name. Studies of other patients reveal that parts of both cerebral hemispheres subserve linguistic and conceptual knowledge in other domains, such as about color (Damasio, Damasio, Tranel, and Brandt, 1990; Damasio and Damasio, 1992).

Studies using PET scanning reveal that when normal people perform certain semantic-processing tasks, neural activity increases in the posterior left-hemisphere frontal lobe (Posner, Petersen, Fox, and Raichle, 1988). However, Tulving (1989) reported preliminary evidence suggesting that regional cerebral blood flow (another presumed indicator of neural activity) is relatively greater in posterior regions of the cerebral cortex during retrieval of impersonal semantic information and is relatively greater in anterior regions during recollection of personally experienced episodes. The apparent discrepancy between these two sets of findings may be attributable to task differences, especially since several brain regions apparently subserve different aspects of semantic memory.

As involved in psychological time, the semantic system contains information about temporal units, concepts, and linguistic expressions. For example, people know that time seems to pass slowly during a boring experience (Block, Saggau, and Nickol, 1983–1984). Semantic memory also enables a person to use temporal metaphors, such as ''the river of time'' (Jackson and Michon, 1992; Michon, 1990). The semantic system, along with the working and episodic memory systems, also apparently mediates temporal perspective, or one's conceptualization of past, present, and future (Williams, Medwedeff, and Haban, 1989; see below). Research has not yet identified the specific cortical areas and processes that subserve linguistic and conceptual knowledge about time, but they are probably similar to those involved in other domains of knowledge such as color.

## Working Memory

Evidence on the working memory system and its role in psychological time comes from several sources. Here I discuss mainly studies of brain damage, neurotransmitters and drug effects, and electrophysiological recording.

### Brain Damage

Patients with frontal lobe damage usually show little or no impairment in remembering that a particular event occurred. Those with damage in the more anterior areas of the frontal lobe, the prefrontal cortex, however, have difficulty performing tasks that require more explicit use of temporal information. For example, Milner and her colleagues found that these patients show serious impairment in judging which of two remembered events occurred more recently (Milner, 1971, 1974, 1982; Petrides and Milner, 1982; Milner, McAndrews, and Leonard, 1990). Their comparative studies of several patients with damage to various areas of the prefrontal cortex suggests that this impairment of temporal memory occurs mainly if there is damage to the dorsolateral prefrontal cortex, specifically in and around Broadmann's area 46 (see Figure 5.1).

Milner and her colleagues also found that encoding external temporal-order information more heavily involves the right prefrontal cortex, whereas encoding internal temporal-order information more heavily involves the left prefrontal cortex (see Milner, 1982). More recently, McAndrews and Milner (1991) found that frontal-lobe patients show normal accuracy in judging the relative recency of two objects that they actively manipulated. One possibility is that the procedural-memory system encodes temporal order information for actions, and if the prefrontal cortex is damaged this information is sufficient to mediate memory for temporal order of specific actions.

Damage in the anterior regions of the left hemisphere (but not of the right hemisphere) frontal cortex also leads to impaired planning in novel situations (Luria, 1966; Milner, 1982; Shallice, 1982, 1988). Damage in these regions does not impair the execution of well-learned, routine plans, perhaps because this function is subserved by the cerebellum rather than by the prefrontal cortex.

### Neurotransmitters and Drug Effects

The neurotransmitter dopamine is found throughout the prefrontal cortex, and a growing body of evidence suggests that one type of dopamine receptor site (D1) plays a critical role in the efficiency of working memory (Sawaguchi and Goldman-Rakic, 1991; Goldman-Rakic, 1992). Some drugs that influence prospective temporal judgments may do so because they influence D1 dopamine receptors in the prefrontal cortex. Dopamine agonists tend to lengthen prospective duration experience (i.e., they increase the subjective time, rate), whereas dopamine antagonists tend to shorten prospective duration experience (Hicks, 1992).

Schizophrenics show various abnormalities in temporal judgment (for reviews, see Johnston [1960], Mo [1990]). Researchers increasingly view psychopathologies like schizophrenia as biological disturbances of specific neurotransmitter systems. Cohen and Servan-Schreiber (1992) reviewed evidence suggesting that schizophrenia involves a reduction of dopamine effects in the prefrontal cortex, resulting in an inability to form and maintain contextual representations. Weinberger (1987) suggested that schizophrenia involves two separate disturbances in dopaminergic pathways, increased activity in the mesolimbic dopaminergic system and decreased activity in the prefrontal cortex. The precise nature of the deficit in schizophrenia remains unclear, but it apparently involves dopaminergic transmission and the prefrontal cortex.

### Electrophysiological Recording

The first electrophysiological evidence that temporal integration of behavior involves the prefrontal cortex arose from the discovery of the contigent–negative variation (CNV) (Walter, Cooper, Aldridge, McCallum, and Winter, 1964). The CNV is a relatively slow surface-negative electrical potential recorded by scalp electrodes. It occurs in the interval between two successive events, beginning about 300 ms after the presentation of the first. It appears if a person anticipates attending to a significant event, such as a temporal interval, a stimulus, or perhaps a movement (Macar and Vitton, 1979). Contingent-negative variation is prominent over the prefrontal cortex, in which it may originate (Borda, 1970; Loveless and Sanford, 1974; Rohrbaugh, Syndulko, and Lindsley, 1976). In duration–judgment experiments, CNV appears when one stimulus starts a duration and another stimulus ends it. Other evidence suggests that CNV amplitude is directly related to prospective duration judgments (McAdam, 1966). If subjects know that they must estimate duration, and perhaps in other situations as well, ''CNV may be an index of a phase of information processing aimed at constituting internal time bases congruent with a particular situation'' (Macar and Vitton, 1979, p. 226). Stated somewhat differently, CNV may be ''an electrophysiological concomitant of attention to time'' (Hicks, Gualtieri, Mayo, and Perez-Reyes, 1984, p. 235). Drugs such as barbiturates that retard activity in the prefrontal cortex decrease both the CNV amplitude and prospective temporal judgments (Hicks, 1992; Hicks et al., 1984); it is possible that they do so specifically because they affect the prefrontal cortex.

Single-neuron recording in the dorsolateral prefrontal cortex reveals that some neurons remain active between the time a stimulus disappears and the time a response is allowed (Kubota and Niki, 1971; Fuster, 1973, 1980, 1985a,b; Niki, 1974; Goldman-Rakic, 1987; Goldman-Rakic, Funahashi, and Bruce, 1990). Because these neurons show sustained activity over short time periods, they could possibly serve as internal clocks, or short-duration timers. Cognitive psychological evidence also suggests that the working memory system subserves short-term memory for an event, on the scale of several seconds (Baddeley, 1986). Goldman-Rakic (1987) proposed that the prefrontal cortex contains several working memory centers, each dedicated to a different information-processing domain. Attending to time may involve one or more of these centers. Contingent-negative variation may therefore frequently reflect the activity of attending to time, subserved in part by the dorsolateral prefrontal cortex.

### Working Memory and Psychological Time: Conclusions

Psychological time, especially constructing and maintaining a temporally and contextually defined present, critically depends on the dorsolateral prefrontal cortex. This area apparently processes temporal contextual information, thereby enabling a person to remember the order of recent events and to prepare or plan for future events (Fuster, 1984). In other words, the prefrontal cortex is a critical component involved in strategic and organizational control of behavior across time, which is why Moscovitch and Winocur (1992a,b) recently proposed calling this system *working with memory* rather than simply *working memory*. There is some lingering controversy about whether the prefrontal cortex processes both temporal and spatial contextual information. Schacter (1987) proposed that the prefrontal cortex is involved in both temporal and spatial context. Lewis (1989) argued that the prefrontal cortex processes temporal contextual information but that the hippocampus (see below) plays a more critical role in processing spatial contextual information.

Both older theories (e.g., James, 1890) and recent research (Block, 1992) distinguishes between prospective and retrospective duration judgment. The working memory system of the prefrontal cortex appears mainly to subserve prospective temporal judgment, or the experience of time in passing. Short-duration judgments, as well as subsequent recency and temporal-order judgments, presumably require the functioning of this system. Milner et al. (1990) proposed two hypotheses on how the prefrontal cortex may subserve temporal-order encoding: (1) "If the frontal lobes parse and organize the temporal contexts of events, one outcome of such operations could be thought of as a direct encoding of temporal tags for events in memory" (p. 991), and (2) the frontal lobes "develop appropriate encoding and retrieval strategies for the reconstruction of temporal order" (p. 992). Although they favored the second hypothesis, the first hypothesis is also tenable, and the two functions are not necessarily mutually exclusive.

## Episodic Memory

Evidence on the episodic-memory system and its role in psychological time comes from several sources. Here I discuss studies of brain damage, neurotransmitters, and drug effects, and electrophysiological recording.

### Brain Damage

The medial temporal lobe of the brain contains several structures, underlying the temporal lobes but apparently also involving parts of the temporal cortex, that are needed to form explicit long-term episodic memories (Squire, 1987, 1992; Squire and Zola-Morgan, 1991). The hippocampus apparently is the most essential structure (see Figure 5.1), and in the interest of brevity I will occasionally use the term *hippocampus* to refer collectively to these several structures. Damage to the hippo-campus produces the most common type of amnesia, called anterograde amnesia, which is characterized by a severe impairment in the episodic memory system: A person permanently loses the ability to encode new personal experiences so that they may be explicitly retrieved at a later time. Hippocampal damage largely spares the other memory systems; a temporal-lobe patient typically can use existing motor skills and learn new ones (procedural memory), can use existing general knowledge and acquire new knowledge (semantic memory), and can display normal short-term memory for recent events (working memory).

Several patients have received operations involving the hippocampus, usually in attempts to relieve epilepsy or remove tumors. Scoville and Milner (1957) tested the memory of 10 such patients. The most well-studied patient, known as H.M., became amnesic following a bilateral medial temporal lobe resection that included the hippocampus. Richards (1973) studied H.M. on a task requiring reproduction of 1 to 300 s durations. H.M.'s reproductions were normal for durations less than 20 s. Thus, H.M. was able to maintain a working memory context for events in the psychological present, presumably because his prefrontal cortex was intact. However, for durations longer than about 20 s, his reproductions were abnormally short: "one hour to us is like 3 minutes to H.M.; one day is like 15 minutes; and one year is equivalent to 3 hours for H.M." (Richards, 1973, p. 281). The most likely explanation is that damage to H.M.'s hippocampus produced a condition in which he cannot permanently encode personal experiences so that they may be explicitly retrieved later. H.M. described his condition as "like waking from a dream" (Milner, Corkin, and Teuber, 1968, p. 217), suggesting a state of consciousness in which present events pervade consciousness and (postoperative) past events do not exist. Kinsbourne and Hicks (1990) reported that Korsakoff patients show a similar, although less dramatic, deficit.

The most common explanation for this type of amnesia is that damage to the hippocampus (and possibly also related medial temporal-lobe structures) impairs the encoding of new episodic memories. There are two main sources of controversy about this conclusion (for useful discussions, see Mayes [1988], Shallice [1988]). First, in addition to anterograde amnesia, or a deficit in encoding new episodic memories, H.M. and many other temporal-lobe patients also show retrograde amnesia. Retrograde amnesia involves a deficit in retrieving episodic memories that were encoded before the hippocampal damage. Most temporal-lobe patients display retrograde amnesia for events that occurred during the several months or, at most, several years before the operation. But Zola-Morgan, Squire, and Amaral (1986) reported the case of R.B., who developed memory impairment following an ischemic episode. Memory testing revealed that R.B. showed extensive and typical anterograde amnesia but little, if any, retrograde amnesia. They subsequently conducted a thorough histological examination of his brain, which revealed a circumscribed bilateral lesion involving the CA1 field of the hippocampus. Thus, damage restricted to only a small portion of the hippocampus is sufficient to produce anterograde amnesia without appreciable retrograde amnesia. It seems that the hippocampus is, therefore, necessary for encoding new episodic memories. The hippocampus and related medial temporal-lobe structures do not directly store episodic memories; instead, storage of them probably occurs throughout all cortical regions that are active in specific information-processing tasks.

A second source of controversy involves semantic memory and the episodic-semantic distinction. In the past, neuropsychologists typically thought that hippocampal damage also impairs semantic memory encoding, and some still do (Squire,

1992). The recent case of K.C. may question this conclusion (Tulving, Hayman, and MacDonald, 1991). K.C. experienced profound episodic (anterograde) amnesia following an automobile accident that damaged several areas, mostly in the left hemisphere, including the left medial temporal lobe and part of the right medial temporal lobe. In spite of this damage, Tulving et al. successfully taught K.C. a large number of three-word sentences, such as "reporter sent review" and "student withdrew innuendo." K.C. implicitly remembered much of this semantic information up to a year later, even though he had no episodic memory for the experience: He could not remember anything about the circumstances in which the learning had occurred. Just as the hippocampus is not required for retrieving existing knowledge, it is apparently not essential for encoding new knowledge (semantic memories).

McAndrews and Milner (1991) presented temporal-lobe amnesic patients with a series of stimuli and then tested their memory by presenting test stimuli in pairs and asking them to judge which of the two occurred more recently. When these amnesic patients were able to remember both stimuli, they performed normally on the recency-judgment task. Thus, although patients with medial temporal-lobe damage show deficits in encoding new episodic memories, when they are able to acquire and explicitly retrieve an episodic memory, they usually can remember temporal contextual information (such as approximately when they experienced the events). As noted earlier, patients with frontal-lobe damage show an opposite kind of performance: impaired memory for temporal information but normal memory for event information per se. This double dissociation suggests that the hippocampus and the prefrontal cortex perform separate but interrelated functions: As the hippocampus encodes information about the *content* of an episode, the prefrontal cortex may supply it with information about the *context* of the episode (for a slightly different view, see Moscovitch and Umilta [1992]).

### Neurotransmitters and Drug Effects

Several neurotransmitters are found in the hippocampus, including acetylcholine, glutamate, and NMDA. Evidence suggests that any drug which interferes with acetylcholine-based neurotransmission in the hippocampus will influence retrospective duration judgments. Hicks (1992) reported that acetylcholine antagonists shorten the remembered duration of a time period. In addition, patients suffering from Alzheimer's disease frequently show severely impaired memory for personal experiences. They typically have damage in several brain areas, including the hippocampus. Their brains also show decreased acetylcholine synthesis (Khan, 1986), which would impair hippocampal neurotransmission. Although researchers have not adequately studied their temporal judgments, one would expect Alzheimer's patients to make abnormal retrospective duration judgments for recent events that they can remember.

### Electrophysiologial Recording

People maintain dynamic internal models of the environment, and events that deviate from the current model require that it be updated. If a person is attending to the performance of some task, presenting a relatively unexpected, but task-relevant stimulus will trigger a positive event-related potential (i.e., a time-locked voltage shift reflecting a change in brain activity), which begins about 300 ms after stimulus onset. Early research (e.g., Halgren, Squires, Wilson, Rohrbaugh, Babb, and Crandall, 1980) suggested that the medial temporal lobe (i.e., hippocampus and related structures) may generate at least some portion of P300. Polich and Squire (1993), however, found that an intact hippocampus is not required for P300 to occur. Activity in several brain regions, including areas of the frontal and parietal cortex, may contain P300-generators that summate to produce the scalp-recorded P300 amplitude (Johnson, 1993). Because P300 is larger following a novel stimulus than an expected one, it may reflect a process of schema- or context-updating (Donchin and Coles, 1988). Thus, P300 may reflect contextual information processing required for the formation of new episodic memories, which then critically requires the hippocampus. Alternatively, P300 may be generated, at least in part, by "an ancillary monitoring and gain-control system that assesses and controls modulation of the basic hippocampal memory system as a function of novelty of the incoming events" (Metcalfe, 1993, p. 333). In Metcalfe's view, firing of hippocampal neurons ordinarily contributes to P300 but is not the sole cause of it.

### Episodic Memory and Psychological Time: Conclusions

The prefrontal cortex and the hippocampus are fairly directly connected and play a conjoint role in the processing of working memory and episodic information (Goldman-Rakic, Selemon, and Schwartz, 1984; Goldman-Rakic, 1987; Olton, 1989). As discussed earlier, the working memory system of the prefrontal cortex apparently generates encodings of information concerning temporal context. This information, which is perhaps in the form of time-of-occurrence of event relative to other events (Hintzman, Summers, and Block, 1975) is critically important for the episodic system. Schacter (1989) proposed "that remembering of temporal order constitutes one component of episodic memory, subserved by the frontal regions, and that remembering of recently presented items constitutes another component of episodic memory, likely subserved by the medial temporal regions" (p. 704).

To the extent that retrospective duration, order, recency, and other similar temporal judgments rely on event information no longer represented in the working memory system, they require the hippocampus for the permanent encoding of events. Temporal memory judgments concerning past events depend on retrieving encoded contextual changes, including changes in process context, environmental context, emotional context, and other contextual associations (Block, 1982, 1990b, 1992; Block and Reed, 1978). Encoding these contextual changes relies heavily on the hippocampus, perhaps using temporal context information generated and supplied by the prefrontal cortex.

**Summary and Conclusions**

Table 5.1 summarizes four memory systems of the brain that subserve psychological time. Any single technique to study the brain necessarily provides limited information. Converging evidence from multiple sources suggests these conclusions about the four systems:

**Table 5.1**
**Memory Systems of the Brain and Associated Characteristics**

| Memory System | Type of Information | Time-Related Behavior/Judgment | Major Brain Area(s) |
|---|---|---|---|
| Procedural | Movement | Movement Timing, Motor Skills | Cerebellum; Cortical Motor Areas (supplementary motor, premotor cortex, and motor cortex) |
| Semantic | Factual (linguistic) | Temporal Concepts | Mainly Left Temporal, Parietal, and Frontal Cortex |
| Working | Temporal Contextual | Prospective Timing; Recency and Order Judgment | Dorsolateral Prefrontal Cortex |
| Episodic | Personal Experience | Retrospective Duration Judgment | Hippocampus; Various Cortical Areas |

1. Procedural memory critically involves the cerebellum, although some structures in the cerebral cortex (e.g., the supplementary motor cortex) are also involved in controlling certain kinds of movement timing.
2. Semantic memory involves widespread areas of the left-hemisphere temporal, parietal, and frontal lobes in understanding and expressing temporal facts and concepts.
3. Working memory involves the dorsolateral prefrontal cortex in constructing an ongoing temporal context; this contextual information is used in tasks such as prospective duration timing, order judgment, and recency judgment.
4. The permanent encoding of episodic memories requires intact functioning of the hippocampus and other medial temporal-lobe structures, apparently working in conjunction with temporal contextual information supplied by the prefrontal cortex; retrospective duration judgments and other long-term temporal memory judgments depend on this episodic information.

# References

Baddley, A. D. (1986), *Working Memory*. London: Oxford University Press.

Block, R. A. (1979), Time and consciousness. In: *Aspects of Consciousness*, Vol. 1, ed. G. Underwood & R. Stevens. London: Academic Press, pp. 179–217.

——— (1982), Temporal judgments and contextual change. *J. Experiment. Psychol.: Learn., Mem., & Cog.*, 8:530–544.

——— (1989), A contextualistic view of time and mind. In: Time and Mind: Interdisciplinary Issues, *The Study of Time*, Vol. 6, ed. J. T. Fraser. Madison, CT: International Universities Press, pp. 61–79.

——— ed. (1990a), *Cognitive Models of Psychological Time*. Hillsdale, NJ: Lawrence Erlbaum Associates.

——— (1990b), Models of psychological time. In: *Cognitive Models of Psychological Time*, ed. R. A. Block. Hillsdale, NJ: Lawrence Erlbaum Associates, pp. 1–35.

——— (1992), Prospective and retrospective duration judgment: The role of information processing and memory. In: *Time, Action and Cognition: Towards Bridging the Gap*, ed. F. Macar, V. Pouthas, & W. J. Friedman. Dordrecht, Netherlands: Kluwer Academic, pp. 141–152.

——— Patterson, R. (in press), Simultaneity, successiveness, and temporal-order judgments. In: *Encyclopedia of Time*, ed. S. L. Macey. Hamden, CT: Garland.

——— Reed, M. A. (1978), Remembered duration: Evidence for a contextual-change hypothesis. *J. Experiment. Psychol.: Hum., Learn., & Mem.*, 4:656–665.

——— Saggau, J. L., & Nickol, L. H. (1983–1984), Temporal Inventory on Meaning and Experience: A structure of time. *Imagin., Cog., & Personal.*, 3:203–225.

Borda, R. P. (1970), The effect of altered drive states on the contingent negative variation (CNV) in rhesus monkeys. *Electroencephalogr. & Clin. Neurophysiol.*, 29:173–180.

Carr, C. E. (1993), Processing of temporal information in the brain. *Ann. Rev. Neurosci.*, 16:223–243.

Cohen, J. D., & Servan-Schreiber, D. (1992), Context, cortex, and dopamine: A connectionist approach to behavior and biology in schizophrenia. *Psychol. Rev.*, 99:45–77.

Damasio, A. R., & Damasio, H. (1992), Brain and language. *Sci. Amer.*, 267:89–95.

——— ——— Tranel, D., & Brandt, J. P. (1990), Neural regionalization of knowledge access: Preliminary evidence. *Cold Spring Harbor Symp. on Quant. Biol.*, 55:1039–1047.

Donchin, E., & Coles, M. G. H. (1988), Is the P300 component a manifestation of context updating? *Behav. & Brain Sci.*, 11:357–373.

Fuster, J. M. (1973), Unit activity in prefrontal cortex during delayed-response performance: Neuronal correlates of transient memory. *J. Neurophysiol.*, 36:61–78.

——— (1980), *The Prefrontal Cortex*. New York: Raven Press.

——— (1984), Behavioral electrophysiology of the prefrontal cortex. *Trends Neurosci.*, 7:408–414.

——— (1985a), The prefrontal cortex and temporal integration. In: *Cerebral Cortex*, ed. A. Peters & E. G. Jones. New York: Plenum Press, pp. 151–177.

——— (1985b), The prefrontal cortex, mediator of cross-temporal contingencies. *Hum. Neurobiol.*, 4:169–179.

Ghez, C. (1991a), The cerebellum. In: *Principles of Neural Science*, 3rd ed., ed. E. R. Kandel, J. H. Schwartz, & T. M. Jessell. New York: Elsevier Science, pp. 626–646.

——— (1991b), Voluntary movement. In: *Principles of Neural Science*, 3rd ed., ed. E. R. Kandel, J. H. Schwartz, & T. M. Jessell. New York: Elsevier Science, pp. 609–625.

Goldman-Rakic, P. S. (1987), Circuitry of primate prefrontal cortex and regulation of behavior by representational memory. In: *Handbook of Physiology: The Nervous System. V*, ed. F. Plum. Bethesda, MD: American Physiological Society, pp. 373–417.

——— (1992), Working memory and the mind. *Sci. Amer.*, 267:111–117.

——— Funahashi, S., & Bruce, C. J. (1990), Neocortical memory circuits. *Cold Spring Harbor Symp. on Quant. Biol.*, 55:1025–1038.

——— Selemon, L. D., & Schwartz, M. L. (1984), Dual pathways connecting the dorsolateral prefrontal cortex with the hippocampal formation and parahippocampal cortex in the rhesus monkey. *Neurosci.*, 12:719–743.

Halgren, E., Squires, N., Wilson, C., Rohrbaugh, J., Babb, T., & Crandall, P. (1980), Endogenous potentials in the human hippocampal formation and amygdala by infrequent events. *Science*, 210:803–805.

Hart, J., Berndt, R. S., & Caramazza, A. (1985), Category-specific naming deficit following cerebral infarction. *Nature*, 116:439–440.

Hicks, R. E. (1992), Prospective and retrospective judgments of time: A neurobehavioral analysis. In: *Time, Action and Cognition: Towards Bridging the Gap*, ed. F. Macar, V. Pouthas, & W. J. Friedman. Dordrecht, Netherlands: Kluver Academic, pp. 97–108.

——— Gualtieri, T., Mayo, J. P., & Perez-Reyes, M. (1984), Cannabis, atropine, and temporal information processing. *Neuropsychobiol.*, 12:229–237.

Hintzman, D. L., Summers, J. J., & Block, R. A. (1975), Spacing judgments as an index of study-phase retrieval. *J. Exp. Psychol.: Hum., Learn., Mem.*, 1:31–40.

Ivry, R. B., & Keele, S. W. (1989), Timing functions of the cerebellum. *J. Cog. Neurosci.*, 1:136–152.

——— Keele, S. W., & Diener, H. C. (1988), Dissociation of the lateral and medial cerebellum in movement timing and movement execution. *Exp. Brain Res.*, 73:167–180.

Jackson, J. L., & Michon, J. A. (1992), Verisimilar and metaphorical representations of time. In: *Time, Action and Cognition: Towards Bridging the Gap*, ed. F. Macar, V. Pouthas, & W. J. Friedman. Dordrecht, Netherlands: Kluver Academic, pp. 349–360.

James, W. (1890), *The Principles of Psychology*, Vol. 1. New York: Henry Holt.

Johnson, R., Jr. (1993), On the neural generators of the P300 component of the event-related potential. *Psychophysiol.*, 30:90–97.

Johnston, H. M. (1960), A comparison of time estimation of schizophrenic patients with that of normal individuals. *Psychol. Bull.*, 57:213–236.

Kandel, E. R., Schwartz, J. H., & Jessell, T. M., eds. (1991), *Principles of Neural Science*, 3rd ed. New York: Elsevier Science.

Khan, A. U. (1986), *Clinical Disorders of Memory*. New York: Plenum Press.

Kinsbourne, M., & Hicks, R. E. (1990), The extended present: Evidence from time estimation by amnesics and normals. In: *Neuropsychological Impairments of Short-Term Memory*, ed. G. Vallar & T. Shallice. Cambridge, U.K.: Cambridge University Press, pp. 319–330.

Kolb, B., & Wishaw, I. Q. (1990), *Fundamentals of Human Neuropsychology*, 3rd ed. New York: W. H. Freeman.

Kosslyn, S. M. (1992), *Wet Mind: The New Cognitive Neuroscience*. New York: Free Press.

Kubota, K., & Niki, H. (1971), Prefrontal cortical unit activity and delayed alternation performance in monkeys. *J. Neurophysiol.*, 34:337–347.

Leiner, H. C., Leiner, A. L., & Dow, R. S. (1991), The human cerebro-cerebellar system: Its computing, cognitive, and language skills. *Behav. Brain Res.*, 44:113–128.

Lewis, R. S. (1989), Remembering and the prefrontal cortex. *Psychobiol.*, 17:102–107.

Loveless, N. E., & Sanford, A. J. (1974), Slow potential correlates of preparatory set. *Biol. Psychol.*, 1:303–314.

Luria, A. R. (1966), *Higher Cortical Function in Man*. London: Tavistock.

Macar, F., & Vitton, N. (1979), Contingent negative variation and accuracy of time estimation: A study on cats. *Electroenceph. & Clin. Neurophys.*, 47:213–228.

Mayes, A. R. (1988), *Human Organic Memory Disorders*. Cambridge, U.K.: Cambridge University Press.

McAdam, H. W. (1966), Slow potential changes recorded from human brain during learning of a temporal interval. *Psychonomic Sci.*, 6:435–436.

McAndrews, M. P., & Milner, B. (1991), The frontal cortex and memory for temporal order. *Neuropsycholog.*, 29:849–860.

Melges, F. T. (1989), Disorders of time and the brain in severe mental illness. In: Time and Mind: Interdisciplinary Issues, *The Study of Time*, Vol. 6, ed. J. T. Fraser. Madison, CT: International Universities Press, pp. 99–119.

Metcalfe, J. (1993), Monitoring and gain control in an episodic memory model: Relation to the P300 event-related potential. In: *Theories of Memory*, ed. A. F. Collins, S. E. Gathercole, M. A. Conway, & P. E. Morris. Hillsdale, NJ: Lawrence Erlbaum Associates, pp. 327–353.

Michon, J. A. (1990), Implicit and explicit representations of time. In: *Cognitive Models of Psychological Time*, ed. R. A. Block. Hillsdale, NJ: Lawrence Erlbaum Associates, pp. 37–58.

Milner, B. (1971), Interhemispheric differences in the localization of psychological processes in man. *Brit. Med. Bull.*, 27:272–277.

———— (1974), Hemispheric specialization: Scope and limits. In: *The Neurosciences: Third Study Program*, ed. F. O. Schmitt & F. G. Worden. Cambridge, MA: MIT Press, pp. 75–89.

———— (1982), Some cognitive effects of frontal lobe lesions in man. In: *The Neuropsychology of Cognitive Function*, ed. D. E. Broadbent & L. Weiskrantz. London: Royal Society, pp. 211–226.

———— Corkin, S., & Teuber, H. L. (1968), Further analysis of the hippocampal amnesic syndrome: 14-year follow-up study of H.M. *Neuropsycholog.*, 6:215–234.

———— McAndrews, M. P., & Leonard, G. (1990), Frontal lobes and memory for the temporal order of recent events. *Cold Spring Harbor Symp. Quant. Biol.*, 55:987–994.

Mo, S. S. (1990), Time reversal in human cognition: Search for a temporal theory of insanity. In: *Cognitive Models of Psychological Time*, ed. R. A. Block. Hillsdale, NJ: Lawrence Erlbaum Associates, pp. 241–254.

Moscovitch, M., & Umilta, C. (1991), Conscious and nonconscious aspects of memory: A neuropsychological framework of modules and central systems. In: *Perspectives on Cognitive Neuroscience*, ed. R. G. Lister & H. J. Weingartner. New York: Oxford University Press, pp. 229–266.

———— Winocur, G. (1992a), Frontal lobes and memory. In: *The Encyclopedia of Learning and Memory*, ed. L. R. Squire. Hillsdale, NJ: Lawrence Erlbaum Associates, pp. 182–187.

———— ———— (1992b), The neuropsychology of memory and aging. In: *Handbook of Aging and Cognition*, ed. F. I. M. Craik. Hillsdale, NJ: Lawrence Erlbaum Associates, pp. 315–371.

Nielsen, J. M. (1958), *Memory and Amnesia*. Los Angeles: San Lucas Press.

Niki, H. (1974), Differential activity of prefrontal units during right and left delayed response trials. *Brain Res.*, 70:346–349.

Olton, D. S. (1989), Inferring psychological dissociations from experimental dissociations: The temporal context of episodic memory. In: *Varieties of Memory and Consciousness: Essays in Honour of Endel Tulving*, ed. H. L. Roediger III & F. I. M. Craik. Hillsdale, NJ: Lawrence Erlbaum Associates, pp. 161–177.

Petrides, M., & Milner, B. (1982), Deficits on subject-ordered tasks after frontal- and temporal-lobe lesions in man. *Neuropsycholog.*, 20:249–262.

Polich, J., & Squire, L. R. (1993), P300 from amnesic patients with bilateral hippocampal lesions. *Electroencephalogr. Clin. Neurophysiol.*, 86:408–417.

Posner, M. I., Petersen, S. E., Fox, P. T., & Raichle, M. E. (1988), Localization of cognitive operations in the human brain. *Science*, 240:1627–1631.

Richards, W. (1973), Time reproductions by H.M. *Acta Psycholog.*, 37:279–282.

Rohrbaugh, J. W., Syndulko, K., & Lindsley, D. B. (1976), Brain wave components of the contingent negative variation in humans. *Science*, 191:1055–1057.

Sawaguchi, T., & Goldman-Rakic, P. S. (1991), D1 dopamine receptors in the prefrontal cortex: Involvement in working memory. *Science*, 251:947–950.

Schacter, D. L. (1987), Memory, amnesia, and frontal lobe dysfunction. *Psychobiol.*, 15:21–36.

———— (1989), Memory. In: *Foundations of Cognitive Science*, ed. M. I. Posner. Cambridge, MA: MIT Press, pp. 683–725.

Scoville, W. B., & Milner, B. (1957), Loss of recent memory after bilateral hippocampal lesions. *J. Neurolog. & Neurosurg. Psychiat.*, 20:11–21.

Shallice, T. (1982), Specific impairments of planning. *Philosoph. Trans. Royal Soc. London*, B298:199–209.

———— (1988), *From Neuropsychology to Mental Structure*. New York: Oxford University Press.

Squire, L. R. (1987), *Memory and Brain*. New York: Oxford University Press.

———— (1992), Memory and the hippocampus: A synthesis from findings with rats, monkeys, and humans. *Psychol. Rev.*, 99:195–231.

———— Zola-Morgan, S. (1991), The medial temporal lobe memory system. *Science*, 253:1380–1386.

Tulving, E. (1972), Episodic and semantic memory. In: *Organization of Memory*, ed. E. Tulving & W. Donaldson. New York: Academic Press, pp. 381–403.

———— (1985), How many memory systems are there? *Amer. Psychologist*, 40:385–398.

———— (1989), Memory: Performance, knowledge, and experience. *Europ. J. Cog. Psychol.*, 1:3–26.

———— (1991), Concepts of human memory. In: *Memory: Organization and Locus of Change*, ed. L. R. Squire, N. M. Weinberger, G. Lynch, & J. L. McGaugh. New York: Oxford University Press, pp. 3–32.

———— Hayman, C. A. G., & MacDonald, C. A. (1991), Long-lasting perceptual priming and semantic learning in amnesia: A case experiment. *J. Experiment. Psychol.: Learn., Mem., & Cog.*, 17:595–617.

Vidal, F., Bonnet, M., & Macar, F. (1992), Can duration be a relevant dimension of motor programs? In: *Time, Action and Cognition: Towards Bridging the Gap*, ed. F. Macar, V. Pouthas, & W. J. Friedman. Dordrecht, Netherlands: Kluwer Academic, pp. 263–273.

Walter, W. G., Cooper, R., Aldridge, V. J., McCallum, W. C., & Winter, A. L. (1964), Contingent negative variation: An electric sign of sensori-motor association and expectancy in the human brain. *Nature*, 203:380–384.

Weinberger, D. R. (1987), Implications of normal brain development for the pathogenesis of schizophrenia. *Arch. Gen. Psychiat.*, 44:660–669.

Weiskrantz, L. (1987), Neuroanatomy of memory and amnesia: A case for multiple memory systems. *Human Neurobiol.*, 6:93–105.

Williams, J. M., Medwedeff, C. H., & Haban, G. (1989), Memory disorder and subjective time estimation. *J. Clin. & Exp. Neuropsychol.*, 11:713–723.

Zola-Morgan, S., Squire, L. R., & Amaral, D. G. (1986), Human amnesia and the medial temporal region: Enduring memory impairment following a bilateral lesion limited to field CA1 of the hippocampus. *J. Neurosci.*, 6:2950–2967.

6

# Consciousness and the Individual Event in Scientific Theory

*David Park*

Abstract   In science, the 20th century may be called the century of the atom; it seems that the next one will be the century of the brain, in which scientists begin to discover how it works. If "how it works" means an explanation in terms of physical processes, one might expect to apply the same intellectual discipline that produced the century of the atom. This paper attempts to show why that cannot be done, why today's criteria for good science are only partly applicable to a science that hopes to unite brain with mind. First, good science depends on studying events that can be repeated, but events in the conscious mind can happen only once. Scientific theories offer to be refuted by predicting the result of an experiment, but the same circumstances that prevent replication also prevent prediction of what a person will think or say or do. And second, since we are conscious only of a single moment called the present, and physical science contains no such entity, it will require radically new thinking, perhaps a new science, to bridge the two.

At the most fundamental level of description, what happens in a thinking brain that makes a thinking mind? At first glance, one might hope to approach an answer to this question using a model which represents the brain as a parallel computer of immense power and so turn loose on it the theory of computation, about which at present we know quite a lot. Of course such a model is useful at certain points, but what specially interests researchers is consciousness, and it is not easy to see even what questions should be asked of a theory of computation that would help one in thinking along that line. Most cognitive psychologists, therefore, have adopted the sensible strategies of experimental science rather than forcing their ideas into an existing but alien framework. Following this approach one might hope slowly to learn more and more about what happens in a conscious mind and so approach an understanding in terms of fundamental physical processes in a mass of brain tissue.

In this way, exact science might find something to say about the mind, but what? This paper is concerned not with exactly what it might say but with the difficulties of concept and method which arise before it can say anything. I think they must inevitably lead to some compromises with what we today consider the proper relation between questions, answers, and evidence in the exact sciences.

In a moment I shall begin to explain what I mean, but first I will defend myself against a protest that I have occasionally heard: Why probe the secrets of the human mind with scientific tools? Science has turned our bodies inside out. Is there no aspect of our life that will be spared this scrutiny? Has reticence no rights? I see two answers: First, it is the nature of human beings to explore, and I cannot imagine that such scruples would ever prevent eager and curious people from trying to answer questions that excite their interest. Second, there is the prospect of reducing human suffering. To mention only one disease of the mind, schizophrenia is one of the most painful illnesses known, it is usually incapacitating, and quite often fatal. It affects the area of consciousness. I think that few scientific workers who love humanity would be deterred by an ideological outcry from trying to lessen this pain.

So much for an introduction. Now I shall begin at the beginning.

The ancients were aware of regularities in the processes of nature, aware that if we prepare a fire or a meal, if we mix dyes or metals, similar causes produce similar effects, and out of these observations the concept of natural law was born: Similar causes *must* produce similar effects. According to this way of understanding things, events can be divided into classes: *All* eclipses of the moon are produced by the shadow of the earth; it is probable that *all* thunderstorms are produced by some cause, natural or divine, but thunderstorms happen only in certain kinds of weather and can often be predicted a few hours in advance. This understanding collided with older ways of thinking. The same society that understood the lawfulness and inevitability of the eclipse could also see it as a sign sent from God, a portent of punishments to come; and there is nothing about the experience of a thunderstorm—lightning, noise, violence—that suggests the automatic and impersonal and inevitable operation of law.

The sense that nature is lawful is not by itself enough to make real science; one must understand what a law of nature is, what it says, and what it leaves unsaid, and in this sense exact science began with Isaac Newton. It was he who first distinguished the law from the conditions in which it acts, so that exactly the same laws are understood to govern processes ranging from what happens on a billiard table after a ball has been struck to the motions of the planets; the laws are the same, only the conditions are different. In this case size doesn't matter, only the external circumstances that there is no sun on a billiard table and no table in the sky. Natural law explains thunderstorms in general and does not speak of *this* thunderstorm. This one may or may not be typical but that is not the point. Natural law does not refer to *any* specific event, typical or not. It speaks in terms of timeless generalities. The statement of such a generality may mention time, but the statement itself sits timelessly on the printed page. The statement *is*; nature also *happens*, and this is where the problems begin, for consciousness happens; it refers only to particulars. It is an event—it happens and is gone—like the sudden flash of a bluebird it leaves nothing behind except a memory.

From Karl Popper's writings on the nature of scientific knowledge (1968), scientists have learned that a properly formulated scientific claim is accompanied by a commentary that tells how it can be proved wrong. The commentary says you must measure or observe something, under controlled conditions. You ought to try it again and again to be sure. If your answer disagrees with the theory then the theory has failed its own test, unless it can be shown that something is wrong with the experiment.

So much for physics and chemistry; so much for experiments on animals or on human reactions to certain stimuli. Pavlov's dogs must have grown weary of that bell. But each experience in our conscious life happens just once. It cannot be repeated, for the memory of earlier experiences changes later ones, and all the sanitary scientific procedures for verifying what anyone might say about it are lacking. Suppose then that someone comes forward with a theory which claims to explain consciousness, what it is like to be alive and having experiences and thinking. I am not sure what kind of words such a theory would use; that is for the future. But if it did make some claim, how could anyone know whether it was true? According to the usual recipe for scientific verification it would have to predict what a person, under given circumstances, would think and say and do. Precisely because one cannot "give" all the interior mental circumstances that determine what someone is going to think at a particular moment, I do not see how this could happen.

If exact science does not aim at the individual event but demands proofs rooted in individual events, it is hard to see how this science can have anything, even in ideal circumstances, to say about consciousness, which exists *only* in the individual event: a wish, a hope, a memory, an interpretation, and which changes at every moment.

I come now to a second difficulty that must finally face those who seek to reduce consciousness to exact science. There is a buzz-word that is often heard when people discuss these questions, *reductionism*; it is usually spoken by people who know little about science, and is usually spoken with a sneer. It refers to efforts to state and solve biological and medical problems by reducing them to the physics and chemistry of atoms and molecules. Scientists rarely use the term; they take the process for granted because they know how successfully it has transformed biology and medicine in our time and has even, as with the synthesis and administration of neurotransmitters, made life easier for some people with problems of the mind. But now, along with those detractors, I am saying that the process of reduction seems to have a limit. Still, one cannot stop here. To quote Werner Heisenberg (1958): "The exact sciences start from the assumption that it will in the end always be possible to understand nature, even in every new field of experience, but that we make no a priori assumptions about the meaning of the word 'understand' " (p. 28). This is the point. If "understand" means understand in terms of the familiar categories of the exact sciences, I fear that human consciousness is beyond their reach, but these sciences are certainly not in their final form, and problems like this challenge us in a very healthy way to extend them. For if our own century might be called the century of the atom, the next one will surely be the century of the brain and the mind.

Understanding the idea of now is obviously crucial to understanding consciousness. Now is when we live our lives; it is when we think what we think and do what we do. The past and future are mental constructions; we cannot observe them. And yet the idea of now is totally lacking in physics. I have often illustrated the point with the simple formula which states that for motion at constant velocity $v$, the distance $d$ traveled in a time $t$ is given by

$$d = vt$$

Our consciousness explores the dimension of time in the direction of advancing events (first we see a bird, then we think for a moment and say it is a crow), and if we think of the numerical value of the time $t$ as continuously getting later and later, then also the distance gets continuously greater and greater. But this act of interpretation already assumes that we understand the exploratory mode that any theory of consciousness must try to explain. Take it away and the formula says nothing about continuous motion. Give me $v$ and a value for $t$ and I will give you $d$; that is all it says, and except for the entropy law of thermodynamics, which specifies the order in which changes tend to occur, there is no principle of change or motion in all of physics that does any more.

Think of it: The physicist's description of the world, the triumph of centuries of inspired labor, does not, even in principle, represent the most elementary datum of our experience. Physicists do not usually mention this gap in their ability to explain, though I know it bothered Einstein very much. We haven't had to fill it. We work in the same narrative mode as novelists or historians ordinarily do, with time spread before us, free to pick up the thread at an earlier or later moment, any moment we wish, but that is not the mode of experience.

I have mentioned two holes in the tent that physicists live in: that lawlike statements do not describe our experience of individual events, and that we seem to sense that our consciousness moves forward in time. Most probably I have just described one big hole. Theory works in the narrative mode; consciousness lives in the experiential mode; a theory of consciousness must do both at once. For us, events happen now and at no other time. The classical physics of Isaac Newton and his followers' physics, lacking now, also lacks the concept of an event, but in quantum mechanics, through what is often described as a defect in the theory, it comes back in again. Briefly, quantum mechanics does not, as Newton did, describe nature as it *is*, but rather as it is known to us from experience. (For example, it predicts what will happen in experiments involving electrons but does not ever say whether an electron is a wave or a particle.) Even a momentary observation of some physical object can tell me something I did not know before; my description of the object jumps to a new description that may be quite different. Is this jump a physical process? One can argue the question either way. It is hard not to think that this is a rather strange way to do physics. After all, if I think that the world has some kind of existence independent of myself, glancing at a thing cannot change it very much; what changes sharply is the state of my own consciousness. But at least *something* changes with the quickness of thought, and a number of authors (d'Espagnat, 1976; Penrose, 1989) have argued that our brains make use of this discontinuity in order to report on a single moment of experience with that vividness that distinguishes the present from all other moments. The brain is a big thing, and quantum mechanics was invented to explain little things like atoms. Big things are made of atoms, and so one might think that having solved the problem of little things we have only to add them together to make a theory of big ones. But it has not worked out that way. Our experience of the world of big things generates a philosophy quite different from the philosophy of little things, and the theory that relates them, known as the theory of measurement, remains full of mysteries, 65 years after quantum mechanics was invented and the philosophy of little things was first worked out.

It is often convenient to claim that truth lurks in the shadows. But generally, study of some particular shadow reveals that nothing ever lurked in it, and my instinct is to be highly skeptical of any claim that quantum theory contains the mystery of consciousness. It just can't be that simple. Still, quantum theory, though it seems to be correct as far as it goes, is not in its final form and probably it never will be, since the form of a theory depends on what is assumed and what questions are asked of it, and both of these will change. Conceivably the demand for a theoretical psychology could coalesce with the demand for a theory of measurement to produce ideas useful in both, but it will take many years of patient work before we are wise enough about how the brain actually thinks so that we even know what questions to ask. And finally let us remember Heisenberg's words quoted above. Consciousness is a phenomenon entirely different from anything the exact sciences have ever studied so far. I am suggesting that if some day we feel able to say that we understand it in material terms, we shall be using the word *understand* in a sense rather different from the way we use it today.

## References

d'Espagnat, B. (1976), *Conceptual Foundations of Quantum Mechanics*, 2nd ed. Reading, MA: Benjamin.

Heisenberg, W. (1958), The idea of nature in contemporary physics. In: *The Physicist's Conception of Nature*, tr. A. J. Pomerans. New York: Harcourt, Brace, & World.

Penrose, R. (1989), *The Emperor's New Mind*. New York: Oxford University Press.

Popper, K. R. (1968), Science: Conjectures and refutations. In: *Conjectures and Refutations*, 2nd ed. New York: Harper.

# II.
# Experiences of Time and Life

# Reflections

The papers of this section deal with the varied experiences of time and life and the ways by which temporal reality is generated, conceptualized, and conveyed. The topics range from the exploration of the technologies employed to measure our living time, to its expressions in word, sound, and image.

> The tolling bell
> Measures time not our time, rung by the unhurried
> Ground swell, a time
> Older than the time of chronometers. . . .
>
> ["The Dry Salvages," Eliot, 1943, p. 36]

Technologies of time measurement have come to play a dominant role in the way we order and direct our lives; yet life as experienced weaves its own temporal fabric, follows its own drummer, possesses a quality which resists quantification. That quality of lived time demands metaphor for its description because it is the symbolic rather than the concrete that speaks to the subtleties of human time experience.

> Words move, music moves
> Only in time; but that which is only living
> Can only die. Words, after speech, reach
> Into the silence. Only by the form, the pattern,
> Can words reach or music reach
> The stillness of the violin, while the note lasts. . . .
>
> ["Burnt Norton," Eliot, 1943, p. 19]

The human awareness of transiency in life has frequently inspired the desire to create objects or ideas which suggest timelessness. When human imagining thus gives lasting form to the fleeting moments of life, it saves those moments from oblivion: they can be relived in imagination as examples of the eternal return, or generate new rhythms and pace or combine the new with the familiar. The modulation by music of the sense of passage, the cinematic play of images, the evocation of simultaneous perceptions in poetry, and weighty silences in music, word, and action give life to forms when those forms are actualized in experience. It is thus that the tapestry of time experience, woven of many elements, enriches the present and expands the store of experience upon which the innumerable expressions of human life depend.

# Reference

Eliot, T. S. (1943), *Four Quartets.* New York: Harcourt, Brace & World, 1971.

7

# Time and Life: An Evolving Relationship*

*Samuel L. Macey*

*Abstract*   The purpose of this paper is to show that the relationship between human life and the time by which its passage is measured has been undergoing a dialectical development which is still evolving. In the past, conceptions and measurements of time were based on the human body and on human activity within the immediate environment. The breakdown of this integration, which began to accelerate in the late 17th century, was a necessary evolution in the story of our dynamic Western civilization.

Increasingly accurate measurements—especially the measurement of time—are an essential prerequisite for the progress of science and technology. The paper discusses how the increasingly accurate measurement of time evolved, and how it has been instrumental in creating or energizing virtually all of the modern academic disciplines. Using figures drawn from my recently published *Time: A Bibliographic Guide* (1991), I discuss the extent and nature of these disciplines. I also give figures for the remarkable increase in scholarly publications related to time, particularly during the past 10 years.

Finally, the paper points to a complicating factor. Although we in the West may sometimes be overcome by a romantic yearning to own 5 acres and a cow, to join a green party, or to throw away our watches in favor of a simpler life, the rest of the world hopes to emulate our material successes—and therefore risks destroying their integration with their own local environments. Nevertheless, this study should help to clarify our choice of goals. In the thesis of our ongoing dialectic, primitive time was integrated into life; in the antithesis, the separation of time and life has been an essential prerequisite for developing Western civilization; in the synthesis, we shall be obliged to seek ways to control the genie that our material and intellectual greed has let out of the bottle. Inevitably, we must discover how to reintegrate life, all lives, within the new and more complex form of environment that we have created.

I propose to demonstrate that the relationship between human life and the time by which its passage is measured has been undergoing a dialectical development which is still evolving. In the past, conceptions and measurements of time were based on

---

*Presidential Address, July 4, 1992.

the human body and on human activity within a limited environment. The destruction of that integration, however, which began to accelerate in the late 17th century, was a necessary evolutionary step in the story of our dynamic Western civilization. This change was accompanied by a sense of loss, and I shall return before the end of the paper to a discussion of related yearnings for a reintegration of time and life in a setting of more human dimensions.

Increasingly accurate measurements communicated through standardized numbers are an essential prerequisite, particularly in respect of time, for developing the science and replicating the technology of modern Western civilization. Earlier forms of measurement were closely related to the human body. For example, our present base 10 numbering system is derived from the biological fact that human hands possess 10 digits. Also, the archaic base 20 numbering system of "Three score years and ten" was based on the number of digits on human hands and feet. Less obvious is the connection between the structure of human hands and the base 12 and base 60 numbering systems strongly entrenched in our months, hours, minutes, and seconds. Georges Ifrah has demonstrated convincingly that these numbering systems are also based on the human body. There is a duodecimal finger-counting method which is still employed in certain Eastern countries: using the thumb of the right hand one counts to 12 by beginning at the tip of the little finger and pointing in sequence to the three obvious segments on each of the four fingers of that hand. The complementary sexagesimal finger-counting method employs the left hand to indicate the dozens from 12 to 60 by respectively closing down each of the fingers and proceeding from the little finger to the thumb (Ifrah, 1985, pp. 65–66).

As early as the 8th millennium B.C., the Sumerians recorded their possessions by using physical representations such as cone-shaped tokens for each small measure of grain and lenticular discs for each sheep. But it took 5000 years for human beings to make the colossal leap into abstract numbers. By 3100 B.C., the Sumerians—who started to write on tablets of clay at this time—began using a small wedge for 1, a small circle for 10, a large wedge for 60, and a large circle for 360 (*Discover*, 1987, pp. 7–8). Many further developments sprang from this system, but the divorce of numbers from a specific human or material reference had begun.

In other forms of measurement, man continued to remain the measure of all things. For example, spatial measurement was based on the human body. The digit or nail—measured across the base of the middle finger—was about three-quarters of an inch; the inch or *unce* was measured across the thumb. The palm—or four fingers—represented 3 inches, the hand 4 inches, and the foot 12 inches. The cubit, which measured approximately 1½ feet, stretched from the elbow to the tip of the middle finger. The yard, which derived from man's girdle or girth, stretched from the nose to the middle finger. The fathom, which measured approximately 6 feet, represented the distance between the extended fingers of a man's outstretched arms.

Longer linear measures were also based on human dimensions. The Roman mile, or *mille*, was 1000 paces (we would call them double paces). The stadium, linked with the length of the chariot race, was 125 paces, or one-eighth of a mile. (Our *furlong*, or furrow-long, reflects a distance related to agriculture.) Ratios between such distances were fairly well fixed, but distances themselves varied between one locality and another to an extent that would be quite unacceptable today. Even as late as 1800, the foot varied between 14.01 modern inches in Frankfurt and 9.84 inches in Darmstadt (Zupko, 1977, p. 171). Man might be the measure of all things, but the measure depended greatly on the individual person and on his specific locality.

In square measure, human activity based on time provided the equally inexact criteria from which standards were derived. For example, an acre was a measure of the amount of land a man could plow in one working day. Its German equivalents of the *Tagwerk* and the more common *Morgen* (or morning's plowing) make the derivation of the words self-evident. As late as 1800, the *Morgen* varied all the way from the modern English equivalent of 2.38 acres in Hamburg to .50 acres in Frankfurt (Zupko, 1977, p. 175). Such standards would hardly satisfy modern surveyors, let alone geophysicists.

The standards by which time was measured in the earlier Western world were equally dependent on human beings, human activities, and the immediate local environment. Although time was ostensibly measured by the relationship between the earth, the moon, and the sun, the human realm influenced many systems of time division and measurement. The calendar was calibrated by priests and subservient to their holy days and sabbaths. In the Judeo-Christian chronology, the story of creation reflected a Chain of Being dominated by man and his anthropomorphic God. Also the Old and New Testaments record the progress of time with lists of human generations from Adam to Noah, from Noah to Abraham, and from Abraham to "Joseph the husband of Mary" (Genesis 5, 11, and Matthew 1). Even historic periods were measured from the wars of man, the accession of his kings, the founding of his games, or the building of his cities. These kinds of subjective measurements would hardly encourage the broader temporal perspectives necessary to evolutionary biologists, geologists, or cosmologists.

Not even concepts of the day and its divisions were divorced from human values. In modern terms, the measured day lasts for 24 equal hours. But in a society without artificial light, when men were tied to the land, the hours of daylight restricted the working day and virtually defined the day itself. One-twelfth of the seasonally variable day was called an hour, which is to say a temporal or unequal hour. Even its parts retain vestiges of seemingly human derivation. John Trevisa, a contemporary of Chaucer, lists a series of divisions, which is similar to the table of Papias in Du Cange (found under *atom* in the *Oxford English Dictionary*). Trevisa divides the "temporal" or "unequal" hour as follows: an hour "conteyneth" 4 points, a point 10 moments, a moment 12 ounces, and an ounce 47 atoms; and an atom can be no further divided (1975, pp. 529–535). The words *ounce* and *inch* are identical. Both mean one-twelfth and are therefore related to the system of base 12 numbers.

What is remarkable about the terms that Trevisa uses is that the *point, moment,* and *atom*—as well as the *minute* which appears in the table of Papias—are all periods which at one time referred to the most minute or tiny or indivisible moments by which time could be measured. By today's standards the point represented around 15 minutes, but this was then considered so minute a moment of measured time that it has given us such terms as *punctuality* or on the "point of death." The *Oxford English Dictionary* defines *atom* as "the twinkling of an eye and that which cannot be divided." It notes further that the Latin *atomus* meant "twinkling of an eye." Apparently this meaning already existed in the Greek *atomos*, for which the *OED* gives a translation from 1 Corinthians 15:22 (actually 15:52): "in a moment, in the twinkling of an eye."

Curiously, the *moment*, which Trevisa lists as a temporal measurement between the *point* and the *atom*, derives from *movement* and refers to a physical movement of the shortest duration. Thus both the moment and the atom are measurements of time that refer to minute human movements. The word *moment* is still with us as a measurement of time, and so is "in the twinkling of an eye," which translates directly into the German *Augenblick*, the French *en un clin d'oeil*, or the Italian *in un batter d'occhio* (Nilsson, 1920, p. 42).

In the Western world there has been a progressive rationalization of time during the past six centuries. In the 14th century, as human beings were drawn to the towns, unequal hours, which were one-twelfth of the daylight or of the night, were among the first measures to be rationalized. At the latitude of London and according to the time of the year, unequal hours could vary in length between 40 and 80 modern minutes. In the whole of his canon, Chaucer (d. 1400) never uses the terms *minute* or *second* as a measure of time. Yet two centuries later, Shakespeare's references to time assume the use of equal hours. He refers to minutes more than 60 times. The second, however, was still elusive, and was never mentioned in his canon.

The great progress in clock technology took place 40 years after Shakespeare's death. The verge and foliot escapement, which controlled mechanical clocks during their initial three centuries, was hardly isochronous. Such clocks varied by up to 15 minutes each day. By the third quarter of the 17th century, however, the requirements of astronomers and the navigators' urgent need to measure the longitude brought us, through Christiaan Huygens, the Dutch mathematician, astronomer, and physicist, first the pendulum escapement for clocks and then the balance-spring escapement for watches. Clocks were now, for the first time, as accurate as was necessary for urban purposes. And because clocks were now some 60 times more accurate than before the discovery of the pendulum escapement, minute and even second hands came into general use.

From then on we have maintained an exponential rate of progress in the separation of time from human life. By concerning ourselves with increasingly larger and increasingly smaller periods of time, we have taken its measurement far beyond the limits of a human scale. In the 18th century geological time was conceived of on the one hand, and on the other hand chronometers and stopwatches were invented. In the 19th century humankind envisioned the large sweeps of time required for the evolution of all forms of life and time was also divided into increasingly smaller units and synchronized through telegraphs, time zones, and electric clocks. Our own Einsteinian century has developed not only the model of a universe that we think has been expanding outward for some fifteen billion years, but also has produced cesium clocks that now measure time to an accuracy of one second in one million years.

One clue to the growing consciousness of time when the pendulum clock was being invented is indicated by the increasingly scientific rather than man-oriented words that were being incorporated into our language. In the 17th century we started using the word *progress* to designate a movement through time. Before then the word *progress* was used in the spatial sense of a progression of queens or kings. Similarly, the word *century* had referred to a number of soldiers. With the increasing awareness of chronology during the 17th century this word also came to represent a unit of time. Also, by inventing the word *anachronism* we demonstrated that we had now begun to acquire an abstract sense of temporal order (Ebeling, 1937).

In the book *Clocks and the Cosmos: Time in Western Life and Thought* (1980), I demonstrated how virtually all major philosophers and a host of minor ones employed clockwork metaphors to illustrate their central arguments during the period 1660 to 1760 (Macey, 1980, p. 70). I have called those years between the inventions of the pendulum escapement and the marine chronometer the British horological revolution. That was the period when horology was the recognized handmaiden of astronomy, and was also spearheading the advances in navigation and technology without which Britain's empire and industrial revolution could hardly have been envisaged. In no small measure, the increasing separation of human life from the human criteria through which the passage of time had been measured had begun to let out of the bottle the genie that represents our lust for power through scientific and material advancement.

In addition, horology took on quite a different role in the service of religion. What came later to be known as Paley's argument from design demonstrated with all the force of logic that the clockwork universe of Newton and Leibniz could have been created by no one other than a Watchmaker God. Understandably, man's major metaphors derive from his current interests. The Watchmaker God implies a clockwork universe—a metaphor one would expect during the British horological revolution. Hitherto, when time was described in human terms, the predominant metaphors had virtually always been animate. Gregory, in "The Animate and Mechanical Models of Reality," argues that "Descartes finally convinced the seventeenth century that physical nature was mechanical like a clock, and did not participate in animal qualities" (Gregory, 1927, p. 302).

The separation of man from the criteria by which time is measured has permitted the quantification of periods of time previously too long or too short to be measured. This has spawned some remarkable consequences in the past 300 years. As I have shown at length in *The Dynamics of Progress: Time, Method, and Measure* (1989), most of the philosophies and disciplines of the modern world are virtually permeated by a concern with progress through time. Consider some of the seminal works that have inspired modern disciplines. Adam Smith argues that wealth will increase over time without government interference; Buffon sets out geological stages; and Condorcet outlines the 10 epochs of moral and social progress. Similarly, the idea of progress and growth through time are essential to the Hegelian and Marxian dialectics; the Darwinian premise of the survival of the fittest; Thomsen's progressive ages of stone, copper, and iron; and the new discipline of anthropology inspired by L. H. Morgan's *Ancient Society; or Researches in the Lines of Human Progress from Savagery Through Barbarism into Civilization* (1877).

A complementary idea of progressive strata is the basis of William Smith's tabulation of geological ages; Lyell's inorganic physical processes, which inspired Darwin; Schliemann's digging in Troy, which gave us archaeology; and Freud's digging in the individual mind, like Jung's digging in the archetypal mind, which gave us psychoanalysis. In our own century, of course, the Einsteinian expanding universe flows directly from the idea of progress through time, and directly or indirectly affects much of our thinking. On earth, as in the heavens, we now seem unable to operate without a vastly more accurate measure of time.

I have shown in *Dynamics of Progress* (1989) how other elements of human measure and endeavor—weight, volume, area, length, numbers, language, labor, and production methods—have also been extensively rationalized during the past 300 years. But time dominates and encompasses them all. As Samuel Alexander wrote of the "so-called theory of relativity," during those heady days of physics earlier in this century, "Every point has four co-ordinates, the time co-ordinate being the fourth. Hence it follows, as Minkowski writes, that geometry with its three dimensions is only a chapter in four-dimensional physics." Furthermore, "Time does with its one-dimensional order cover and embrace the three dimensions of Space, and is not additional to them" (Alexander, 1920, p. 58).

R. V. Jones has much more recently observed that "The millionfold or so improvement in precision of time determination over the last fifty years far exceeds the improvement in precision of length measurement in the same period and so it has been worth substituting time for length as the property to be measured wherever possible" (Jones, 1985, p. 8). For example, the standard meter has been redefined as the distance that light travels in a vacuum during one-299, 792, 458th of a second. One should add that radar has been developed during the same period, and vast distances are now measured in light years. Navigating with a sextant and the most accurate chronometers of the 18th century, one could determine one's position at sea to within one half a degree or some 30 miles. We are now able to measure to the nearest inch the 154,680,381 inches between the Owens Valley and the Haystack observatories. But we do so by training their radio telescopes on a quasar several billion light years away and measuring with atomic clocks the difference in the two arrival times of the quasar's radio emissions (Bartusiak, 1981, pp. 82–83). Without increasingly accurate time measurement, voyages to the planets would be as unthinkable in our own day as was the accurate mapping of the world before the advent of the chronometer.

As a result of the developments at which we have been looking, time and time measurement have been widely introduced into existing disciplines during the past three centuries and have engendered many disciplines and subdisciplines that did not previously exist. Before the increased specialization of our post-Faustian era, university disciplines were divided into the four faculties of divinity, medicine, law, and philosophy, in each of which the Faust of both Marlowe and Goethe qualified for the doctorate at Wittenberg. Today, all four of the original faculties, including religion, are permeated with concerns regarding time on this earth.

Bishop Ussher even capitalized on the incipient interest in time and chronology in the 17th century to demonstrate when the world was created. The time calculated, and at first very widely believed, was 9:00 A.M. on Sunday, October 23, 4004 B.C. That would eventually prove a little too presumptuous for the new temporally oriented discipline of geology, though religion continued to be concerned with creation, eschatology, and the argument from design.

In medicine, time management has proved essential for controlling hospitals, surgeries, psychiatric wards, dental practices, and medication, and precise time measurement for assessing time perception, pulse rates, and a whole host of biological periodicities. The law, for its part, has become increasingly aware of the fees to be earned through what is known as "limitation of actions." These are restrictions through which the law imposes a wide range of time controls over the activities of individuals and corporations.

Faust's doctorates in law, divinity, and medicine are still granted today in the very universities that existed in his own time. But the fourth faculty, namely that of philosophy, has exploded into a whole spectrum of disciplines. Many of these disciplines have been engendered or at least inspired by the subject of this essay, namely, by an accurately measured earthly time divorced both from man's immediate environment and his concerns regarding eternity. There are now scores of new disciplines and subdisciplines in which we may receive what is called a doctorate in philosophy. The first new offspring was, of course, natural philosophy, which was the 17th-century term for science.

Music, mathematics, and astronomy, which were once part of the medieval educational system, have become increasingly associated with temporal control and measurement. Mathematics, after a long apprenticeship to astrology, has proved remarkably adaptable (*mathematicus* means both a mathematician and an astrologer, and *horoscope* means observing the time or hour). It now serves the time-oriented disciplines through such new strategies as calculus, dynamics, and more recently time series analysis. During very recent years, the combination of mathematics and computers has become one of the most potent tools of the modern world.

In some of the modern subjects, the time-oriented nature of the discipline is immediately evident. These include aging, archaeology, history, horology, navigation, time management, and time measurement. In other subjects, important subdisciplines have developed in which the time component is equally obvious. This is true of film and TV as well as the iconography of time and death in art; of evolution and biological clocks in biology; of demography in geography; of paleontology and geochronology in geology and geophysics; of utopian, futuristic, and science fiction as well as prophecy and alternate history in literature; of rhythm, meter, tempo, and the metronome in music; of quantum and relativity theory, thermodynamics and entropy, time and space, and time's arrow in physics; and of memory and time perception in psychology.

I was occupied for several years with producing *Time: A Bibliographic Guide* (1991), which comprises references to some 6000 books and articles, broken down into 25 disciplines and about 100 subdisciplines. The exercise has convinced me that the study of time is now taking on a life of its own. The disciplines and subdisciplines whose time-oriented nature is immediately evident from their titles have provided just over half of the subject headings in *Time: A Bibliographic Guide*. Since they parallel the related subject headings in the main periodical indexes, these disciplines have also provided the basis for my attempt to arrive at a numerical estimate of time-related literature at large (Macey, 1991).

A physical count of such time-related entries in the main periodical indexes for the years 1980 to 1989 yielded a total of 50,293 published articles. After making percentage adjustments to allow for multiple listings, entries not under the headings used, and what J. T. Fraser has called indexers "untutored in the study of time," this figure translated into an estimate of some 90,000 periodical publications related to the study of time. Two methods were employed to extrapolate from the figure of 90,000 during the last decade to the current century as a whole. The first produced a total figure of 155,000 articles by adding the 90,000 directly to J. T. Fraser's estimate of 65,000 articles for 1900 to 1980. The second method produced a total figure of 205,000 by extrapolating from the articles in *Time: A Bibliographic Guide*, which had been chosen on a relatively random basis insofar as dates were concerned. Since the true figure will probably lie on the broad band between these two estimates, we may assume that some 180,000 time-related articles have been published during this century.

By using similar methods with the *Library of Congress Subject Catalog*, which lists some 4 million volumes even during its first 25 years of publication, I have been able to estimate for the first time that about 95,000 books on time-related subjects have been published since the beginning of the current century. These quite remarkable figures suggest something of the size and ramifications of the subject with which we have thus far been concerned. But the most recent increase in the growth of time-related studies that has taken place since man became separated from the temporal scales of his immediate environment is even more remarkable. My calculations in *Time: A Bibliographic Guide* indicate that approximately one-third of all time-related books have been published during the past seven years, and some 45 percent of all time-related articles have first appeared during the past decade. These figures are also supported by J. T. Fraser's estimate in 1980 that out of a total of 65,000 time-related articles which had appeared since the beginning of the century some 40,000 had been published during the previous 15 years (Fraser, 1981, p. 267).

The study of time is a multifaceted discipline. It has permeated and energized virtually all of the other academic, industrial, and social concerns that lie behind the modern phenomena we have come to know as Western science and technology. There are, of course, fashions even in these developments. Leading the field at the moment are the very recently constituted disciplines of aging, time series analysis, temporal perception, chronobiology, chronopharmacology, and perhaps less obviously telecommunications and computer-controlled temporal allocation. Other once essential temporal disciplines, such as the dialing of sundials, the production of hourglasses, and the study of comparative chronology, have long since been consigned to the Burgess shale of temporal activity. But the extent to which virtually all disciplines are energized by temporal considerations continues its strong upward trend.

This essay has outlined the extent to which the increasingly accurate measurement and consciousness of time during the past three centuries has permeated the Western psyche, dominated its science, and allowed technology to progress. But the 500 million watches now being sold annually to the 5 billion men, women, and children on this earth signal an even more dramatic change. Clearly the West is no longer alone in having become separated from inexact but personally significant time measures.

We may sometimes be overcome by a Romantic yearning to own 5 acres and a cow, to join a green party, or to throw away our watches. But very few of us would be prepared to accept the much lower standard of living that a return to the less accurate time measures of the past would inevitably entail. The genie is now out of the bottle and there is little we can do but modify the effects of the dangerous hegemony that Western technology has given us over this earth's other forms of life. In the thesis of our ongoing dialectic between time and life, primitive time was integrated into life; in the antithesis, the separation of time and life has brought us Western civilization; in the synthesis, we must seek some way to control the genie that our material and intellectual greed has permitted to escape, and we must reintegrate ourselves with the new and more complex environment we have created.

For good or ill, what the Western world has achieved today the rest of the world will have emulated 60 years from now. What better reason can there be for encouraging our best young minds to become engaged in every aspect of the study of time, and thereby involve themselves with that most vital force in our intellectual lives, in our material lives, and in the life of our increasingly time-compact globe?

## References

Alexander, S. (1920), *Space, Time, and Deity: The Gifford Lectures at Glasgow, 1916–1918*. London: Macmillan.

Bartusiak, M. (1981), The ultimate timepiece. *Discover*, May:79–83.

*Discover* (1987), Editorial: Lumps of clay that gave birth to numbers. March:7–8.

Ebeling, H. L. (1937), The word *anachronism*. *Mod. Lang. Notes*, 52:120–121.

Fraser, J. T. (1981), A report on the literature of time, 1900–1980. In: *The Study of Time*, Vol. 4, ed. J. T. Fraser, N. Lawrence, & D. Park. New York: Springer.

Gregory, J. C. (1927), The animate and mechanical models of reality. *J. Philosoph. Studies*, pp. 301–314.

Ifrah, G. (1985), *From One to Zero: A Universal History of Numbers*, tr. L. Blair. New York: Viking Penguin.

Jones, R. V. (1985), Time and distance. *Horolog. J.*, 128:6–8, 14.

Macey, S. L. (1980), *Clocks and the Cosmos: Time in Western Life and Thought*. Hamden, CT: Archon Books.

——— (1989), *The Dynamics of Progress: Time, Method, and Measure*. Athens: University of Georgia Press.

——— (1991), *Time: A Bibliographic Guide*. New York: Garland.

Morgan, L. H. (1877), *Ancient Society; or Researches in the Lines of Human Progress from Savagery Through Barbarism into Civilization*. Chicago: C. H. Kerr, 1907.

Nilsson, M. P. (1920), *Primitive Time Reckoning*. Lund, Sweden: C. W. K. Gleerup.

Trevisa, J. (1975), *On the Properties of Things*, tr. of *Bartholomoeus Anglicus De Proprietatibus Rerum*. Oxford: Oxford University Press.

Zupko, R. E. (1977), *British Weights and Measures: A History from Antiquity to the Seventeenth Century*. Madison: University of Wisconsin Press.

8

# The Separation of Time and Nature

*Lennart Lundmark*

*Abstract*  The basic argument of this paper is that clocks and other technical devices have molded our concept of time to a much greater extent than we are generally aware of. Science has not been excluded from this influence.

Mechanical clocks started to push the conception of time in an absolute direction as early as the mid-14th century when hours of equal length were introduced. With the arrival of pendulum clocks in the late 17th century the irregularities of the apparent motion of the sun became a problem and true solar time was changed to mean solar time (= clock time) in civilian life. This led to absolute time being conceived of as "natural." The technical innovations in the communications sector then made standard time necessary and finally the need to conserve energy by utilizing more daylight has led to the introduction of summertime. All these interventions have made everyday time a self-contradictory mixture of absolute and conventionalist concepts.

Solutions to problems regarding time are often sought among philosophers, whether in advanced science or profound existential and psychological deliberations. In this paper I shall take another route and point to everyday technological change. I shall argue that it has had a greater impact on our conception of time than has deep thinking behind wrinkled foreheads. I shall, of course, not be as reductionistic as to claim that technical change is the *only* factor molding our conception of time. But in my opinion it has not been given the consideration it deserves.

Before the invention of the mechanical clock, time was held to be relational and based on nature. It was relational in the sense of being directly dependent on events and processes and having no independent character. Time was natural in the sense that it was based on processes not deliberately constructed by human beings for the purpose of measuring time; that was true of sundials, for example, because they were constructed only in order to divide up a natural process.

The hour was also a natural unit in those days, and it was therefore not a uniform measure. Whether the days were long in summer or short in winter, they

97

were divided into 12 parts. That meant that the length of the hours varied with the season. In the Middle Ages they were called *horas inequales*. Sundials, water clocks, the early astronomical clocks, and the monasterial *alarii* were all constructed to show horas inequales (Bilfinger, 1892; Landes, 1983; pp. 76–77).

We can be fairly sure that the mechanical clock was invented somewhere in West Europe around the year 1300 or maybe a decade or two before that. The mechanical clock did not fill any general needs at the time except maybe in monasteries and with astronomers. It was not given a special name, but was called horologium like the other devices for telling the hours. It seems as if the 14th-century Europeans did not see mechanical clocks as a revolutionary invention, probably because the clocks were so inaccurate that virtually all other means of telling the time were better. The minimum error seems to have been about 1 hour a day (von Bertele, 1953).

But the early mechanical clocks created a revolution in one respect. With their spread during the 14th century the unequal hours were abolished. There have been some claims that this was a result of the burghers needing equal hours for their early capitalistic calculations. This seems to be a typical example of anachronistic thinking, that is, imposing the values of our own time on the past. Except for maybe the astronomers nobody had any need for equal hours in the 14th century, especially not in the West European heartland where variations in the length of the day are in no way extreme over the seasons.

The reason why unequal hours were abolished was that it was virtually impossible to construct automatic striking mechanisms adapted to unequal hours. Having a striking clock was important for the prestige of the medieval city.[1] That was also the reason why clocks spread so quickly. David Landes (1983) has rightly emphasized that their function was mainly "totemic." He has also claimed that "equal hours announced the victory of a new cultural and economic order." That might be a fair conclusion as well. But Landes only emphasizes that the equal hours made it possible for the new commercial and administrative layers of society to develop new numerical skills and a new calculating mentality (pp. 77–78).

There is, however, another aspect which I find more important. With the equal hours the first step was taken toward a separation between time and nature. The unequal hours were closely related to the natural variations of the length of the day. With the equal hours the abstraction of time starts. The reason was a technical coincidence. It was of benefit for the ability to calculate time in arithmetical terms.

---

[1]This is basically Gustav Bilfinger's conclusion (1892). Other scholars have given mercantile calculations a greater role in this change (Drummond Robertson, 1931; LeGoff, 1960; Edwardes, 1965; Landes, 1983). Cipolla (1977) provides a more composite explanation listing both prestige, interest in mechanics, and the need to calculate and delimit daily activities. Still those advocating mercantile explanations have not been able to convincingly pinpoint exactly what activities needed equal hours. For the working day the differences were insignificant, and on a yearly basis it simply meant that people had to work longer hours in the summer and shorter in the winter, which, incidentally, does not seem to be a bad idea. No calculations concerning production and interest were made on an hourly basis in those days.

But that does not necessarily mean that it was beneficial for our ability to understand time.

The second step in alienating time from nature was taken when "absolute" time got its breakthrough. Absolute time flows of itself, and from its own nature, "equably without relation to anything external...," as Newton phrased it. To investigate that breakthrough it is necessary to start with the medieval background to Newton's absolute time.

During the 14th century most philosophers accepted that there could be a potential time, a time that could be filled with events even if nothing happened. But if there were a potential time, how could it be measured? There were no clocks, either in heaven or on earth that could measure the time that proceeds when nothing is happening. One way of solving the problem was to claim that such a clock only had to exist as an abstract mathematical concept. But the philosophers were not ready to take that step. To quote Pierre Duhem's (1985) conclusion: "They did not dare declare that the movement intended to mark time for all other movements is also a pure concept not realized in nature—that the absolute clock is an abstract clock existing only in the mind" (p. 351).

That was the position all the way up to the 17th century. What gave the late 17th-century scientists the confidence to state that an absolute time was a viable concept? Maybe they finally found something that they thought could function as an absolute clock? In my opinion the fact that the pendulum clock was introduced in England only a short time before Newton first developed his theory of an absolute time has not been given enough consideration. The London clockmakers were busy making pendulum clocks as early as 1660 and Newton did not explicitly state his concept of absolute time before 1668, maybe even some years later (Westfall, 1980, pp. 301–303).[2]

The new type of regulator in clockworks, the pendulum, constituted a virtual revolution in the precision of timekeeping. Before the pendulum was introduced, the best clocks had a minimum error of about 15 minutes in 24 hours, so mechanical clocks were still unfit for serving as "absolute" clocks. With the pendulum as regulator, clocks of a fairly good quality had an error of less than 20 seconds a day (von Bertele, 1953).

Newton was explicit in his reference to pendulum clocks. In the *Principia* (1683) he states that the necessity to distinguish absolute and relational time in astronomy "is evidenced . . . from the experiments of the pendulum clock" (p. 8). In my opinion it is highly probable that one of Newton's reasons for introducing the concept of absolute time in the *Principia* was the hitherto unseen uniformity of time indications made possible by pendulum clocks. They gave Newton and others the confidence they needed to accept an absolute time. The pendulum clock could

---

[2]Pierre Gassendi formulated a concept of absolute time in 1640, using almost the same words as Newton, that is, before the first pendulum clock was made (see the excerpts in Capek [1976, pp. 195–196]). The question here is, of course, not who first mentioned it, but when it got its breakthrough. Speculations about something similar to an absolute time had been going on at least since the Middle Ages, so neither Gassendi nor Newton was original in that respect.

be seen as filling the role of an absolute clock. And even if it still has its shortcomings the new technological optimism of the day surely made it plausible that the minor irregularities would be overcome shortly. Once again we can discern a down-to-earth technological influence behind a fundamental shift in the conception of time.

Generally Newton's physics is considered to be the cause of the definitive breakthrough of absolute time. But was Newton really so important when it came to the acceptance of a concept of absolute time? (By that I mean the everyday concept of an irreversible absolute time that is dominant today.) In my opinion the pendulum clock's influence on everyday life was a much more important factor than Newton's natural philosophy.

Let us look at the world outside theoretical science. In Newton's age it had been known for a long time that the apparent motion of the sun is irregular. Even the ancient Greeks had some inkling of it, but in the 16th century astronomers started to take it into account on a more systematic basis. For their calculations they needed a time which was as uniform as possible and therefore they devised a mean solar time where the irregularities of the sun were leveled to an equalized time, a time that was basically the same as the clock time we use today. True solar time can be up to 16 minutes ahead of mean time, and up to 14 minutes behind it. The difference between them is called the equation of time.

Mean solar time was strictly for astronomers. But the pendulum clocks were so uniform in their measurements that the equation of time became obvious also outside the ranks of astronomers. In everyday life the sun was still considered as the decisive regulator of time, and when the discrepancy between true solar time and fairly uniform pendulum-clock time became obvious, the problem could be tackled in two ways: One was to build clocks that showed true solar time or both true and mean solar time. This was a very complicated task but some "equation" clocks were made. The second, and common way of solving the problem was to devise tables giving the difference between mean and solar time for each day or week of the year. Such tables were often pasted on the inside of the door of clock cases (King, 1978, pp. 118–131; Landes, 1983, pp. 122–124).

The tables posed new problems. They were not devised to verify mean time, but to convert it to the solar time people lived by. To comply with them clocks had to be reset at frequent intervals, not a good thing to do to a precision instrument. How often should the adjustments be made, for example, and with what precision?

The debate over true solar time versus mean solar time went on for over a century. There were many clocks around in the 18th century and most clock owners were discussing whether their clocks showed the correct time or not. In those debates the relation between true and mean solar time must have been a major issue.

Successively, clocks won the battle against nature. Mean time became standard, even sundials were built with conversion scales for the equation of time. Big cities decided to leave true solar time and follow the mechanics of the clock instead. Mean solar time was made standard in Geneva in 1780, in London in 1792, in

Berlin in 1810, in Paris in 1816. More peripheral areas came slightly later. In the Swedish official almanac mean time was made standard in 1841.[3]

The shift to mean time exerted a major influence on how the notion of time was tied to natural processes. With the introduction of mean time as the civilian standard, absolute time became "common sense." Time was now embodied in clocks and "without relation to anything external. . . ," to quote Newton once more. The uniform flow of mechanized mean time certainly made its influence felt in all spheres of society, not excluding philosophers and scientists. This was the second major step in separating time and nature.

The third step was standard time (Thrift, 1977; Howse, 1980; Zerubavel, 1982; Bratky, 1989; O'Malley, 1990). In most countries the changeover to mean solar time had not been accompanied by a standard time like the one we have today. Still people clung to nature wanting their time to be exact in the sense that midday should occur when the sun was at its highest in the sky. As the sun appears to be constantly moving, every city, town, and village had its own time, and communications and other activities had to be organized in accordance with that. Mechanical clocks were set to local time whether one used true or mean solar time.

When national standard times were introduced in the late 19th century it was not an easy task to convince people in general that it was necessary. In both England and the United States there were protests of different kinds. Some clergy argued that the local time of their region was God's time and that the new time was an abomination. A more profane opposition found the new time imprecise and unnatural, which was exactly what it was from a local point of view. In Sweden a proposal for standard time was rejected by the government in 1871 because such a time was

---

[3]The implementation of this measure varied. In France it was suggested by the Bureau des Longitudes and in 1816 the Prefect of Paris decided that mean solar time was to be civil time in that city. In Berlin the Academy of Sciences started to keep its official clock at mean solar time. In Sweden mean solar time was made standard in the official almanac issued by the Royal Academy of Sciences. But was mean solar time ever officially introduced in England before Greenwich Time was made the national standard? There have been claims that such a measure was taken in 1792, but the sources are somewhat dubious.

The French historian of science M. G. Bigourdan mentioned the year 1792 in *Annuaire pour l'an 1914 publié par le Bureau des Longitudes, Paris 1914*, p. B 8. He did not further substantiate it. Bigourdan's statement was repeated by eminent scholars like J. Drummond Robertson and Derek Howse, but it has never been checked so far as I can see. There is, however, some material indicating that no "official" decision to use mean time as the standard was made. *The London Almanac*, issued by the Stationer's Company, used true solar time as late as 1840. It is proved by the fact that there are notices like "Clock 2 minutes fast" or "Clock 5 minutes slow" printed at various dates during the year (I am aware that the *Nautical Almanac* introduced mean solar time in 1834, but that was, of course, a matter only for navigators and other specialists.)

So this is where I stand at the moment and I shall probably have to go back to London to do some further research on the problem. At the moment it seems to me that there was no decision to change to mean solar time in England, while still keeping local times. It just spread along with Greenwich Time as the national standard. If that is correct there was no "second step" in the standardization of time in England. It simply merged with the third step, that is, the introduction of a national standard overruling the multitude of local times.

But I am in no way sure of this and I would, of course, be extremely thankful for any suggestions that may help me solve the problem.

considered too inexact in relation to "natural time." After a boost in industrialization and improved communications during the 1870s, standard time was introduced by law beginning January 1, 1879. Sweden was the first country to introduce standard time *by law*. In England most activities were already in accordance with Greenwich Mean Time by the 1850s, but a law regarding standard time wasn't passed until 1880.

The introduction of standard time was another intervention in natural relational time made necessary by technological change, mainly the spread of railways. Natural processes were no longer determining factors, but rather the contingencies of technological innovation or other demands that people were confronted with in everyday life.

The three steps mentioned here have probably molded our concepts of time to a higher extent than we usually realize when we try to think "freely" and "objectively" about it. This gradual accommodation to a concept of time even more independent of nature continues to this day.

We can see this clearer in respect to the fourth step, summertime, which has been virtually unopposed since the mid-1960s when it was introduced in the United States, and around 1980 when it was coordinated throughout Western Europe. There were lots of discussions before summertime was introduced, but the protests have been rather limited since. They have almost exclusively consisted of some individual grumbling when the clock has to be reset.

We seem to have accepted as a convention the location of time-scales and their relative independence from the bonds of nature. But that seems to have happened only during the last few decades. When summertime was introduced during World War I the reactions were different. There had been proposals early in this century in both England and the United States to make better use of daylight by changing the clock, but they were invariably rejected. Then in 1916 Germany introduced summertime, or daylight saving time (DST) as it is also called. Within a few months Austria-Hungary, England, France, Italy, Holland, and the Scandinavian countries all embraced the measure, advancing their clocks by 1 hour. The suspicion was, of course, that Germany had found a "secret weapon" to further its war effort. When DST was introduced in the United States in 1918 it was promoted by posters carrying such slogans as: "Set the clock ahead one hour and win the war!" and "Mobilize an extra hour of daylight and help win the war!"

In spite of that the reactions in the United States were fierce. Pressure groups soon formed and started lobbying for a repeal. Hundreds of petitions were sent to members of Congress urging them to abolish the law. One of the petitions was signed by more than 120,000 people. A straw vote in a Midwest town indicated that 90 percent of its 12,000 inhabitants were against DST (Patrick, 1919; Movahedi, 1985). Three types of arguments were commonly raised. First, there were farmers claiming that the cows refused to come home while the sun was high in the sky, and that the milking schedule was disrupted. The second type of argument was that sleep was disturbed, especially for workers and children. The third proclaimed that DST was an unholy interference with God's time.

Protests like this were voiced in all countries that experimented with DST during the First World War. In most cases it was abolished after only one summer. In Sweden a thorough investigation of the reactions showed that more than half of the population had simply neglected to change their clocks in spite of summertime being determined by law (Lundmark, 1986).

The reactions to DST during World War I show that nature still played a far greater role in deciding time indications than is the case today. What happened when people accepted summertime was that the general everyday conceptions of time became an even more confusing mix of the absolute and the conventional. Time became somehow conceived of as proceeding at a steady uniform pace, but the nametags of time indications could still be changed according to contingent needs.

Now an even more conventionalist approach than summertime is gaining ground; stopping clocks in political assemblies is beginning to spread. If a decision has to be made before a certain date—according to law or other regulations—the politicians simply stop the clock in the assembly when they are about to exceed the time limit. They go on until they have reached a conclusion, then they start the clock and sign the document with a date that may be many days old. I first heard about this when it was done at the U.N. conference on peace initiatives in Stockholm in 1986. But now the practice is apparently spreading and I have been told that it is a rather common practice in France. Of course everybody knows that this is just a fictitious stopping of time, but to us it seems a bit awkward and we do not really feel at ease with it. My guess is that in two or three decades this will be a normal procedure and nobody will care. The separation of time and ordered natural scales will have proceeded one step further.

From this paper somebody might conclude that I propose that we should return to unequal hours, true solar time, and local times; that is, of course, not my intention. Nor is it my aim to uncover some kind of "essential" time behind the confusing conceptual shrouds that have covered it over the centuries. My aim has simply been to emphasize that the concept of time is molded by practical procedures and demands in everyday life to a higher extent than is normally realized. Our understanding of time has constantly been adapted to the practical problems at hand. Against that background one might ask if it is meaningful to envisage time as a concept with some kind of universal qualities. The most fruitful way to conceive of it seems rather to be as a tool for our attempts to order a surrounding reality filled with a confusing multitude of change.

With such a point of departure one can also consider all questions about possible topological properties of time as void of meaning. In the perspective taken here, time has neither a beginning nor an end, it is neither linear nor cyclical, neither continuous nor discontinuous, neither reversible nor irreversible. But we can create temporal concepts with such properties and they can serve us more or less efficiently in different situations, but they cannot be assigned any universal qualities. This is an instrumentalist view of time, and to me that seems to be the most fruitful approach.

Some of those doing scholarly work related to the concept of time are prone to stick to the absolutist approach, being afraid that their object of study in some way would disappear if they abandoned it. They need not worry. An instrumentalist approach aims at the increasingly important question of how concepts are applied to reality, and even more important, what the outcome of different strategies will be for our adaptation to the cultural and physical environment. To shed light on such questions is to me the most important task for scholarly activities today.

## References

Annuaire pour l'an 1914 publié par le Bureau des Longitudes, Paris, 1914.

Bertele, H. von (1953), Precision timekeeping in the pre-Huygens era. *Horolog. J.*, 95:795–805.

Bilfinger, G. (1892), *Die Mittelalterlichen Horen und die Modernen Stunden.* Stuttgart: Verlag von w. Kohlhammer.

Bratky, I. R. (1989), The adoption of standard time. *Technol. & Culture*, 30:25–56.

Capek, M., ed. (1976), *The Concepts of Space and Time.* Boston: Boston Studies in the Philosophy of Science, Synthese Library.

Cipolla, C. (1977), *Clocks and Culture.* New York: W. W. Norton.

Drummond Robertson, J. (1931), *The Evolution of Clockwork.* London: Cassell.

Duhem, P. (1985), *Medieval Cosmology.* Chicago: University of Chicago Press.

Edwardes, E. (1965), *Weight-Driven Chamber Clocks of the Middle Ages and Renaissance.* Altrincham, U.K.: John Sherratt & Son.

Howse, D. (1980), *Greenwich Time and the Discovery of the Longitude.* New York: Oxford University Press.

King, H. C. (1978), *Geared to the Stars. The Evolution of Planetariums, Orreries, and Astronomical Clocks.* Toronto, Canada: University of Toronto Press.

Landes, D. S. (1983), *Revolution in Time.* Cambridge, MA: Harvard University Press.

LeGoff, J. (1980), *Time, Work and Culture in the Middle Ages.* Chicago: University of Chicago Press.

Lundmark, L. (1986), Soltid, standardtid, sommartid. Naturens klocka och samhällets rationalitetskrav. *Historisk tidskrift*, 4:457–483.

Movahedi, S. (1985), Cultural perceptions of time: Can we use operational time to meddle in God's time? *Compar. Studies in Soc. & Hist.*, 27:385–400.

Newton, I. (1683), *The Mathematical Principles of Natural Philosophy.* Los Angeles: University of California Press, 1962.

O'Malley, M. (1990), *Keeping Watch. A History of American Time.* New York: Viking Penguin.

Patrick, G. T. (1919), The psychology of daylight saving. *Scient. Monthly*, Nov.:385–396.

Thrift, N. (1977), The diffusion of Greenwich time in Great Britain. *Working paper 188*, Leeds, U.K.: School of Geography, University of Leeds.

Westfall, R. S. (1980), *Never at Rest. A Biography of Isaac Newton.* Cambridge, U.K.: Cambridge University Press.

Zerubavel, E. (1982), The standardization of time: A sociohistorical perspective. *Amer. J. Sociol.*, 88:1–23.

9

# Cross-Language Universals in the Experience of Time: Collocational Evidence in English, Mandarin, Hindi, and Sesotho

*Hoyt Alverson*

*Abstract*   Collocations—stock phrases, idioms, aphorisms, or other expressions that gain their relative fixity from culturally patterned conditions of language use—reflect usage-induced attachments of words for one another and express the ideological and institutional bases of communication and knowledge. Provided one can control for linguistic borrowing, collocations that appear and reappear in different languages, independent of cultural contingency, may reflect or express universal categories of experience.

In the research reported here, a corpus of collocations, which are about or which predicate properties of time, has been obtained from native speakers of English, Mandarin, Hindi, and Sesotho. A categorization of these collocations for each language is made using the kinds of metaphoric predications employed in the collocations. This exercise has yielded five basic, metaphorically characterized, universal categories of temporal expression that appear in all four languages, despite the independence of these languages from one another and despite the very different conditions of belief, knowledge, and institutional structure that exist among the cultures and communities that natively speak these languages.

## Collocation

The structure of process called "collocation" is neither a strictly syntactic-level nor a lexical/morphological-level form; it exists, rather, between the fixed form of phrases that compose the lexical meaning of words and the open, largely generative

Support for this study was provided by a grant from the Claire Garber Goodman Fund for Anthropological Research.

syntax of ordinary sentences. For a person who is not a native speaker of English, phrases like (1) stubborn ox, stone deaf, scared rabbit, sly fox, old maid, proud father, Jewish mother, dirty tramp, dumb blonde, Commie symp, or (2) smokes like a chimney, swears like a trooper, stares like a zombie, drinks like a fish, runs like a rabbit, may seem like simple phrases. But a native speaker knows they are clichés, idioms, aphorisms, or other stock phrases used as epithets. The words ''go together'' in a more intimate way than the grammar would normally indicate. In general, collocations reflect cultural patterning in the repetition (use, acquisition, and transmission) of language and the cultural patterning of knowledge, situation, and purpose, of which language is always—at least tacitly—expressive. The repetition occurs within texts or discourse, so that much of what a collocation expresses is derived from its context of repeated use (see Tannen [1989] for further discussion).

Collocation explains why, in modern English, we *bake* break, cake, apples, fish, and ham, while we *roast* nuts, beef, lamb, and pork, and can either bake or roast chicken and turkey. These collocations are evidence of cultural history—the cuisine of Saxon England, the Norman invasion and its culinary impositions on courtly life, the shared techniques for preparing of chicken, and the late arrival of turkey (a New World fowl) into the English world. Because collocation or prepatterning is created in and reflects its context, it forms a linguistic index of culture. Without the game of poker, how could we account for idiomatic expressions like: ''He has an ace up his sleeve,'' ''She gave me a fast shuffle,'' ''She's dealing from the bottom of the deck,'' ''You'd better not get lost in the shuffle or else have an ace in the hole,'' ''You four-flusher!'' ''Let's lay our cards on the table,'' ''What do we do when the chips are down?'' ''Stay with the blue chips,'' ''How do they stack up against the penny-ante stuff?'' ''Well, they sweeten the pot by upping the stakes,'' ''I'll stand pat,'' ''Just play above board and don't pass the buck,'' ''Otherwise you might wind up in hock''? Working back from such phrases, we are led ineluctably to a cultural practice—poker.

## The Language of Space and of Time

From the above flow two intriguing questions: (1) if there are transcultural universals of temporal experience, would these not be expressed in universal, translanguage collocations involving time? and (2) what kind of experience(s) would such universal collocations of time express? There is a school of thought within linguistic semantics that argues that semantic meaning can be understood as based upon or derived from our ''crude'' experience of space or spatiality. In this view, the basis for meaning invariance in natural language lies in (1) the way in which humans, irrespective of culture, experience space; (2) how language expresses that experience; and (3) abducts it to constitute other meanings/experiences.

Ray Jackendoff (1983) has provided a cogent statement of this spatialization thesis:

> The semantics of motion and location provide the key to a wide range of further semantic fields. . . . The significance of this insight cannot be overemphasized. It means that in exploring the organization of concepts that, unlike those of physical space, lack perceptual counterparts; we do not have to start de novo. Rather, we can . . . adapt . . . in so far as possible the independently motivated algebra of spatial concepts to new purposes. The psychological claim behind this methodology is that the mind does not manufacture abstract concepts out of thin air. . . . It adapts machinery [the experience of space] that is already available [Jackendoff, 1983, p. 189].

If these claims are correct, then from them flows a hypothesis: if (1) the experience and therefore the linguistic expression of crude space is universal (invariant across language/cultures); and if (2) the process (i.e., collocation) of typifying nonspatial experiential domains (e.g., time) in terms of spatial ones is invariant across language/cultures; then (3) the linguistic expression of the experience of time will also have a universal component across languages and cultures.

Four methodologic requirements must be met to falsify this claim.

1. One must identify, in as culturally/linguistically unbiased a way as possible, a formal object, experience of time, into which specific time-experiences of different cultures/languages can be translated, thereby rendering them theoretically cognate; that is, as instances of a theoretically justified metacultural category time.
2. The cultures selected for comparative study should be characterized by very different beliefs, knowledge, and institutions that compose temporal experience to minimize the chance of mistaking for a universal traits which arise from accidentally similar beliefs or institutions.
3. A set of language-specific, yet universally occurring, linguistic data must convincingly represent or express those metaculturally justified temporal experiences.
4. An analysis of those language-specific data must reveal the degree to which the intra- and cross-language/culture variation in the expression of temporal experience can be associated with (1) linguistically constant (universal) expressions/experiences of space, and with (2) linguistically/culturally variable and particular expressions/experiences, whether spatial or not.

## Collocational Data

Data from four languages/cultures were collected to test this hypothesis—English, Mandarin, Hindi, and Sesotho—to provide a large and random sample of collocations that express temporal experience. Examples of these data are given in the appendix (space limitations preclude listing the entire corpus, which runs to more than 200 in each of the four languages). The experiences of time in the cultures of the Judeo-Christian/West European world, China, North India, and Bantu-speaking Africa are as different from one another as any four such cultures can be (see Alverson [1994] for relevant background). By inspection and analysis of these data, one can determine to what extent these collocations derive from more basic spatial expressions. The collocations should also make clear the extent to which such spatial concepts form a universal core or whether they vary among languages/ cultures and in what ways temporal experience among languages is alike or uniquely different because of culturally particular beliefs and institutions.

The data were elicited from two native speakers each of Mandarin, Hindi, and Sesotho. Each grew up in a community where his or her native language was the dominant common language, and none acquired fluency in English until after adolescence. The data were collected over several interviews. I elicited them by describing for each respondent, purely as a heuristic device, the conceptualization of time contained in certain words in English that might be found in relevant collocations—instant, moment, slow, fast, now, then, later, earlier, again, era, epoch, period, duration, hour, day, month, year, clock, and, of course, time. I asked that they think of ways of expressing in their native language various aspects, dimensions, instances of "temporal experience" using stock phrases, aphorisms, sayings, or figures of speech that they had heard growing up in their home communities. I did everything possible, given the practical necessity of working with multilingual informants, to prevent the respondents from translating English exemplars into their native language.

## Categorization of Collocations

The data displayed in the appendix have been placed in five categories, the same categorization as was used with the corpus. These started out as a rough-andready, intuitive classification based on the metaphoric term in the collocation that predicates of time a property or relationship, but the initial classification proved surprisingly robust in its cross-language applicability. The categories are: (A) Time is a (divisible) entity; (B) Time is its effects; (C) Time is a medium in motion; (D) Time is (I) a linear course, (II) an orbital course; and (E) Time is its method of ascertainment/measurement. (Letter designations correspond to appendix listings.)

To check the reliability of my judgments in assigning collocations to the categories, I gave three students a description of the root metaphor that I deemed to be common to the five categories, plus a sixth category, "don't know/can't decide." I then gave them all the collocations, including English translations of those in the other three languages. I asked them (independently) to assign each collocation to one of the six categories in forced-choice fashion.

For all but a few of the foreign collocations, mainly Mandarin, there was 100 percent agreement on the assignment of collocations to the five basic categories. There were a few expressions in which three of the four of us were in agreement, but this occurred less than a dozen times in over 800 sortings (the phrases in the appendix marked by an asterisk were those *not* placed in the same single category by each rater). While reliability does not insure validity, this extremely high reliability, along with other evidence described below, strongly suggests that the categories have face-validity.

## Discussion of Data Characteristics

Each collocation in the four languages could, of course, be translated into a well-formed phrase in any other language. This is not at issue here, nor is this idea any longer very controversial. But the investigation of the spatialization hypothesis raises different questions: Which collocations occurring in any one language are found *as collocations* in any and all other languages, irrespective of the culture(s)' native-speaking population(s)? Which collocations are unique to a particular language because they express demonstrable, culturally particular background conditions of belief and institutionalized language use?

Here we have collocational evidence from four languages/cultures that reveals how time is metaphorically schematized in language and therefore, I argue, how time is experienced. There appear to be five basic, metaphorically constituted experiences predicated of time. This does not mean that time is strictly included in those domains. Rather, something inchoate—expressed, for example, by time, "*shi*," "*samay*," and "*nako*"—has been the argument typified by such predicates as entity, linear course, orbital course, medium in motion, method of ascertainment, and deictic space (space defined by one's physical body and its attitude and orientation).

On the face of it, the Sesotho collocations present one incontrovertible piece of evidence against the hypothesis that "time" collocations exist merely as a function of the cumulation of conscious, critical, philosophical thought concerning time. The Sotho do not have such thought, never have had, yet their language exhibits the same categories of temporal collocation as English, Mandarin, and Hindi. Moreover, as a quick perusal will show, many of the particular metaphors that occur within specific categories, whether English, Mandarin, or Hindi, also appear in Sesotho.

On the other hand, space–time "deixis"[1] in Sesotho is identical to that of English and Hindi and recognizably similar to Mandarin. The term *nako* shows wide and similar privileges of collocation within and between categories as that shown by the counterpart terms—*time* in English, *shi* in Mandarin, and *samay/waqt* in Hindi/Urdu.

Many collocations express culturally particular beliefs and institutions. For example, the Indian cultural preoccupation with the destructive and other causal effects of time is apparent. Similarly, despite a linearly spatialized space–time deixis

---

[1]"Deixis" is a linguistic therm that means the experience of space as a function of one's corporeal self-experience—up, down, here, there, right, left, before, behind, etc.

and ample experience of "linear" clock time, Hindi, uniquely in this sample, resists extrapolation of time as an indefinite line, either straight or orbital; thus *samay/ waqt* do not collocate with scalar or vectoral metaphors of quantity. Furthermore, neither Hindi nor Sotho collocationally predicate of time the many monetary, market, or commodity values that we have in English, and there is a pervasiveness of spatial and locomotor metaphors for temporal experience in Sesotho. These entail actual experiences of walking, traveling, seeing, and so on. The Sotho also do not express an absolute mechanical clock time and its attendant metrication. In Sesotho there is little collocational expression of time as an orbit, though they have numerous collocations predicating a linear or scalar quantity to time, the Chinese method of valuating goods—"inch of, foot of"—is applied to time as a measure. The Chinese "deictic" coordination of space and time differs uniquely from the other three languages in its reversal of facing-direction and time movement; also, its use of the vertical dimension of space to describe time is unique in this sample.

Let me elaborate the last point briefly. Mandarin collocations of space–time deixis express an experiential perspective in which time is a medium in motion bearing events (like English), but unlike English, the experiencer is always stationary, facing the direction of the past (i.e., that which has passed) with one's back to the future. The past is before one; the future is/comes from behind. The endless course of time flows from behind the speaker, moving forever before/in front of one into the past. The future, coming from behind the experiencer, moves closer and closer. Once the time medium *cum* events have passed the experiencer, they become past events and are before or in front of one. Events have, like the experiencer, a front (further past), back/behind (more recent past/future) orientation. Unlike English, events that are before or have passed the experiencer continue to move farther and farther away from the experiencer, who remains "now" but still in the ordinal sequence. In Mandarin, the before (past), behind (future) orientation applies to the relative temporal positioning of all events, irrespective of the "now" of the experiencer. Future and past events alike show what one respondent calls a "sequencing" of "scheduling."

Mandarin has a vertical (up-down) dimension in its space–time deixis expressed collocationally. (In English we have this dimension, but it is not expressed in collocationally rich fashion. Despite "down through the ages," we don't have "up through the past"; the "Descent of Man" doesn't have as sequel or converse the "Ascent of Apes.") In Mandarin, upper (*shang*, toward or from the past), down (*xia*, toward or from the future), is the genealogical metaphor of time *par excellence*. Lest the reader hastily conclude that the space–time deixis of Mandarin is uniquely odd, note that the same deictic schematization is found in Latin (see Bettini [1988] for a very convicing demonstration).

Despite the specific differences that each language's collocations show, we have here abundant evidence that temporal collocations in Hindi and Sesotho are—in frequency, category, and in specific metaphors employed—very similar to those of English and Chinese. These similarities persist, despite the supervenience of marked cultural differences in belief, knowledge, and production that have peculiarly informed the respective temporal experiences.

Four striking features are found in the collocations from each of the four languages.

First, in each there is a key lexical item that has wide privileges of collocation across the categories of metaphor that are predicated of time—*time* in English, *shi* in Mandarin, *samay* or *waqt* in Hindi-(Urdu), and *nako* in Sesotho.

Second, for each of the four languages the collocations seem to fall into five (natural?) kinds or types based on the fundamental metaphor which contains time and which interacts either as argument or as predicate, with terms meaning (A) as divisible entity; (B) a causal force or effect; (C) a medium in motion; (D) a linear or orbital course; or (E) an artifact, that is, construction, created by the process of ascertainment or measurement.

Third, within each metaphor-defined category of collocation are a large number of specific lexical items and lexical senses entering into metaphoric predication with the word for *time* which are common to all four languages. Examples within each of the five categories, which correspond to the lists in the appendix, include:

(A)    time: wasting, saving, having, using, filling, ripening;
(B)    time: destroying, wearing, revealing, healing, befalling, happening, making happen;
(C)    time: flowing, moving, passing, coming, waiting, fleeting, bearing events with it, conducting events in it;
(D)    time is: ahead, behind, in front of, looked into/back upon; time has: location, accompanies, comes round;
(E)    time is: lost, kept, minded, discerned, marked, gained, divided, told.

Fouth, in each language there are what appear to be a large number of collocations in each language for most of the five categories. (Hindi's lack of extrapolation of time as a line and Sesotho's lack of orbital time or mechanical clock-based metaphors, are excepted here.)

There is a tantalizing indication in ethnographic and linguistic data that the conscious philosophical elaboration of thought concerning the nature or experience of time in any culture seems to be confined to a selection and embellishment of one or more of the metaphoric categories that appear to be experiential universals (see Alverson, [1994], for elaboration). Culturally specific philosophies of time, including time's tacit apperception among the Basotho, seem to consist in propositionally expressing the metaphoric content of one or more of the categories of collocations; for example, the Greek idea of time as a medium of motion in a circular orbit; the Roman (Christian) idea of time as a linear course with a beginning and end; the Chinese idea of time as a medium in motion bearing already scheduled events; the Indian idea of time as an all-encompassing entity without beginning, end, or even linearity, a kind of n-dimensional hypervolume, filled with all kinds of powers and purposes; and finally the Sotho idea of time as a three-dimensional space in which culturally patterned events take place and through which the individual moves.

These observations suggest that culturally particular constructions of time are based on elaborate propositional expressions of aspects of the five universal categories. Even modern scientific treatments of relativistic time (e.g., time is one dimension of a four-dimensional space–time manifold) trade heavily on those universal categories and metaphors of time experience described above.

## Conclusion

Even if one cannot finally and determinately state what time is, these collocational data allow us to state that, across languages and cultures, the same predicates have been used in the same collocational structures to predicate something—often the same thing—of time. And if time is what is said of it, if it is how it is described, and if statements and descriptions in language are the expression of experience, we can say that time certainly seems to have been similarly experienced in diverse settings and places.

These data and their discussion substantiate, I believe, Merleau-Ponty's claim that "to have a body is to possess a universal setting, a schema of all types of perceptual unfolding and of all those inter-sensory correspondences which lie beyond the segment of the world which we are actually perceiving. A thing is therefore not actually given in perception, it is internally taken up by us, reconstituted and experienced . . . in so far as it is bound up with a world, the basic structures of which we carry with us" (1962, p. 326). "In language . . . man superimposes on the given world the world according to man" (Merleau-Ponty, 1962, p. 188).

## References

Alverson, H. (1994), *Semantics and Experience*. Baltimore: Johns Hopkins University Press.

Bettini, M. (1988), *Anthropology and Roman Culture: Kinship, Time, Images of the Soul*. Baltimore: Johns Hopkins University Press.

Jackendoff, R. (1983), *Semantics and Cognition*. Cambridge, MA: MIT Press.

Merleau-Ponty, M. (1962), *Phenomenology of Perception*, tr. C. Smith. London: Routledge & Kegan Paul.

Tannen, D. (1989), *Talking Voices: Repetition, Dialogue, and Imagery in Conversational Discourse*. Cambridge, U.K.: Cambridge University Press.

**Appendix:**
**Selected Examples of Relevant Collocations of Time from**
**English, Mandarin, Hindi/Urdu, and Sesotho**

**English**

List A: Time is a divisible entity or substance

| | |
|---|---|
| spend time | kill time |
| divide time | outlay of time |
| seize/catch the time/moment | time is money |
| waste (of) time | spare (one) time |
| save time | do something with one's time |
| have time | good/bad/fine/troubled time(s) |

List B: Time as known from its effects

| | |
|---|---|
| father time | time-worn |
| time destroys | time wears away |
| ravages of time | time conquers all |
| time is on our side | time heals |
| time reveals/discloses | time and tide befall/betide (one) |

List C: Time is a medium in motion

| | |
|---|---|
| flow of time | time moves on |
| time is coming (when) | time is past |
| time has passed (one by) | time flies |
| time goes/flows/slips/slides by | on the wings of time* |
| time is coming/approaching | time marches on |

List D: Time is a course
(I) A linear course

| | |
|---|---|
| look ahead (in time) | look into the future |
| look back in time | look into the past |
| X follows/precedes/succeeds Y in time | time is short |
| here and now* | then and there* |
| into infinity | time without end (eternity) |
| beginning of time | length of time |
| a long time | a short time |

(II) An orbital/circular course

| | |
|---|---|
| life is a circle | the round of life |
| a wheel of fortune | ashes to ashes/dust to dust |
| time will come again | the seasons come around |
| over and over again | the swift seasons roll |

(III) Space–Time Deixis (space–time as experienced from perspective of speaker's body)
events that are before me, I will come to
events that have already happened, are behind me

---

*Indicates that the expression was not placed in this category with 100 percent reliability.

events that come/arrive late, will follow me
events that are later, I will come to
events that happened before, happened earlier, are behind me/I leave behind
what lies before/has yet to happen/is later/has yet to come to us/we face

---

List E: Time is its ascertainment or measurement

| | |
|---|---|
| lose time | gain time |
| keep time | tell time |
| give time | run out of time |
| time to burn | divide time |
| mind time | time weighs heavy on one's hands |

---

List F: Some nonspatial expressions of time
fast, rapid, slow, now and then, concluding, often, seldom, beginning, changing, again, earlier, waiting, moment, instant, sooner, later, delaying

---

## Mandarin

List A: Time is a divisible entity or substance

| | |
|---|---|
| *shijian* | time-partition/space |
| *yishi* | lifetime |
| *ji* | twelve years, century, year |
| *jiyuan* | year-beginning/first of era |
| *yibeizi* | lifetime |
| *cunyin shijing chibi feibao* | *a foot jade is not treasure, an inch of shade—a unit of time—is fought for |
| *he shiyi* | fit/suitable time |
| *shiji chengshu* | time-opportunity ripe |
| *zhuajin shiji* | grasp tight time-opportunity |
| *shiji* | lose gone-moment |

---

List B: Time is its effects

| | |
|---|---|
| *shiling* | time ordering/causing |
| *shishi* | time condition/tendency |
| *shibi* | time-evil |
| *shiyu* | time-praise/fame |
| *shiyun* | time-fate/luck/fortune |
| *nianlao* | time-worn |
| *shibuwoyu* | time not me wait (time does not wait for me) |

---

List C: Time is a medium in motion

| | |
|---|---|
| *congqian* | follow from* |
| *quri* | gone day |
| *wangri* | one day |
| *lairi* | come day |
| *shijian liushi* | time-flow-gone |
| *shijiande jiaobu* | time's footstep (pass away) |

---

List D: Time is a course
(I)  A linear course

| | |
|---|---|
| qian | before/in front of (time) |
| hou | after/behind (time) |
| xianzai | appear present (to eye) |
| xianshi | now/present to eye (time) |
| muqian | eye-before/front (now) |
| qianshi | before time (formerly) |

(II)  An orbital/circular course

| | |
|---|---|
| yinian you yinian | one year again one year |
| yitian you yitian | one day again one day |
| yiri you yiri | one sun again one sun |
| niansui diegeng | year-year in turn exchange/substitute* |
| yiyang laifu | one spring come return |
| yiyuan fushi wanxiang geng xin | one beginning again begin myriad image renew |
| riri yeye | sun/day sun/day night night* |
| xingshuang lüyi | star front repeatedly move |
| xingyi douzhuan | star move constellation turn |

(III)  Space–Time Deixis

| | |
|---|---|
| shijian | time-partition (space) |
| guangyin | play of light-shade and shadow = time |
| cong . . . qi/kaishi | from . . . beginning (space and time) |
| cong . . . dao | from . . . to (space and time) |
| qian | before/previous/earlier/in front (space and time) |
| qiantian | before-day (day before yesterday) |
| qiansheng | previous lives |
| qianbian/tou/mian | front-side, end, face (space only) |
| hou | after/later/behind (time and space) |
| houtian | day after tomorrow |
| houshi | later life/generation |
| houbian/tou/mian | behind, back-side, end (space only) |
| shang | above/up/top/before/previous (time and space) |
| shangwu | forenoon, morning |
| shanggeyue | previous/upper month |
| shang yi beizi | previous/former life |
| xia | below/down/later/future/next/after (time and space) |

Events that have already happened are those which are before (*yiqian*) or have passed (*guoqu*) the experiencer/speaker. Of the events that are before, have already passed the experiencer/speaker, those which were experienced earlier (*xian*) are before/in front of (*qian*) those that were experienced later/after (*hou*).

Events that have not yet happened to the experiencer are those that will come (*jianglai*) or are yet to come (*weilai*).

Events that will come or are yet to come are all later or after/behind (*yihou*) the experiencer.

Persons of the upper (i.e., former, previous times [*shang*]) are members of one's before or earlier (*xian*) generations.

Persons of the lower (i.e., later times [*xia*]) are members of one's after/behind (*hou*) or later (*wan*) generation.

## List E: Time is its method of ascertainment or measurement

| | |
|---|---|
| *shiling* | time-ordering |
| *shijie* | time segment* |
| *nianling* | age-teeth* |
| *anshi* | according to time |
| *zhunshi* | accurate time |
| *shoushi* | keep on time |
| *wushi* | mistake/miss time |

## Hindi-Urdu

### List A: Time is a divisible entity or substance

| | |
|---|---|
| *samay bitaana* | pass time |
| *waqt lagaana* (Urdu) | devote/take time |
| *waqt baant'naa* | allocate time (not a collocation) |
| *aek pal* | a moment (borrowed from English) |
| *waqt barbaad kayrna* (Urdu) | destroy time |
| *samay nasht' karnaa* | destroy time |
| *makkhee maarnaa* | kill flies ("kill time") |
| *samay kee bachat* | save time |
| *samay kaatnaa* | to get through time |

### List B: Time is its effects

| | |
|---|---|
| *jab samaye aataahai, tabhi woh cheez hoti haey* | when its time comes, only then does it happen |
| *waqt tabaah kartaa haey* (Urdu) | time destroys; levels/wastes time |
| *samay naashak haey* | time is a destroyer |
| *samay sab khatam kar detaa haey* | time destroys/ends everything |
| *samaye kaa koyee mukaablaa nahin kar saktaa* | no one can compete with time |
| *waqt kay nishaan* (Urdu) | marks (ravages) of time |
| *samay/waqt hamaaray saath haey* | time is with us/on our side |
| *waqt kay saath sacchchaaye saamnay aa jaatee haey* | time reveals/will unfold the truth |
| *samay main utaar chadaav* | time raises and lets fall its events |

### List C: Time is a medium in motion

| | |
|---|---|
| *samay kay beetnay say* | with the passing of time |
| *waqt kay guzarnay say* (Urdu) | with the passing of time |
| *waqt kee dhaaraa* | stream (flow) of time |
| *samay kaa bahaav* | flow of time |
| *samay/waqt kay saath bayhnaa* | to go with the flow of time |
| *samay kay badhtay huay kadmon ko rok daynaa* | stop the advancing steps of time |

List D: Time is a course
(I) A linear course [Note: there are very few, if any, collocations that predicate of time a spatial or orbital linearity. Below are some collocations which are translatable into English expressions of linearity which make use of other metaphors.]

| | |
|---|---|
| *aagay daykhnaa* | to see ahead (''think of the future'') |
| *bhavishya vichaarnaa* | to ponder the future (= previous phrase) |
| *beetay samay ko daykhnaa* | to look at the time that has passed |
| *beetay huay kal ko daykhnaa* | to see/look into the past |
| *beetay huay samay kaa smaran karnaa* | to study the time past |
| *beetay samay par drishtee daalna* | vision cast on time past |
| *kaafee samay* | long time (not a collocation) |

[no collocations for beginning or end of time]


II. An Orbital/Circular Course: There are likewise few phrases that predicate of ''time'' an orbital course.

| | |
|---|---|
| *zindagi aeyk chakkar haey* | life is a circle |
| *zindagi ghoom phir kar phir waheen* | life roams around and comes back to the same place again |
| *waqt/samay phir aayegaa* | time will come again |


(III) Space–Time Deixis
*jo mayray saamnay haey, us tak main pahoonch jaaoongaa*
what lies before me (in space) I shall reach

*jo ho chukaa, woh mayray peechay rah gaya*
what has happened/done is behind me (in time)

*jo kuch baad main hogaa/aayegaa, woh mayray peechay rahayga* (Hindi-Urdu)
whatever happens/comes in the after shall follow me/will keep behind me (this collocation implies time as a schedule of events already laid out—as in ''those who have gone before in time are ahead of me, those who follow me in time are behind me'')

*jo kuch baad main hogaa/aayegaa, main us tak pahoonch jaaoongaa*
whatever comes later/afterward I shall reach.

*Y kay baad X* (Hindi-Urdu)           X follows Y in time
The phrase is commonly used in cases where one needs to say ''X happened after Y'' or ''X came after Y''

| | |
|---|---|
| *Y kay paeyhlay X* (Hindi-Urdu) | X precedes Y in time/X comes before Y |
| *abhee aur iss-hee wakt* (Hindi-Urdu) | here and this very time |
| *yaheen aur abhee* (Hindi-Urdu) | here and now |
| *paeyhlae* (Hindi-Urdu) | before/earlier (in time) |
| *iskay poorva* (Hindi) | this before |
| *hamaaray saamnay* (Hindi-Urdu) | in front of us (in space) |
| *peechay rah jaana* (Hindi-Urdu) | behind in time/space ''afterwards'' |
| *samay say aagay* (Hindi-Urdu) | ahead of/before time |

List E: Time is its method of ascertainment or measurement

| | |
|---|---|
| ghad'ee | clock; moment/point in clock-time |
| aarambh main (Hindi) | in the beginning (of any event/period) |
| shuru main (Hindi-Urdu) | in the beginning (of any event/period) |
| samay/waqt kaa andaazaa lenaa | estimate or idea of the time |
| vaha ghad'ee nikat aa rahee haey, jab | that clock time is drawing near when |
| ghad'ee aa rahee haey | the clock-moment is coming/approaching |
| waqt/samay kaa andaazaa rakhnaa | keep tabs on time (while doing X) |

## Sesotho

List A: Time is a divisible entity or substance

| | |
|---|---|
| selebisa nako | cause time to be used |
| fetisa nako | cause time to pass |
| qeta nako | finish time |
| arola nako | separate (out) time |
| arorela nako | distribute/apportion (one's) time amongst |
| nka nako | seize/take time |
| hapa nako | snatch time |
| t'soara nako | catch, get hold of, time |
| monono o ka moso | fatness (of animals—i.e., riches/reward is tomorrow—i.e., in the future) |

List B: Time is its effects

| | |
|---|---|
| nako tse thabileng | time which gladdens* |
| nako ke ntat'a tsohle | time is the father of all (things) |
| nako ke mookameli | time is the dominating one |
| nako ke sesinyi | time is a destroyer |
| nako ke sebatalatsi | time is a leveler |
| lesupi la nako | ravages of time |
| re it'soarelelitse ho nako | we hold onto ourselves by (means of) time |

List C: Time is a medium in motion

| | |
|---|---|
| nako ha entse e ya | as time was going on |
| nako e t'soanetseng ha e fihla | as the necessary time arrives |
| nako e fihlile | the time/moment has arrived |
| ha nako e ntse e ea | as the time was passing/going |
| ha nako e feta | as time passed |
| ho tsamaea ha nako | the going/marching of time |
| ho lelemetshea ha nako | the flow of time |
| ho nyolosa le ho theosa ha nako | the rise and ebb of time |

List D: Time is a course
(I) A linear course

| | |
|---|---|
| nako e khut'soane | time is short |
| hloelisa nako tse tlang | to look from above at time that comes |
| sheba tse tlang | to look ahead to those things coming |
| sheba bokamoso | to look at things of tomorrow |
| ha sesa feleng | that place which does not give out (infinity) |
| ile boilatsatsi | gone to the place of no day/sun (infinity) |

(II) An orbital/circular course
[There are no collocations in this category; the expressions below are ordinary sentences.]

| | |
|---|---|
| *nako e (li) ea potoloha* | time revolves (i.e., events recur) |
| *nako li it'supa moliqalileng teng* | time points (directs) itself to where it started |

(III) Space–Time Deixis

| | |
|---|---|
| *X e latela Y* | X it follows Y (space and time) |
| *X e tla ka morao ho Y* | X it comes at the back of (behind, later) Y (space and time) |
| *x tsamaea mehlaleng ea Y* | X goes in the tracks (i.e., footprints) of Y (space and time) |
| *X e etella Y* | X precedes Y |
| *X e tla pele ho Y* | X (is seen by an observer) to come in front of (before) Y |
| *X e ka pele ho Y* | X is (absolutely) at/with/to the front of Y (space and time) |

List E: Time is its method of ascertainment or measurement

| | |
|---|---|
| *lahleheloa ke nako* | to lose time (literally, to be lost by time) |
| *fetoa ke nako* | to be passed by time |
| *bolela nako* | tell one the time |
| *bala nako* | read/count the time (contemporary usage) |
| *fetisa nako* | make time pass |
| *liehisa nako* | to cause time to be delayed |
| *ka nako e baliloeng* | at the time agreed upon |

# 10

# The Light of Time: Einstein and Faulkner

*Paul A. Harris*

*Abstract*   Light bears a fundamental, intriguing relation to time in the human imagination across many fields of investigation. In this essay I explore and compare the relation of light to time in Einstein's special relativity theory and William Faulkner's novel *Absalom, Absalom!* In his reconception of time in the special theory, the motion of light enables Einstein to correlate quantitatively the proper times of different observers; as the reference with respect to which time measurements may be transformed, light functions as such an indispensable explanatory device that it begins to play the role of a ground for time in rhetorical expositions of the special theory. Conversely, while light is one of the predominant metaphors for time in Faulkner's prose, his use of light in *Absalom, Absalom!* goes beyond metaphor. I argue that in some sense he seems to have taken the metaphor of light literally, as almost a physical basis for time. The relation between light and time in Faulkner is further enriched by the hovering presence of death and the function of memory. Unlike relativity theory, where the motion of light helps to fix events precisely in space-time, Faulkner's novels depend on light as a medium in which memory can refigure events.

Among the many reasons that time has remained such an enduring subject of contemplation is the simple fact that it remains nestled somewhere between the ideal and perceptual domains of our being. Time "itself" eludes perception: The contents of the past or future can be neither directly registered by the senses, nor derived from sensory experience; while we apprehend things in the present, we never lay hands on the present "itself." Yet this view of time does not do justice to the richness of our experience of time, for in certain momentary flashes time seems to materialize for us; we grasp fleetingly some quality or essence of time. These events take place at an intuitive level, where the boundary between perception and conception becomes fuzzy or effaced completely. Light is frequently the means by which such epiphanies occur; through light we gain access to something like a mental contact with time. Kant held that time is the form of the inner sense; in my view light embodies the external form of time. In this paper, I explore how light harbors what I call the texture of time in the more intuitive conceptual dimensions of two very different modes of expression: Einstein's special theory of relativity and William Faulkner's *Absalom, Absalom!* (1936).

Because such a wide gulf separates these intellectual creations, let me make a brief statement regarding method. More than making a comparison between physics and fiction, this essay seeks to filter the views of each domain through the lenses of the other. Instead of reducing one field to the other, or demonstrating strict analogies between the two, I try to align them so that we can see the work of Einstein and Faulkner as complementary materials that relate light and time in provocative ways. Thus while I do offer brief technical explications of certain tenets of Einstein's special theory of relativity, I also *interpret* its physical and mathematical definition of light; I bring a literary sensibility to bear on the philosophical implications of the physics. In doing so, I extrapolate from a suggestive remark dropped by J. T. Fraser (1971) that "in order to better understand such problems as the time vs. space–time syndrome of Relativity Theory . . . we must turn to an understanding of reality through ways other than the physical sciences" (p. 487). Conversely, my treatment of Faulkner seeks to show how bringing relativity theory's notions of light and time to bear on *Absalom, Absalom!* changes our reading of the novel.

In fact, I would contend that Faulkner's sensibility toward light shares a great deal with that of the physicist, though only speculative evidence of direct influence of Einstein's theory on Faulkner has been documented. This is not to say that Faulkner read and understood how physics correlates light and time, but that he was aware of developments in modern physics and this awareness informs his larger outlook on the world. Gerald Holton resolves this issue gracefully and forcefully: in discussing the physics metaphors Faulkner employs in *The Sound and the Fury* (1929), Holton argues that Faulkner's work exemplifies "a grand fusion of the literary imagination with only dimly perceived scientific ideas. There are writers and artists of such inherent power that the ideas of science they may be using are dissolved, like all other externals, and rearranged in their own glowing alchemical cauldron" (Holton, 1986, p. 15).

My point of departure in aligning Einstein's special theory and Faulkner's novel is simply that in both cases light serves a complicated dual purpose: Light is the physical form that time assumes in its primordial state, serving as a kind of physical source of time. Simultaneously, light provides the governing principle that organizes different experiences and qualities of time. Thus light affords Einstein and Faulkner some hold on time: Time becomes, through light, a medium that can be molded, shaped, and to some extent controlled. Because time for physicist and author alike has both a physical aspect (in its correlation to physical processes or events) and an epistemological function, it occupies a liminal position with respect to the fundamental dualism that haunts the study of time: The essential split between materialist views based on physical time, which does not have an irreversible directedness from past to future, and mental time, consciousness, and its impression of temporal flow. But as we shall see, Einstein and Faulkner's use of light cannot fully cover the breach between these orders of time: This dualism ultimately resurfaces in different ways in special relativity and *Absalom, Absalom!*

For most of us, the opposition between physical time and mental time obtains almost by definition. The biologist Gerald Edelman (1992) asserts that this dichotomy will remain intact until consciousness can be explained in terms of physics (p. 4). Thus the laws of physics concern themselves exclusively with the nature of time in physical processes; they seek to jettison the problem of consciousness. But, as P. T. Landsberg observes, mental time is not so easily gotten rid of. Landsberg (1982) concludes his introduction to *The Enigma of Time* by stating that for physicists, "future advances" in the study of time are contingent on their ability "to remove a problem of self-reference. We are stumbling here because we do not know how to introduce the observer properly" (p. 28).

The problem of self-reference indeed proves an obstinate one, for in trying to remove it, one creates further confusions that reproduce a diametric opposition between the "real," Platonic, immutable universe and an illusion of time produced by consciousness. The long train of controversies surrounding the philosophical implications of Einstein's special theory of relativity exemplify such confusions. On the one hand, the special theory seems to manage and contain temporal dualism: The speed of light, a universal constant, provides an atemporal standard of measure that anchors physical time. On the other hand, this quantity enables the physicist to calibrate the relative motions and times among different observers. The special theory thus seems to resolve the problem of how to introduce the observer in a manner that sustains its control over time: The "observer" is a human perceiving instrument that measures time and space, not a person with a complex psyche. It does seem ironic that relativity theory is frequently cited as proof that all perspectives are just "relative" and equally true, for Einstein's real achievement was to use the motion of light as what he called a "universal formal principle" that upholds a complete *Weltbild*, a static picture representing an atemporal universe.[1]

The special theory does not supplant the absolute world of Newton with a completely relativistic universe—as would be envisioned by a radical constructivist. The special theory rather inverts the grounds on which stability in the universe rests. In typically laconic fashion, Einstein (1905) overturned physics by noting that

---

[1] It is extremely interesting that this control over time originates in a kind of thought experiment predicated not on pure mathematics but on the imagination of a human observer. Einstein's insight into the fact that the motion of light is a universal constant hinges on a rather playful, imaginative reflection on light itself. In his retrospective "Autobiographical Notes," Einstein recollects how this "principle resulted from a paradox upon which I had already hit at the age of 16" (1949, p. 53). The paradox posits an observer traveling near the speed of light, who sees a light beam: The conundrum is, what does the observer see? Does he see light standing almost still (i.e., "a spatially oscillating field at rest")? No, Einstein decided, light would have to look the same to this observer as to a stationary observer. In other words, light will have the same velocity relative to an observer, no matter what velocity the traveler is moving at relative to the light beam. "One sees," Einstein writes, "that in this paradox the germ of the special relativity theory is already contained" (1949, p. 53). One also sees that Einstein resolved this paradox by simply making the constancy of the speed of light a postulate; that he assumed his solution to the paradox was correct, and promptly removed himself as the onlooker at the scene of the observer and the light beam.

"the phenomena of electrodynamics as well as of mechanics possess no properties corresponding to the idea of absolute rest" (p. 37). This "conjecture" he termed the " 'Principle of Relativity,' " and then added his second postulate, "that light is always propagated in empty space with a definite velocity *c* which is independent of the motion of the emitting body" (p. 38). Essentially, the motion of light has replaced Newton's absolute rest as the invariant feature of the universe by which the relations among relative inertial frames of reference are measured. We have shifted from a universe oriented by the unmoving container of absolute Newtonian space and time or the long-sought ether, to one calibrated by the invariable motion of light. An observer's location is given by three spatial dimensions and one time dimension. The time dimension is the observer's "proper time"; mathematical operations called the Lorentz Transformations translate the proper times of different observers, their different measurements of space and time, into one another. Proper times are correlated by an abstract variable termed the coordinate time.

A convenient way to illustrate the contrast between the Newtonian and Einsteinian worlds is to consider how "now" is understood in each one. In Newtonian mechanics, a person experiences a particular moment in absolute space and time, so that "now" extends throughout the universe at that instant. In special relativity, light itself delimits the extension of "now" at a distance: In a graphic rendering called a Minkowski diagram, paths of light extend outwards from the observer's position, creating what is called a "light cone." Events that occur within the cone have a causal (temporal) relation to the observer, while those that lie outside the cone are, as Arthur Eddington (1928) put it, in an "absolute elsewhere."

Looking at a Minkowski diagram brings most of us into the confusions that reintroduce the divorce between physical and conscious time, though nonphysicists almost inevitably form a dualistic interpretation of Einstein's universe through one of two errors. We may unwittingly equate the time dimension with our consciousness. This is incorrect because, as David Park (1971) has argued, we would really need to add a dimension to the diagram to represent consciousness; here, we have only a "now." On the other hand, we may conclude that the static diagram represents an immutable universe, relegating time to something like a side-effect of consciousness. To help rectify this situation, J. T. Fraser (1977) introduced a "structured Minkowski diagram" (Figure 10.1). Without adding any new physics, Fraser aligns different temporal levels within the diagram, from the atemporal path of the light cone to noetic temporality. This modification overcomes the radical dualism between mind and matter by embedding them both within a single ontology. Fraser's diagram reminds us that special relativity is a theory that relates different temporalities with respect to a constant located at the atemporal level of nature. The Minkowski diagram can accommodate the emergence of different temporal levels, but it cannot account for their qualitative differences.

Rather than attempting to resolve any of the logical paradoxes and technical problems engendered by special relativity or its limited representation in the Minkowski diagram, my brief discussion of Einstein's theory is meant to bring out the way in which it raises light to a universal formal principle that, figuratively speaking, shapes time, and places conditions on the transmission of signals among observers

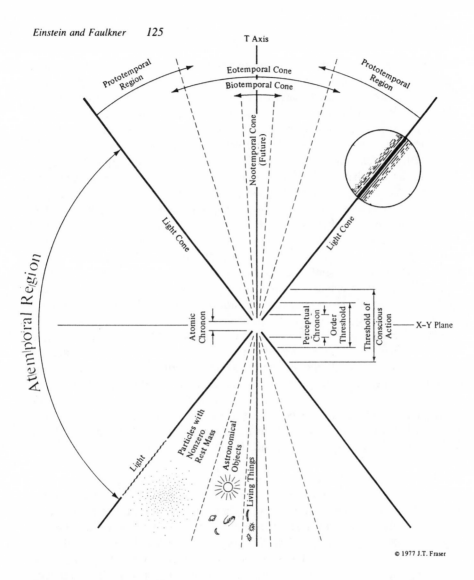

*Figure 10.1.* The Structured Minkowski Diagram.

Motion in the *X–Y* plane of this diagram may be represented by three-dimensional curves in the *X–Y–T* space. The initial reference is to the here-and-now of an inertial human observer located at the origin *O* of the coordinate system. At a particular instant, remembering his past and imagining his future he makes a rapid mental survey of the world. Using his clock as his time reference, he plots his time units along the *T* axis.

Special relativity theory permits him to assume any kind of motion in the *X–Y* plane as possible and hence admit its appropriate worldline as valid, as long as the worldline remains inside the light cone. But nature does not permit him to select freely any object as suitable to move in a manner appropriate to any arbitrary worldline.

The regions limited by and external to the light cone are unreachable from the here-and-now of the observer; he or she can say nothing about worldlines in that region of the diagram. Worldlines in the line cone proper must be those of particles with zero rest mass. Worldlines that pass through the prototemporal regions close to the light cone must be those of elementary particles. Worldlines more or less within the eotemporal cone are likely to be those of astronomical objects. But this region also includes all living things which move at the ordinary speeds of living organisms. Their worldliness would fall within the biotemporal cone. The worldline of the observer herself is the T-axis.

in time. While Bergson did not perfectly understand the special theory (the definitive view on this subject is Capek [1991]) he saw that its revolutionary aspect lay in the consequences of this emphasis on light as the basis for all measurements of time. The axiomatic status of light counters the instinctive preference for solid objects and unchanging matter firmly established in natural philosophy at least since Descartes and Newton. Bergson (1923) observed that with relativity, "it is the figure of light which imposes its conditions on the rigid figure" (p. 50). Bergson points to the fact that the dilation of time and contraction of space measured by different relative observers inevitably follow from the absolute motion of light.

This reversal in natural philosophical perspectives has its corresponding effect in the domain of metaphysics as well. Throughout the Judeo–Christian–Classical intellectual tradition, light has served primarily as the metaphor for truth, for spiritual insight (for a historical overview of this phenomenon [Blumenberg, 1957]). Gilles Deleuze (1986) asserts that the theory of light in special relativity "breaks with the whole philosophical tradition which placed light on the side of spirit and made consciousness a beam of light which drew things out of their native darkness" (p. 60). In relativity, the universal speed of light is rather a first principle, the ontological condition on which time and consciousness alike are predicated. Light is a "condition" both in the sense of a material state, and a necessary premise or antecedent proposition. In the apt phrase coined by Arthur Eddington (1928), the light path in special relativity represents "the grain of world-structure" (p. 54) in any region of space–time—so long as one realizes that this is a grain only, not something as substantial as wood. It is with this image of a universe whose space–time fabric is the grain of light that I want to turn to Faulkner, as if this grain were the imaginary field populated by his characters and traversals in time.

I believe that it is fruitful to filter our view of Faulkner through Einstein's physics because, more than any writer I know, Faulkner imputes to time an elusive but inescapably physical quality. Critics usually treat time in Faulkner as an existential theme, or analyze how he manipulates narrative time through disrupting the chronology of events in the plot by telling the story out of order. But I think relativity sensitizes us to the way that time permeates the landscapes of Faulkner's fiction as something inextricably bound to space, an encompassing medium or overarching atmosphere sensible to consciousness. I have the feeling that Faulkner felt time in his innermost physiological makeup, that he experienced his Southern heritage not as distant history but as something like a form of body-memory. The past saturates settings in Faulkner novels in the form of sensuous images—time floats around as the smell of wisteria, the tolling of old bells, the dust on a country road. Faulkner's primary metaphor for the physical medium of time, though, is light: Specific forms

---

In this diagram the observer and the furnishings of the universe are not ornaments introduced to help people understand the scheme but determinants of the world representable by the diagram.

Reproduced from J. T. Fraser (1977), *The Genesis and Evolution of Time*. Amherst: University of Massachusetts Press.

of light express different qualities of time. As the medium of time, light also becomes the vehicle through which Faulkner can bend time to specific purposes: Light collapses events throughout time into one another, and propagates stories from one period instantaneously into a present. Faulkner uses phrases to describe time that sound borrowed from physics, as when he terms future events in *Absalom, Absalom!* "still undivulged light rays."

Faulkner called his fictive world of Yoknapatawpha County "a cosmos of my own," and *Absalom, Absalom!*, published in 1936, represents his most ambitious attempt to map this universe, for it brings a sweeping historical tableau to bear on a contemporary situation (Sundquist [1983] argues this point in detail). The book recounts Thomas Sutpen's "design" to build a plantation and bear progeny—to create a Southern aristocratic family at one stroke. The design self-destructs from within his own divided house. His white son Henry shoots Charles Bon, Sutpen's son by an octoroon mistress, because Bon is engaged to Judith, Henry's sister. Thus there is a crossing of paradigmatic Southern themes: incest and miscegenation overlap against the backdrop of the Civil War. This story is told in 1910. In the first half of the novel, Rosa Coldfield and Mr. Compson tell it to Quentin Compson; then Quentin and his Harvard roommate Shreve recreate it for themselves.

Broadly speaking, this novel confronts us with a literary version of the definitive dilemma of relativity theory: How do we find a stable frame of reference, when all that exists is a multitude of relative viewpoints, all equally "true"? The problem in *Absalom, Absalom!* could be formulated as follows: How do we reconcile the idea that this novel maps Faulkner's fictive *uni*-verse with the fact that as readers we experience a *multi*-verse—that we only see this universe from multiple, partial narrative perspectives?

I believe that Faulkner, like Einstein, finds in light the grain of time, the ground that stabilizes relations among different observers or narrators. But, of course, there is an irreducible difference between how these relations are expressed in physics and literature. In relativity theory, the Lorentz Transformations calculate an invariant, symmetrical reciprocity between the time measurements of different observers. In literature, instead of quantitative equivalence, we look for qualitative comparisons between the experiences of different characters or narrators. Rather than mathematical equations, the tool for forging links between different experiences is metaphor, and metaphor recognizes qualitative difference even as it creates a basis for comparison. But while a literary critic might allow that in *Absalom, Absalom!* light is the underlying metaphor that connects together the different narrators and their respective narrations, I feel that light is more than a metaphor for time. I think it is the linguistic figure Faulkner found most adequate in his struggle to express in language the prelinguistic, physical sensation of his native Mississippi time-scapes.

In my concluding reflections, I will discuss a few passages to show how light serves Faulkner two purposes in *Absalom*: First, he uses specific images of light to express the different relations characters have to the story and the past. In the novel, there are three central space–time locations where the story unfolds: Rosa Coldfield's office, Mr. Compson's porch, and the dorm room shared by Quentin and Shreve. Each of these scenes of narration is permeated by its own peculiar light, and in each case this light embodies the temporal relations that define the storyteller's narrative time—their relation to the past story in this present moment. But second, even as light defines each character's experience of time, it functions as the hinge that superimposes different narrative sequences onto each other—metaphors of light serve not only as the associative connections among the different narrators and scenes of narration, but as the means to *merge* past and present times.

Let us turn to a passage from the novel, the opening sentence of *Absalom, Absalom!* From it we may already unpack several dimensions of Faulkner's narrative time.

> From a little after two oclock until almost sundown of the long still hot weary dead September afternoon they sat in what Miss Coldfield still called the office because her father had called it that—a dim hot airless room with the blinds all closed and fastened for forty-three summers because when she was a girl someone had believed that light and moving air carried heat and that dark was always cooler, and which (as the sun shone fuller and fuller on that side of the house) became latticed with yellow slashes full of dust motes which Quentin thought of as being flecks of the dead old dried paint itself blown inward from the scaling blinds as wind might have blown them [p. 7].

Like relativity theory, this sentence uses light to measure time: The discrete time interval ("a little after 2 oclock until almost sundown") is measured and embodied by light (the sun shining on the house). Time has a physical correlate: Light assumes visible, physical form as dust motes. "Time" is thus counted not by a ticking clock but by the physical motion of light. Indeed, the only thing that moves in this passage is the sun and the "yellow slashes full of dust motes" in the blinds.

This sentence makes us experience time as a physical presence by interweaving space and time into a single expression. It depicts both a setting and the proper times peculiar to each character. The office sealed for 43 summers embodies the static quality of Rosa Coldfield's life: She remains trapped by her fury at events from half a century before; for her, time has stopped long ago. Rosa's narration that afternoon recounts her experiences from 43 years before. The intriguing relation set up in the sentence is that light becomes the medium that propagates the narration from an atemporal (chaotic) source to an observer. From Quentin's viewpoint, the elapsing time traced as sunlight and dust motes transports the story from Rosa's past to his present. The story originates as light, taking visible form as dust motes for Quentin's listening consciousness.[2]

The way that time must stand nearly still in the present in order to receive this past temporal order is expressed through Faulkner's stylistic techniques: The furious rush of the words tries to pack all these temporal orders in one sentence. This sentence is broken only by a dash, a comma, and a parenthetical clause; it struggles

---

[2]Faulkner's use of the dust motes image convinces me that he knew Lucretius well; in *De Rerum Natura*, this image stands for a primordial ground of time—dust motes are moved by the sway of atoms "at the first-beginnings of things"; from this atemporal level, time and events "by successive degrees emerges upon our senses" (2.128–130).

to contract its spatial extension to a point, and the different time series it depicts to an instant. Its surface is dappled with meticulous repetitions designed to fuse disparate components into a closed whole. The double use of "still"—first to describe the unmoving afternoon and then the fact that Rosa carries on a tradition begun by her father—equates lack of motion with lack of change in time; the pairings of the "hot" day and the "hot" room and the "dead" day and the "dead" paint hold time still and identify the day with the spatial location (the room in the house). Even the passage of time marked by movement of the sun must be slowed by verbal repetitions: the sun shining "fuller and fuller," and the repeated fact that "it should have been later than it was." By bathing the scene of narration in this specific form of light, Faulkner uses light as a metaphor that stitches together physical and mental time series. The imagery implants in our minds an imaginatively construed sensuous medium, that we then continue to hold fixed as the location into which the subsequent narration flows.

Light thus mediates between an atemporal domain where the whole story is, as the narrator terms it, like a dream that arrives "stillborn and complete, in a second," and the level of consciousness, where to hear the story "depends on elapsing time." In other words, as the story bears down on Quentin, time slows down. He experiences a kind of time dilation as he is consumed by Rosa's narration. When he intuits the story as an atemporal whole, he realizes that "the sun seemed hardly to have moved" (p. 22). This process builds throughout the first half of the novel, until it snaps midway, and in the second half, Quentin looks back on the story. That is, instead of the past bearing down on him, slowing down Quentin's experience of the elapsing present, the relation is reversed: Quentin is journeying from the present back toward that past, a domain that is, even as he moves toward it, already there waiting for him.

This reversal occurs in chapter 6, which begins by referring to a letter from Quentin's father lying in the lamplight of a dorm room at Harvard. For Quentin, the "dead summer twilight" of the "dead dusty summer" from the Compson porch—the scene of chapters 2 to 4—is "attenuated up from Mississippi" by this letter. Within this crossing of two time series Faulkner inserts Quentin's memory of the trip he and Rosa make to the Sutpen mansion. The details of this trip are only revealed in the last chapter, so that this halfway point aligns the light-times of both preceding narrations and those that will round out the novel.

The qualitatively different type of light, the lamplight, marks the crucial shift in narrative viewpoint. Quentin transforms from a listener receiving the past to a partner with Shreve recalling and actively reconstructing the Sutpen story. The lamp is a traditional literary figure for memory, and the lamplight embodies the new narrative time in the novel—Quentin and Shreve become the work's memory of itself, supplementing and fleshing out the story for themselves. I use the metaphor "flesh out" nearly literally: What we see happen is that the lamp of memory illuminates the dust mote contents of the story, and makes the ghostly characters in the story come alive.

To show how we may imaginatively yoke together Faulkner's fiction and Einstein's physics in complementary terms, we could configure the levels of time in *Absalom, Absalom!* along the lines of a reversed structured Minkowski diagram. While the Minkowski diagram calibrates time and temporal levels to the atemporal umwelt composed of the chaotic motion of light, here we would orient temporal levels from the viewpoint of the conscious character-narrators. Thus rather than "reading" the structured Minkowski diagram from the atemporal umwelt inward, imagine that because memory looks backward, we follow memory-light rays emitted from the noetic level in the "middle" of the diagram outward toward the atemporal paths of light. In terms of the novel, we visualize memory-light intersecting at some points with the dust motes of the story projected, in the first half of the novel, from primordial light.

If atemporal light issues from the paths along the light cone, from where does memory light originate? It is when the moment of death seems imminent that memory contracts to its fullest tension; the specter of death jarringly reverts our gaze back over our lives and history. Thus in this thought experiment, we add an oblique, shadowy dimension below the noetic level, that of death. Death is in a strange way the point of origin for Quentin's figure in the novel: Faulkner readers know that Quentin has already committed suicide 7 years before in *The Sound and the Fury*. It is as if Faulkner needed to explain Quentin's death in the earlier book by setting it against the larger historical backdrop of the Civil War period.[3] Quentin is already haunted and absorbed by this past, and lives in the present as a fated repetition, a ghost, of the figures from that past: He is described as being "too young yet to be a ghost, yet having to be one for all that" (p. 9).

With these speculations in mind, let us return to the text. Light imagery pervades the opening pages of the middle chapter in *Absalom*. The lamplight illuminates the letter from Quentin's father; the letter brings with it the summer twilight. Imagery links up in chains of the different events and narrations in the novel: The dust motes issuing from the atemporal level in the opening sentence of the novel become the "moonless September dust" (p. 175) on the road during Quentin and Rosa's nocturnal journey to the mansion. While the dust in Rosa's office was kept in the cool darkness, this is "heat-vulcanized dust" which rises:

> [In a] dustcloud in which the buggy moved not blowing away because it had been raised by no wind and was supported by no air but evoked, materialized about them, instantaneous and eternal, cubic foot for cubic foot of dust to cubic foot for cubic foot of horse and buggy . . . the dustcloud moving on, enclosing them . . . bland, almost friendly, as if to say, *Come on if you like. But I will get there first; accumulating ahead of you I will arrive first . . .* [p. 175].

In the opening scene, the dust motes serve as the medium through which the story is conveyed from atemporal light to Quentin. Here, in the light of Quentin's memory, the "galvanized" dust cloud takes on a monstrous life of its own; it becomes a thing in its own right, the embodiment of the story itself. The dust cloud story

---

[3]It is probably a pertinent psychobiographical fact that Faulkner was also mourning and working through the death of his favorite brother Dean, who crashed flying the plane Faulkner had bought him.

precedes Quentin because it is the goal toward which he travels back into the past. As if to suggest that it is Quentin's active memory that brings the story to life, Quentin is depicted "feeling exactly as an electric bulb" (p. 176) as he rides through the dust.

The dust cloud is a foreshadowing figure for the ghost from the past that awaits Quentin at the mansion—as we learn in the last chapter, he will discover Henry Sutpen there, waiting to die. In this final chapter, the narrator recounts how Quentin can still "taste the dust" as he lies in bed at Harvard remembering this trip and his encounter with Henry. Rather than a climactic confrontation or enlightening insight, however, there is simply a static exchange of phrases between the two:

> *And you are—?*
> *Henry Sutpen.*
> *And you have been here—?*
> *Four years.*
> *And you came home—?*
> *To die. Yes.*
> *To die?*
> *Yes. To die* [p. 373].

The way that this passage is italicized, blocked off, and rendered in verse form on the page gives this episode a visual quality similar to the dust cloud; the physical form of the language matches the object it describes. The phrases seem to merge together, as questions are interrupted and answered, and the questioner echoes the answer back. This formal merging effectively collapses the Sutpen story from the past into Quentin's present. The temporal levels stretched out in memory's light from the domain of death collapse back to their origin: the conversation merges Henry ghost with Quentin ghost in a looping repetition around the axis "to die." This is the highly ambivalent spirit in which the novel concludes a few pages later.

While specters of death haunt the pages of Faulkner, and Einstein's special theory does not explicitly involve human lives but time and physical events, the work of both men does, in the end, bear directly on "Time and Life." In very different ways, I believe that Einstein and Faulkner sought to quick-freeze time and life into a form that, brought into the proper light, would free life and time to move again. Einstein's theory recuperated a universal reference frame, creating a timeless picture of the universe, but one that also gave each frame of experience its due recognition. Faulkner seeks through his chains of metaphor to freeze time by fusing disparate time experiences together—even if only for a fragile moment, in the fleeting context of fiction. But behind this attempt to overcome time is an impulse toward life. Faulkner once said "Life is motion," and that his aim as an artist was "to arrest motion, which is life, by artificial means and hold it fixed so that a hundred years later, when a stranger looks at it, it moves again since it is life" (Cowley, 1959, p. 139). I have addressed "Time and Life" only indirectly then, from the domains that shadow it on either side—the chaotic motion of atemporal light and the blank space of death. But from these virtually extratemporal dimensions there shines the eerie light of time.

# References

Bergson, H. (1923), *Duration and Simultaneity*, tr. L. Jacobson. New York: Bobbs-Merrill.

Blumenberg, H. (1957), Licht als Metapher der Wahrheit. *Studium Generale*, 10:432–447.

Capek, M. (1991), What is living and what is dead in Bergson's critique of relativity. In: *The New Aspects of Time*, Vol. 125, Boston Studies in the Philosophy of Science, ed. Boston: Kluwer Academic.

Cowley, M., ed. (1959), *Writers at Work*. New York: Viking.

Deleuze, G. (1986), *Cimena I: The Movement-Image*, tr. H. Tomlinson & B. Habberjam. Minneapolis: University of Minnesota Press.

Eddington, A. (1928), *The Nature of the Physical World*. New York: Macmillan.

Edelman, G. M. (1992), *Bright Air, Brilliant Fire: On the Matter of the Mind*. New York: Basic Books.

Einstein, A. (1905), On the electrodynamics of moving bodies. In: *The Principle of Relativity*. New York: Dover, 1923, pp. 35–65.

——— (1949), Autobiographical notes. In: *Albert Einstein: Philosopher-Scientist*, ed. P. A. Schipp. New York: Tudor, pp. 2–95.

Faulkner, W. (1929), *The Sound and the Fury*. New York: Random House.

——— (1936), *Absalom, Absalom!* New York: Random House.

Fraser, J. T. (1971), The study of time. In: *The Study of Time*, ed. J. T. Fraser, F. C. Haber, & G. H. Muller. New York: Springer, pp. 479–502.

——— (1977), *The Genesis and Evolution of Time*. Amherst: University of Massachusetts Press.

Holton, G. (1986), *The Advancement of Science and its Burdens*. New York: Cambridge University Press.

Landsberg, P. T., ed. (1982), Introduction. In: *The Enigma of Time*. Bristol: Adam Hilger.

Lucretius, *De Rerum Natura*, tr. W. N. D. Rouse. Cambridge, MA: Harvard University Press, 1975.

Park, D. (1971), The myth of the passage of time. In: *The Study of Time*, ed. J. T. Fraser, F. C. Haber, & G. H. Muller. New York: Springer, pp. 110–121.

Sundquist, E. (1983), *Faulkner: The House Divided*. Baltimore: Johns Hopkins University Press.

11

# Real Time, Life Time, Media Time: The Multiple Temporality of Film

*Sabine Gross*

*Abstract*  Film seems to provide the closest approximation of lived, experiential time that any technical medium is capable of. Cinematic time is time reconstituted, however; the camera extracts from time what it records, and then reinscribes it into time through the real-time mechanism of projection. A closer look at the history and technology of the medium not only shows that time had to be dissected before it could be reassembled, it also reveals that film's seemingly natural and immediate impression of time is a simulation based on an unprecedented degree of mediation—it is a ''reality effect'' rather than reality. Time as represented in film is thus fundamentally different both in origin and effect from the time of real life. This is confirmed by stylistic devices that seek to capture ''natural'' time spans on the screen: Within the cinematic medium, time is governed by different laws than in real life, even though the spectator ultimately does translate cinematic time back into experiential time.

> The photographic image is the object itself, the object freed from the conditions of time and space that govern it.... For photography does not create eternity, as art does, it embalms time, rescuing it simply from its proper corruption. Viewed in this perspective, the cinema is objectivity in time. The film is no longer content to present the object enshrouded as it were in an instant, as the bodies of insects are preserved intact, out of the distant past, in amber. The film delivers baroque art from its convulsive catalepsy. Now, for the first time, the image of things is likewise the image of their duration . . . [Bazin (1971), pp. 14–15].

This is how André Bazin, one of the most famous film theorists, situates film with respect to the visual arts and photography. He establishes time as a crucial dimension

The author wishes to thank Kristin Thompson, David Bordwell, Chuck Wolfe, and Edward Branigan.

of the visual arts and visual media in general. Not only does Bazin define the different media in terms of time, he also implies that their purpose, indeed their ambition, is to represent the temporality of human existence while moving beyond the restrictions it imposes. Each medium commemorates and keeps alive human time in its own characteristic fashion. Painting and sculpture "create eternity"; that is, they triumph over time, transcend it, capture the spiritual essence of human existence by removing it from time's reign. In direct contrast to this timelessness, many critics have emphasized the pronounced historicity of photographs. They stop time, mummify it as they do their objects, and speak to us from a past that they keep alive in its pastness. Roland Barthes captures both the relentless precision and the limitations of photographic cameras when he refers to them as "clocks for seeing" (Barthes, 1981, p. 15). He speaks of a photograph as giving the spectator "the absolute past," and when he says of historical photographs that "there is always a defeat of Time in them," two readings of this statement suggest themselves: While photography defeats time by arresting it and recording the moment, time itself moves on and renders the photograph a memento of the past. "There is a superimposition here: of reality and of the past," asserts Barthes (p. 76), and calls this "essence" of the photograph its "That-has-been."

Compared with the archival quality of the photograph, which preserves a particular moment in time, film offers a mechanism for storing and then releasing (i.e., re-storing) time. While it shares the illusions of depth and space with photography, it is unique in creating the illusion of time that makes it so realistic. Film and video provide the closest simulation of lived experiential time that any technical medium is capable of. Where photographs offer us a trace of the past, the most conspicuous feature of cinematic time is the apparent presentness conveyed by both images and spoken sound. Taking up Barthes' formulation, Christian Metz (1991, p. 6) asserts that "the movie spectator is absorbed, not by a 'has been there,' but by a sense of 'There it is.' " Its effect is so much stronger because it combines the certainty of a pro-filmic reality (the knowledge that real bodies executed the movements) with the powerful reality effect of motion in time, and usually, with a real-time sound track.

Given the lifelikeness of events on the screen, the logical assumption is that film lends itself naturally to representing time and the way human beings live it. We experience time as duration, motion, rhythm, and change, and that is what cinema delivers. It approximates temporal reality in a way no other medium can, mobilizes time and liberates it from the representational constraints that other media subject it to. If the photographic truth which fixates time can be seen as its annulment and death, film seemingly reanimates and reinstates time. The extent to which the medium makes time present and embodies it for the perceiver makes film *the* medium of time.[1] Films are not *about* time, they are *of* time, indeed they seem to *be* time.

---

[1]This must not be taken to mean that films have a tendency toward thematizing time explicitly. On the contrary, the cinema rarely offers treatises on the nature of time, as philosophy and literature have frequently done.

The relationship of film and time is, however, rather more complicated than this sketch admits. In fact, perhaps the most striking feature of film as an audiovisual medium is its multiple temporality. This essay attempts to bring historical and technical aspects of cinematic temporality and its representation of time into closer view.

If we turn to the history of photography and film, we find that the supposedly clear-cut dichotomy of embalming time versus bringing it to life does not hold up under closer scrutiny. Interestingly, because of their long exposure times, early photographs inevitably recorded time by depicting its progression in fuzzy outlines or blurred movement across the frame. The supposedly static medium actually chronicled movement and the passage of time. It captured unfolding events instead of isolating frozen moments of time. The oldest extant photograph, a heliograph taken by Nicéphore Niépce in 1826, had an exposure time of 8 hours (Oeder, 1990, pp. 248–249). Because of the rotation time of the shutter, human figures in so-called panoramic photographs could actually appear twice in the same picture, with an infinitesimal difference in age determined by the shutter speed (Oeder, 1990, p. 254). The kind of temporal photographic precision we take for granted nowadays was established only when shorter exposure times increased the degree of segmentation imposed on the continuum of lived temporality. By the 1870s photography had reached a degree of technical sophistication that allowed significant reductions in exposure time and thus the fixation of what could be called a moment, rather than a stretch, of time.

With this, represented time becomes discontinuous, and it is precisely this discontinuity of photography which made it not only a precursor, but a precondition of film. Before the images could be made to move, they had to be arrested. With enormous technical effort—a battery of 24 cameras, each with its own trigger—Eadweard Muybridge in 1878 sliced up the continuity of a horse's gallop into distinct images and thus for the first time made visible what had hitherto been hidden from human eyes in the continuity of movement. Muybridge proceeded to take hundreds of photographic series of a great variety of activities (1887).

Etienne-Jules Marey was another pioneer of what was fittingly called chronophotography or "time-light-writing." His sequence photographs combine a number of short exposures on one disk, arranging them into a sequence of movement and inscribing succession into the photograph in a new way in an attempt to recreate a space–time continuum.[2]

Motion studies such as these carried out the dissection of movement, and thus of time, that was a precondition for making pictures move. Cinematic time is time reconstituted. The camera does not record continuous time; it takes samples or slices of time as static images. The spatial sequence of these frames is then converted into temporal succession through the real-time mechanism of projection, which inscribes them into time again, resulting in the impression of movement.[3] In a sense,

---

[2]Muybridge's work has received considerably more attention than Marey's. However, two recent books on Marey and his work (Braun, 1993; Dagognet, 1993) address this neglect.

[3]This succession in turn limits the spectator's freedom in time. While painting, sculpture, and photography do not restrict the time course of the spectator's viewing, film does so most emphatically.

the projector borrows present time to reanimate the past. Strictly speaking, film does not present time. Paradoxically, it was the earlier photographs with their blurred outlines and lack of definition which actually presented images of continuous time. Film simulates time by taking advantage of the inertia of the visual system. At a minimum rate of about 16 frames per second (fps), the visual apparatus is tricked into filling in the temporal gaps between the slices of time offered by the frames, thereby transforming small differences between static images into continuous movement (see Hochberg [1986] for a detailed description of the process summarized here). Early films actually used much higher rates—of about 50 fps—once that became technically feasible. By the 1890s, frequencies of up to 100 pictures per second were possible. In fact, the number of frames per second directly affected the choice of subject matter. Richardson (1967, p. 24) quotes a letter written to him on January 24, 1925, by Thomas Edison, in which Edison states: " 'With my early pictures the rate at which they were taken varied from 40 to 50 per second. This gave a smooth and beautiful reproduction even though the movements photographed were quite rapid.' "

Edison then points out that at the reduced rate of 16 fps, "sudden and rapid movements" could not be filmed satisfactorily and therefore had to be avoided. Film, the medium of time par excellence, did not admit just any kind of time: Its verisimilitude depended on the self-imposed censorship of filtering out what could not be presented convincingly.

In order to convert static to dynamic images and create the impression of continuous movement and temporal flow, several problems had to be overcome (see Hochberg [1986] for a discussion of the perceptual mechanisms involved, such as masking, flicker fusion, and apparent movement). Initially, the use of photographic plates or disks limited the number of pictures that could be taken. The invention of the film roll by Eastman in 1885 made it possible for the first time to shoot—and project—longer sequences. (Most early film cameras doubled as projectors, with an added source of light.)

The main difficulty in early attempts at chronophotography was how to take and project pictures at sufficiently short intervals of time to suggest a continuous sequence. Not surprisingly, a patent for a camera-projector filed in 1890 emphasized the crucial fact that the images were to be projected "sufficiently near each other in time, to convey the idea of life and motion" (Coe, 1981, p. 60). This posed the problem of getting the roll of film to move steadily, but at discrete intervals, and of making the time it took to move from one frame to the next "invisible" for the spectator.[4]

A brief look at the film mechanism—a strip of film is moved, then stopped for exposure or projection of a single image-frame, then moved to the next

---

Time is prescribed by the process of projection which requires the viewer's continued compliance and unwavering attention.

[4]The Lumière brothers' *cinématographe* pioneered a claw device to move the film (which was eventually superseded by a sprocket mechanism). The patent awarded them on February 13, 1895 states that of the total showing time for each frame, one-third was required to draw the film down (Liesegang, 1986, p. 60).

frame—shows clearly that both the motion and the continuity we see on the screen are illusory. The only actual motion is that of the film being scrolled through the projector. In fact, the spectator sees movement where there is none, and does not perceive the movement that actually takes place.

Film simulates and reconstructs time, rather than capturing it. But this ability to simulate time and provide an illusion of movement and temporal continuity is quite effective. The unprecedented degree of mechanical mediation and the technical (and physiological) complexity of the process in no way detract from the cinematic "impression of reality" (Metz, 1991). The panic of the Lumières' 1895 audience as the train on the screen approached them is one of the legends of cinema history that has been quoted again and again. Musser (1990, pp. 152–155) relates a somewhat similar reaction to an American train film in 1896, and Vaughan (1990, p. 63) describes the audience's reaction to one of Lumières' earliest one-minute films, *A Boat Leaving Harbour*. When it was shown in London in February 1896, "visitors came forward after the performance to poke at the screen with their walking sticks, convinced that it must be made of glass and conceal a tank of water." It is difficult to determine the extent to which such accounts have themselves undergone editing for maximum effect. In recent years, the notion that early audiences were naive victims of a cinematic realism which they mistook for reality has itself been attacked as a bit of myth-making. It would seem to constitute a projection onto the history of cinema of a much more recent claim that film possesses a privileged relationship to reality. Early cinema did not define itself—and was not received by its audience—as a mirror of reality. Indeed, the mere duplication of reality would probably have failed to engage spectators' interests even in the early days of cinema. Part of the power of the new medium was the incompleteness of the illusion. No real tree could match the fascination of the aspen-tree on the screen precisely because its leaves moved in a different space and time. On December 4, 1896, the *British Journal of Photography* published a rather clumsy piece of humorous verse which emphasizes the attraction of film while pointing out the shortcomings of the new medium:

> But it is a wonder really
> How the constant flood of life
> O'er the screen keeps moving freely
> Full of action—stir and strife. . . .
> 'Tis far from perfect in its movements,
> 'Tis very hard upon the eyes;
> The jolly wobble no improvements,
> Smooth running films a surprise.
> Still successful beyond reason,
> Spite of all its erring ways,
> Holding first place in the season
> Is the "Living Picture" craze

[Quoted in Coe, 1981, p. 76].

Both Gunning (1990) and Musser (1990) have placed the cinema in the tradition of vaudeville, magic, and visual illusion. Gunning (1990, p. 96) suggests that the viewers' attitude may well have been one of detachment and critical appraisal rather than identification and naive credulity. This perspective allows a new reading of, for example, the following description of a detail from the 1896 Lumière film *The Forge*: " 'When the hot iron is plunged into the barrel of water, the steam rises in a most natural manner' " (G. R. Baker in a review of the first London presentation of Lumières' Cinématographe published on March 6, 1896, in the *British Journal of Photography*; quoted by Coe, 1981, p. 71). What is expressed here is admiration not so much for the detailed photographic depiction as for a trick successfully executed, with an undertone of amazement that steam in a film would actually behave like real steam, as if in special compliance with the filmmaker's wishes—much like the reaction elicited by the rabbit or dove a magician unexpectedly pulls out of a hat.

It seems, then, that the attraction of film was not so much its depiction of reality, but the construction of a magical world (Mast, 1980), something closer to the successful execution of a trick which one admires without being taken in by it, knowing that what one sees is a deceit meant to entertain.

Another aspect of film history that has been investigated in detail brings us back to a literal dimension of cinematic time: the question of what speed films were actually made at and projected during the early decades of cinema. It is a widely held assumption that before the advent of sound in film led to the general adoption of the 24 fps standard, 16 fps was the norm for silent films. Recent research has shown, however, that the majority of silent films were *not* filmed at 16 fps. In fact, since early cameras had no speed indicator, cameramen had no way of telling, except in very approximate fashion, at what rate they were filming. And even if with extensive practice their cranking became sufficiently consistent, variations were inevitable and at times deliberate. Coe (1981) gives a sketch of the historical facts:

> The vast majority of cameras made up to the general adoption of sound in the late 1920ies were turned by hand. Most cameras were provided with a handle which exposed eight frames per turn, so that it was operated at two turns per second for the "normal" running speed of 16 frames per second. *This speed was an average speed only; there was no standard speed as such.* Cameramen would "undercrank" to speed action up on the screen and "overcrank" to slow it down [p. 84; emphasis added].[5]

Clearly, the time of movements executed in front of the camera will only be reproduced on the screen if the rate of filming is identical to the rate of projection. The impression of real time on the screen, far from occurring naturally, is a function of matching two technical parameters. Our present expectations of verisimilitude

---

[5]The practice of under- and overcranking shows that even in the earliest days of cinema, time was manipulated for aesthetic purposes—a strong historical argument against the "objectivity in time" for which Bazin praised film.

in the representation of the speed of movement on the screen are based on this provision. It is interesting to note how far techniques and technicians fell short of that standardization during the first decades of the cinema. We have come to accept—even though our subjective experience cannot always be brought in line with that knowledge—that time is homogeneous and continuous; it does not stop and start, and it neither speeds up nor slows down. Time in early film did all those things. Neither homogeneous nor continuous, it was subject to a considerable range of both deliberate and unintentional variation.

Under ideal circumstances, it would have been the job of the projectionist to match the variable speed of the hand-cranked camera with his similarly hand-cranked projector. As a handbook for projectionists emphasized in 1915: "One of the highest functions of projection is to watch the screen and regulate the speed of projection to synchronize with the speed of taking" (cited in Brownlow, 1990, p. 286).

In reality, the often arbitrary variation at the camera end was compounded by a similar array of projection speeds. Considerable variations in speed were necessitated by rigid exhibition schedules. Obviously, if a nine-reel film had to fit into the same time slot as a seven-reel film, it had to be projected about 20 percent faster. Brownlow (1990, p. 286) states: "During the nickelodeon period films were projected at whatever speed suited the management" and quotes a projectionist who remembers: " 'running a full 1,000 ft reel in 12 minutes at the eight o'clock show, and in the afternoon I used to project the same reel so slow it took Maurice Costello ages to cross the set. Those were my manager's orders' " (cited in Brownlow, 1990, p. 285).

The projection speed, and with it the verisimilitude of time on the screen, was quite literally in the hands of the projectionist turning the handle. In fact, the ease with which it could be manipulated led to a tug-of-war between film distributors and exhibitors in which the commercial dimension of cinematic time left little room for attempts at temporal plausibility:

> Projected at the "correct" speed of 16 fps, a full 1,000 ft reel of 35 mm film would last 16 minutes. The Essanay Film Company of Chicago tried to beat wily exhibitors by printing the running time of the films on the posters. The exhibitors retaliated by pasting a strip of paper over the line. Some unscrupulous theatre managers could get through a full reel in six minutes! Ten minutes was acknowledged to be "more usual." Yet, even today, on standard 24 fps sound projectors, 1,000 feet takes eleven minutes [Brownlow, 1990, p. 285].

Filmmakers themselves exhibited equal disregard for any standardization of speed. In 1914, Griffith recommended different projection speeds not only for different reels of the same film, but even for different parts of the same reel (Brownlow, 1990, p. 296). The projectionist's handbook already quoted above makes a valiant attempt at upholding a standard of temporal verisimilitude in the view of these rather exasperating circumstances when it states emphatically: "THERE IS NO SUCH THING AS A SET CAMERA SPEED. The correct speed of projection is the speed at which each individual scene was taken, which may—and often does—vary wildly'' (Brownlow, 1990, p. 286).

The reason for the practices outlined here was not only a lack of standardization exploited by greedy entrepreneurs. It also had to do with the technical–perceptual interface of the medium, namely the fact that at a projection speed of about 16 fps, the flicker effect was irritatingly noticeable. The shutter mechanism which would later eliminate it had not been invented yet. An increase in projection speed helped to reduce the effect, and while studio production gradually followed suit and raised the filming speed as well, it kept lagging behind. But the discrepancy did not cause alarm. It seems that the synchronization of speeds that would lead to a realistic depiction of time was not even considered necessary and that, on the contrary, a discrepancy that led to an accelerated and distorted depiction of time was cheerfully accepted. Amazingly enough, Coe (1981, p. 84) tells us, "in 1927 it was proposed that all film should be taken at 16 frames per second and projected at about 21"!

So even if today's retroactively imposed projection standard of 16 fps actually matches the frequency at which a silent film was shot, this may not be what contemporary spectators saw, or even expected. It seems that audiences at the time saw films "faster than life," that the time represented on the screen had a different speed, and that the comic or bizarre effect of speeded-up motion which we associate with the projection of silent films at sound speed, may actually be closer to the historical spectators' experience than a supposedly authentic rendering through projection at 16 fps.[6]

This brief excursion into the history of filmmaking and projection should suffice to cast doubt on film's supposed "objectivity in time." The presumed "naturalness" of cinematic time is predicated upon an exact match between camera speed and projection speed. It requires standardized equipment, and in fact it was only through the advent of sound film in 1927 that the norm of 24 fps was established for both filming and projection. With this standard in place, the simulated time on the screen finally equaled the time of real movements—more than 30 years after the brothers Lumière premiered their first films in public.

But even with standardized speed, film has a degree of freedom in the depiction of time which makes cinematic time much less than an objective measure of time. In fact, the represented duration may deviate strikingly from the real-time duration, and even within a single shot, manipulations of "natural" experiential time are possible. Slow motion, accelerated motion, and time-lapse photography all allow subtle or drastic deviations from temporal verisimilitude. On the one hand, they may approximate our subjective experience of time more closely than an objective rendering of time.[7] On the other hand, they defamiliarize our experience of time and make us aware of the technical intervention of the medium. In fact, in film

---

[6]Another device that is used regularly is that of stretch-printing, where the film is run through an optical printer which prints every second frame twice. While this affects the perceived quality of movement, projection of a stretch-printed film at 24 fps results in the same projection time as projection of the original film at a supposedly correct 16 fps.

[7]In 1916, Hugo Münsterberg was one of the first theorists to point out this ability of film to simulate—and objectify—subjective psychological processes.

alone can time be made to move backwards smoothly and continuously, combining the verisimilitude of movement in time with the bizarre effect of "impossible" movement sequences.[8]

Film's ability to manipulate time is nowhere more evident than in editing. The technique of cutting and then joining together different shots creates a rhythmic texture of different paces in which duration and repetition contribute to the overall meaning, and shape the significance we assign to what we see on the screen. The degree to which films are compositions not only in, but also *of* time is acknowledged by filmmaker Wim Wenders' (1988, p. 13) statement that "in film it is not thoughts that have to fit together, but sequences of time."

Cuts provide a convenient way to establish a continuity without the technical problems of continuous staging and camera activity. The editing process establishes discontinuity at the macroscopic level of cinematic time and allows the selective presentation of a complex, large-scale time. The earliest one-shot films strike modern viewers as somehow strange precisely because they show an action in its entirety without breaks or elisions, so used are we to editing as a means of taking apart and reassembling events and, by extension, forming and reforming temporal continuity. As Bottomore (1990) reports, it was a major discovery when filmmakers realized in the late 1890s and early 1900s that they did not have to show everything they had filmed, and that it was possible not only to stop the camera between events, but also to take out less interesting passages.[9]

There was nothing natural to the reassembling of time through editing. In fact, early film theorists like Arnheim (1933) felt that in some respects film should not even try to simulate a plausible space–time continuum which, after all, was neither inherent nor natural to the medium. A close look at editing principles reveals in detail the extent to which cinematic time is constructed: In changing from one point of view to another, a cut may actually be *less* noticeable if there is either a gap or an overlap between two shots. (Bordwell [Bordwell and Thompson, 1990, p. 85] singles out this device of *overlapping editing*, and the *cross-cutting* of two or more lines of action, as prominent and "canonic" techniques of temporal manipulation in film.) The illusion of continuity in time is certainly not automatically produced by a succession of frames and shots. Unless a number of rules are respected, the result is visual incoherence rather than the desired continuity effect (Hochberg, 1986). Over several decades, classical Hollywood cinema developed to perfection

---

[8]Martin Amis' novel *Time's Arrow* (which actually has a predecessor in a 1948 short story by Austrian writer Ilse Aichinger titled "Spiegelgeschichte" ["mirror story"]) can serve as an example of a written text attempting a similar effect, namely, that of reversing the arrow of time. Amis' narration is highly sophisticated, but that only emphasizes the awkwardness which constrains his ambitious project. The narrative manipulation of time (addressed below) cannot be extended to include a representation of time flowing backwards.

[9]Since "cutting from one shot to another was thought to be visually disruptive," a piece of blank film was inserted between shots (Bottomore, 1990, p. 104), a practice that would probably strike *modern* audiences as disruptive. In fact, in his 1974 film *Effi Briest*, German filmmaker Rainer Werner Fassbinder uses fades to and from white (rather than black) as a way of pointedly setting apart individual scenes and puncturing visual and narrative continuity.

techniques of invisible editing that included the art of condensing and expanding time and masking jumps and gaps (Bordwell, Staiger, and Thompson, 1985). The most conspicuous editing device for accelerating time is, of course, montage. Amazingly enough, the passing of decades can be represented within seconds of screen time. Perhaps even more amazingly, this radical telescoping of time is perfectly comprehensible to the trained viewer to whom it is clear, for instance, that the famous breakfast montage in Orson Welles' 1941 film *Citizen Kane* sums up years of a gradually deteriorating marriage. (For the role of montage in presenting an image of time, see also Deleuze [1985, esp. chapter 2.3].)

The cinematic representation of time utilizes the technical possibilities offered by the medium, but it also relies on the viewer's knowledge of the conventions of storytelling in any medium, including, but by no means limited to, film. In his history of American film up to 1907, Musser (1990) emphasizes that "spectators could not assume that a film story would unfold in simple chronological order" (p. 5). Scenes were arranged in a nonlinear temporality, where filmmakers would pick up a story again at an earlier point in time, and show simultaneous action in succession. Linear progression and the presentation of simultaneously occurring actions through cross-cutting were little used before 1908. In fact, it was often largely up to the viewer to create a temporal structure for what unfolded on the screen. "The filmmakers assembled spectacular images that evoked the story rather than telling the story in and of itself; indeed, images jumped from high point to high point with crucial causal connections left unarticulated" (Musser, 1990, p. 5). If the narratives were stereotypical enough, the audience could be relied on to supply the temporal and causal connections, so the films did not have to show them. Getting the temporal dimension of stories onto the screen required technical sophistication, precisely because film is a sign system, and temporal relations cannot be shown directly, but must be signified in some other way.[10]

Narrative can be defined as the art of arranging and rearranging the elements of a story in time. This acknowledgment that story-telling is a temporal skill informs David Bordwell's (1985) assertion that in classical Hollywood cinema, narration poses "as an *editorial* intelligence that selects certain stretches of time for full-scale treatment (the scenes), pares down others a little, presents others in highly compressed fashion (the montage sequences), and simply scissors out events that are inconsequential" (1985, pp. 160–161). Learning to telescope time, to move from one point in time to another, and to use repetitions, is part of telling a story in any medium. Narrative skill means knowing what to leave out, what to withhold until later, and what to introduce by anticipation. In that sense, narrative time is always different from real time. The presentation of events by a narrator, while it

---

[10]Cf. Christian Metz' (1991) question: "How does the cinema indicate successivity, precession, temporal breaks, causality, adversarial relationships, consequence, spatial proximity, or distance, etc.? These are central questions to the semiotics of the cinema" (p. 98). Some such indicators would be: fades, dissolves, and wipes; camera movement; the relation between image and sound; speeded-up motion or montage; changes effected by using different film stock, filters, or lenses; changes in lighting, scene, costume or makeup (which includes aging the characters).

has to follow established and comprehensible schemata, may deviate markedly from the actual or inferred timeline of those events.

In watching films, we are dealing with three levels of time: the time frame of the events that form the basis of the narrative (the story time); the time of their arrangement (how much is shown and in what sequence—plot time); and the actual time of the showing (the screen time—about 2 hours for a fiction film).[11] The three time spans can be drastically different. In Orson Welles' *Citizen Kane*, the six or so decades of Kane's life are covered within a week of plot time beginning with his death, during which a reporter investigates his life and listens to a number of recollections about him; all this is presented in little over 2 hours of actual viewing time. If such a compression of time is possible, clearly its representation is highly conventionalized. In fact, the temporal structure of film has very little in common with the real-life time of events. Unabridged real-life duration of time is employed only for the most crucial or spectacular scenes. We are so accustomed to the rhythm and temporal selectivity of editing that long takes and long held shots which attempt to capture ''natural'' time spans, often strike the viewer as unnatural. Spectators quickly lose their patience when asked to sit through the actual duration of an action. In one of his films, Wim Wenders includes a drive through a tunnel in real time—about a minute of blackness with occasional lights. What seems a natural time span in real life becomes interminable in the context of film time, and Wenders has recorded spectators' aggressive and furious reactions and their frustration at having to wait until the image actually changes (Wenders, 1988, pp. 12–13). Even in a continuous scene, plot time can be accelerated and compressed into a shorter span of screen time without the spectator noticing the deception.[12] Most often, it is experimental filmmakers who utilize real-time depiction of events, and they do so precisely because it deviates from the conventions and jolts viewers. In his experimental film *Nostalgia*, the first installment of his 1971–1972 series *Hapax legomena I–VII*, Hollis Frampton repeatedly shows photographs burning to ashes. Each time, the shot is held until even the last slight movement of ash particles has ceased. Those last 30 seconds or so seem excruciatingly long, and the result is either aggressive boredom (as Wenders observed) or an unusual attention and sensitivity for the slightest movement. In other words, the natural time span itself becomes unnatural when presented on the screen. The perception of temporal duration as presented by films, once again, has little to do with the temporal structure of events in real life.

While this article has largely been concerned with visual aspects of cinema, it has to be emphasized that film is an audiovisual medium, even though synchronized

---

[11]For a model of the various possible relationships between the three times see Bordwell (1985, chap. 6, esp. pp. 76–88). Bordwell also provides a number of detailed examples. Genette's (1980) analysis of temporality in narrative is indispensable and has become a classic. For a detailed and thorough account of temporality in film and how we understand it, see Branigan (1992).

[12]Howard Hawks' 1939 film *His Girl Friday* offers examples of this technique, as Bordwell and Thompson (1990, pp. 304–305) point out; in the first scene, a clock actually shows that 31 minutes of continuous story time have been compressed into 12 minutes of screen time without any ellipses or gaps in time.

sound in film as we know it only dates back to the second half of the 1920s. The role of sound in establishing cinematic time is, however, usually underestimated. It is striking how much more flexible visual time is compared with auditory time. Even montage sequences with their radical elisions and juxtapositions are usually readily understood by spectators. In contrast, spoken language and dialogue cannot be similarly manipulated without deteriorating into incomprehensibility. Jean Epstein's experiments with slow-motion sound (Weis and Belton, 1985, pp. 143–144) do not extend to speech, which is very resistant to any kind of distortion such as acceleration or slowing down.

Sound also contributed significantly to the "reality effect" by making films more naturalistic (which was one of the reasons why formalist theorists advocated the use of asynchronous sound). Film is still widely considered a visual rather than an audiovisual medium, however, and it may come as a surprise to find critics in film history who consider film as "illustrated phonography" (Altman, 1992, p. 121), such as this 1929 voice: " 'When it comes right down to it, what is a talking picture but a phonograph record with plenty of amplification behind it?' " (quoted by Altman [1992, p. 121]). Sound has been largely subordinated to images both in cinematic practice and in film theory, as Altman points out in a recent volume on sound which tries to redress that imbalance. The very term *synchronized sound*, while it attests to the importance of sound in coestablishing a temporal continuum, also subordinates it to the image track; the generally used categorization of sound as either "simultaneous" or "displaced" (Bordwell and Thompson, 1990, p. 259) vis-à-vis the images reinforces the same hierarchy. There is no expression such as "sound time" to correspond to the established concept of "sound space."

Sound permits a lot of manipulation in terms of space, but not in time. We have seen above that the introduction of synchronized sound was instrumental—and highly effective—in bringing about a rapid standardization of filming and projection speed. The introduction of synchronized sound also affected the representation of time in cinematic narratives. Williams (1992, p. 132) states that "dissolves within continuous scenes [were] accepted by audiences and filmmakers before recorded sound but virtually prohibited afterwards." Chion (1992, pp. 104–105) points out how much narrative flexibility was lost when intertitles were replaced by sound, which in its standard form meant, and still means, "dialogue spoken in the present tense by the characters." The visual track of a film is much less bound to real time than the sound track. No matter how high the degree of visual temporal manipulation, sound can usually be relied on—except in avant-garde films—to stabilize the temporal continuity. Not surprisingly, Sesonske (1980, p. 425) calls sound "perhaps the major device for establishing temporal structure in film."

Finally, let me resituate the argument of this paper by briefly sketching a more comprehensive cognitive–experiential perspective. Cinematic time is created for and presented on the screen through the interplay of conventions, viewing expectations, and technical processes. Does the high degree of mediation, then, make it entirely artificial? Certainly on one level, cinematic time is an illusion, a way of manipulating the visual system into synthesizing continuous motion, and therefore time. As a rule, the viewer is not aware of this perceptual manipulation, since it remains below the threshold of consciousness. Is cinematic time, then, a construct that has nothing in common with natural time? While there is nothing natural about the process of its creation, yet the result—and the fact that we recognize it as an image of time—is not independent from the time that we live every day, but is based on the same mental processes. In a larger sense, therefore, cinematic time is not so much a construction as a *re*construction, since it relies on human concepts and knowledge of time. The measurable temporal structure of film has little in common with the real-life time of events. It is all the more amazing that this highly sophisticated and multileveled representation is so readily translated back into lived time. The fact that temporal verisimilitude or plausibility is not identical with natural time, real time, makes it all the more intriguing to discover how it is brought about and how it fits into the way our mental and physiological processes work. It also underscores the complexity of what we might be tempted to take for granted, namely our everyday experience of time. Our inferences about screen temporality are based on both our world knowledge and our narrative practices which enable us to assign time to events. While time does underlie all human perception, developing a sense of time and acquiring the necessary mental operations, as Jean Piaget's work (1971) has shown, takes years, during which we gradually learn what to expect of time and which kinds of temporal patterns "fit." It is this knowledge and experience that is brought to bear on cinematic time. *We* take the images on the screen and provide a temporal framework in which they make sense. Ultimately, our recognition of time presented in any medium is based on the way we experience time; it is the cooperation between the technical medium and human perception and cognition that projects time on the screen.

## References

Aichinger, I. (1987), Spiegelgeschichte. In: *Österreichische Erzählungen des 20. Jahrhunderts*, ed. A. Brandstätter. München: dtv.

Altman, R., ed. (1992), *Sound Theory, Sound Practice*. New York: Routledge.

Amis, M. (1991), *Time's Arrow*. New York: Harmony.

Arnheim, R. (1933), *Film as Art*. Berkeley & Los Angeles: University of California Press, 1957.

Barthes, R. (1981), *Camera Lucida. Reflections on Photography*, tr. R. Howard. New York: Farrar, Straus & Giroux.

Bazin, A. (1971), The ontology of the photographic image. In: *What Is Cinema*, Vol. 1. Berkeley & Los Angeles: University of California Press, pp. 9–16.

Bordwell, D. (1985), *Narration in the Fiction Film*. Madison: University of Wisconsin Press.

—— Staiger, J., & Thompson, K. (1985), *The Classical Hollywood Cinema: Film Style and Mode of Production to 1960*. New York: Columbia University Press.

—— Thompson, K. (1990), *Film Art. An Introduction*, 3rd ed. New York: McGraw-Hill.

Bottomore, S. (1990), Shots in the dark. The real origins of film editing. In: *Early Cinema: Space-Frame-Narrative*, ed. T. Elsaesser. London: British Film Institute.

Branigan, E. (1992), *Narrative Comprehension and Film*. London: Routledge.

Braun, M. (1993), *Picturing Time. The Work of Etienne-Jules Marey*. Chicago: University of Chicago Press.

Brownlow, K. (1990), Silent films—What was the right speed? In: *Early Cinema: Space-Time-Frame*, ed. T. Elsaesser. London: British Film Institute, pp. 282–290.

Chion, M. (1992), Wasted words. In: *Sound Theory, Sound Practice*, ed. R. Altman. London: Routledge, pp. 104–110.

Coe, B. (1981), *The History of Movie Photography*. London: Ash & Grant.

Dagognet, F. (1993), *Etienne-Jules Marey. A Passion for the Trace*. Cambridge, MA: Zone/MIT Press.

Deleuze, G. (1985), *L'image-temps. Cinéma*, Vol. 2. Paris: Les Éditions de Minuit.

Elsaesser, T., ed. (1990), *Early Cinema: Space-Frame-Narrative*. London: British Film Institute.

Genette, G. (1980), *Narrative Discourse. An Essay in Method*, tr. J. Lewin. Ithaca, NY: Cornell University Press.

Gunning, T. (1990), ''Primitive'' cinema. A frame-up? Or the trick's on us. In: *Early Cinema: Space-Frame-Narrative*, ed. T. Elsaesser. London: British Film Institute, pp. 93–103.

Hochberg, J. (1986), Representation of motion and space in video and cinematic displays. In: *Handbook of Perception and Human Performance. I. Sensory Processes and Perception*, ed. K. R. Boff, L. Kaufman, & J. P. Thomas. New York: Wiley, pp. 22–51.

Liesegang, P. (1986), *Dates and Sources. A Contribution to the History of the Art of Projection and to Cinematography*, tr./ed. H. Hecht. London: Magic Lantern Society of Great Britain.

Mast, G. (1980), Kracauer's two tendencies and the early history of film narrative. In: *The Language of Images*, ed. W. T. Mitchell. Chicago: University of Chicago Press, pp. 129–150.

Metz, C. (1991), *Film Language: A Semiotics of the Cinema*, tr. M. Taylor. Chicago: University of Chicago Press.

Münsterberg, H. (1916), *The Film. A Psychological Study*. New York: Dover, 1970.

Musser, C. (1990), *The Emergence of Cinema: The American Screen to 1907*, Vol. 1. New York: Scribner.

Muybridge, E. (1887), *Muybridge's Complete Human and Animal Locomotion*. New York: Dover, 1979.

Oeder, W. (1990), Vom Traum Zenons zu Cantors Paradies. Das fotografische Reglement von Zeit, Sichtbarkeit und Bewegung. In: *Zeit-Zeichen*, ed. C. Tholen & M. Scholl. Weinheim: Acta humaniora, pp. 247–263.

Piaget, J. (1971), *The Child's Conception of Time*. New York: Ballantine.

Richardson, F. H. (1967), What happened in the beginning. In: *A Technological History of Motion Pictures and Television*, ed. R. Fielding. Berkeley & Los Angeles: University of California Press, pp. 23–41.

Sesonske, A. (1980), Time and tense in cinema. *J. Aesth. & Art Crit.*, 38:419–426.

Vaughan, D. (1990), Let there be lumière. In: *Early Cinema: Space-Frame-Narrative*, ed. T. Elsaesser. London: British Film Institute, pp. 63–67.

Weis, E., & Belton, J., eds. (1985), *Film Sound. Theory and Practice*. New York: Columbia University Press.

Wenders, W. (1988), *Die Logik der Bilder*. Frankfurt am Main: Filmverlag der Autoren.

Williams, A. (1992), Historical and theoretical issues in the coming of recorded sound to the cinema. In: *Sound Theory, Sound Practice*, ed. R. Altman. London: Routledge.

12

# The Expressivity of Tempo and Timing in Musical Performance

*Judy Lochhead*

*Abstract*   As a temporal art, music engages its listeners in the temporal unfolding of its sounds; as a performed art, it engages us through the control performers *exercise* over the sounding successions. This study presents an argument for correlating the temporal differences between recorded performances of the same piece with differences in the expressive qualities the performances project, and in making this correlation, I seek to demonstrate how the specific, performed instance of a musical work engages listeners in temporally unique expressive qualities.

The study first outlines a theory of musical expression based on recent research in the areas of emotions, cognitive science, and philosophy. Specifically, I adopt Maurice Merleau-Ponty's ideas of perception and Mark Johnson's theory of cognition as "metaphorical projection" to develop an idea of music's expressive qualities as a cognitive engagement with the world. I employ theories of emotion from Ronald de Sousa and Richard Lazarus to support the idea that musical expressivity is determined through a repertoire of responses that are constituted intersubjectively and that have a cognitive function within a social framework. The study second defines how tempo and timing are measured and considered.

The correlation of performance differences and expressive quality is exemplified in three works from the Western, classical tradition: J. S. Bach's "Goldberg" Variations, Beethoven's Piano Sonata No. 30, Op. 109, Finale, also a set of variations, and Roger Sessions's Third Piano Sonata, first movement. While a written format cannot directly demonstrate the sounding correlations for which I argue, it can suggest it. I refer interested readers to the discography of performances considered, and I encourage all to test my argument with other pieces and performances.

To observe that music is a temporal art is in part to recognize that it is performed. While music's essential temporality may be located in the sounds constituting a

*Acknowledgment.* The discussion following the oral presentation of this paper at the meeting of the International Society for the Study of Time in July 1992 was most gratifying and helpful to subsequent revisions. I wish to acknowledge in particular the insightful comments of Lewis Rowell.

musical presentation, in the Western classical tradition[1] a score—the written speci-fications for sound combinations—supports another dimension of temporality: Scores entail the possibility of multiple, distinct realizations of "a piece." These realizations differ with respect to a number of musical factors, among which are tempo and timing; that is, the speed and variations in speed at which musical events occur in any given performance. This study invites readers to consider the temporal differences between recorded performances and to correlate them with musical ex-pressivity.

Three works are considered: J. S. Bach's "Goldberg" Variations, Beethoven's Piano Sonata No. 30, Op. 109, Finale, also a set of variations, and Roger Sessions's Third Piano Sonata, first movement. Before considering the music in detail, the terms of the comparisons must be established: First, musical expression needs defi-nition, and second, the facets of musical tempo and timing need specification.

To assert the expressivity of music is simple enough, and despite the fact that many find the assertion obvious, there are many prominent critics and composers who find it problematic.[2] To argue over whether music is or is not expressive *is* to engage in questions of essences. A more fruitful path of inquiry leads us not to such ontological issues but rather to the question, If people hear music as or as not expressive, how does this occur? The framing of this question locates the possibility of musical expressivity not in the "sounds themselves" but rather, in either the "listener herself" or some relation between the "listener herself" and the "sounds themselves."

If expressivity is located in the "listener" as the subjective opposite of the objectivity of musical sound, then discussion of musical expressivity rests on shaky, "subjectivist" grounds. This has been the implied stance of many music scholars who, while not denying that music has expressive characteristics, choose not to discuss these features on the basis of their subjective instability. Since evidence for expressive character is available only through verbal report, and since any two listeners may not agree on the character of a given musical passage, musical expres-sivity has been considered subjective in the sense that it has relevance only for an individual: In other words, it has no universal meaning.

If expressivity is located in the relation between "listener" and "musical sound," then one must specify the nature of the relation and how expressive mean-ing is generated from it. Two related difficulties emerge: the lack of agreement about the naming of expressive character in specific instances and a failure to correlate unproblematically specific musical structures or types with particular ex-pressive qualities. Both difficulties rest on the assumption of a causal relation run-ning from music to expressive character. The lack of evidence for a one-to-one,

[1]The issue of performance arises particularly in Western classical music for which a notated score exists. The questions of performance addressed here are not an issue for electronic works (even if a score has been devised) and in improvised or "oral tradition" music.

[2]Eduard Hanslick and Igor Stravinsky are notable adherents of an antiexpressive, formalist position. Peter Kivy (1989) discusses the historical and aesthetic context against which to understand their posi-tion; see also Stravinsky (1942) and Hanslick (1854).

causal relation between some specifiable musical type and some specifiable expressive type is sufficient to demote expressivity to the realm of the "unanalyzable."

Since I wish to demonstrate that specific temporal features of a performance are correlated with the apprehension of expressive character, it is necessary to spell out briefly an approach to expressive meaning in music that transcends these various difficulties.

Recent research in the areas of emotions and cognitive science provides a conceptual framework for conceiving and discussing musical expressivity in satisfactory terms (Johnson [1987], Lakoff [1987], de Sousa [1987], Dennett [1991], Lazarus [1991], Varela, Thompson, and Rosch [1991]). Either implicitly or explicitly, a theory of perception resembling that articulated by Merleau-Ponty in his *Phenomenology of Perception* (1962) underlies this recent research. Merleau-Ponty argued that perception is itself a kind of knowing or a broadly conceived cognition, asserting that to perceive is to apprehend a world as significant. Arguing against a perception conceived as a neural response to stimuli, which is the basis for "higher level" cognitive acts, Merleau-Ponty's theory posits both a relation between world and subject and the intersubjectivity of perceptual meanings. Merleau-Ponty further demonstrated that we perceive our world in this cognitive sense through our bodies. Thus, concepts are not "pure" mental representations and the lived body is the site of understanding where all the senses interpenetrate.

Merleau-Ponty's notion of perception sheds light on the work of Roger Scruton, a philosopher–aesthetician, who writes: "in our most basic apprehension of music there lies a complex system of metaphor, which is the true description of no material fact" (Scruton, 1983, p. 85). To say we hear a pitch as high or low is to use metaphorically terms referring to spatial relations. Through metaphorical projection, relations by which we orient our bodies in space are used to orient our bodies in sound. Another philosopher, Mark Johnson, working in the area of cognitive science and language, applies the idea of metaphorical projection to all forms of thought. In his *The Body in the Mind* (1987), Johnson argues that we are cognitively engaged with our world through "schematic structures salient in our most mundane experience (e.g., structures of containment and force) [which] can be extended and elaborated metaphorically to connect up different aspects of meaning, reasoning, and speech acts" (Johnson, 1987, p. 65). Johnson shows further how these schematic structures have a basis in the lived body, and how the structures are imaginatively extended to a wide variety of experiential domains. In other words, he articulates a view of cognition, a view close to Merleau-Ponty's perception, that is creative and essentially metaphorical.

Recent research on emotions is based on a similar notion of cognition. Ronald de Sousa writes that "emotions are best regarded as a kind of perception . . ." (de Sousa, 1987, p. 45), by which he means perception in the Merleau-Ponty sense: The emotions are a means by which we understand the world through our bodily interactions with it. Another author, Richard Lazarus, offers a "cognitive–motivational–relational" theory of emotions which, among other things, entails the idea that "emotion is largely a learned response . . ." (Lazarus, 1991, p. 129) that plays a role in how we relate to and act in our world, or to use Sartre's terms, how we "apprehend our world" (quoted by Lazarus, 1991, p. 131). Both authors understand emotions not as standing in a one-to-one relation with an object or stimulus, but rather as a repertoire of responses that are constituted intersubjectively and that have a cognitive function within a social framework.

The ideas of authors just discussed establish a foundation for a theory of musical expressivity. Attentive listening to music is a perceptual act and, as such, a cognitive act in the sense of Merleau-Ponty and Johnson. Through the creativity of metaphorical projection that Johnson posits as a fundamental component of cognition, listners may choose to engage their intersubjectively constituted repertoire of expressive responses as a means of apprehending a musical presentation; they may also choose not to engage these responses.[3] Appropriate expressive responses are constrained by the sounds themselves through metaphorical similarity and through social conventions that define a signifying relation between sound and expressive meaning. (Mark Johnson [1987, chapter 5] discusses the issue of constraints on metaphorical meaning. De Sousa discusses appropriate emotional responses in terms of "paradigm scenarios" [1987, pp. 181–184 and passim].) While the theory can only be sketched here, two issues need further discussion: those of "appropriate expressive responses" and of the relation between a particular expressive quality and the sound structure with which it is correlated.

The issue of "appropriateness" recalls the question of whether a one-to-one correspondence can be established between specifiable sound combinations and specifiable expressive characters. Unlike earlier authors who make claims for a limited number of emotions which through combinations generate a wider range, recent emotion scholars recognize the diversity of distinguishable emotions but argue that these may be clustered together according to certain types. Such clustering of emotions within a larger repertoire proves suggestive for musical expression (Lazarus [1991, pp. 66–67] cites the cluster analysis of 135 emotion names by Schwartz, Kirson, and O'Connor [1987]). If, through metaphorical projection, an emotion enters into the perception of a musical presentation, then a number of distinct emotions within a cluster would correlate meaningfully with a particular musical instance. Whether a perceived emotion–music correlation is appropriate must then be judged on the basis of possibility, not a one-to-one definition of correctness.[4]

---

[3]The decision to hear expressively or not may be conscious or unconscious. On the one hand, a music scholar interested in syntactic and formal structures may consciously choose to listen solely to those features; this is clearly a perceptual possibility. The same scholar may in another instance feel free to hear expressive features as a musical constituent. On the other hand, a listener who believes that music is not expressive hears music as relations between sounds. As I suggested earlier, the more fruitful line of inquiry is not whether music is or is not expressive but rather how it may or may not be expressive.

[4]Another related but slightly different way of dealing with differences in the naming of expressive character in music may be developed from the notion of "basic level category" articulated by Lakoff (1987). He argues that cognitive language uses works first at a conceptual level between those of greater generality and specificity. At this basic level there is intersubjective agreement and thus conceptual ease with respect to categorical naming.

In attempting to relate expressive characteristics to musical presentations, we could argue that differences in naming expressive character are due to cognitive engagement at a level that is not basic. Thus, if one person identified a musical passage as morose and another the same passage as dejected,

**Example 12.1**
**Expressive Terms Applied to Pieces Considered**

| Bach | Beethoven | Sessions |
|------|-----------|----------|
| Elegant | Gentle | Gently |
| Black despair | Stable | throbbing |
| Melancholy | Fantasy | Powerfully |
| Tragedy | Intensity | lyrical |
| Comedy | Magical | Restraint |
| Grace | Static | Passionate |
| Grief | Stillness | Dark and |
| Delicate | Tensile strength | uneasy |
| Vigorous | Filigree | Hammered |
| Stormy | Gentle | Unfathomable |
| Noble | Mystery | calm |
| Docile | Sturdy | |
| Aloof | Vigorous | |
| Tranquil | Stormy climax | |
| Composure | Harmonic surprise | |
| Command | of a dark instead | |
| Virile | of a bright color | |
| Rusticity | Purple patch | |
| Boisterous | | |
| | | |
| (Program Notes from | (Tovey [1931], and Program | (Program notes from LaBrecque |
| Rosen and Gould 81*) | notes from Rosen, Goode, | by J. McCalla, and from |
| | and Rangell) | Helps by J. Boros) |

*Program notes are included with recorded performances. Recording information is listed on the following examples.

Example 12.1 lists some expressive words applied to the music considered here. Quickly surveying the list we note words such as *vigorous, aloof, filigree*, and *hammered* that are not simply emotion terms; that is, words used to describe expressive quality in music emerge from virtually all aspects of our lived experience. Through metaphorical projection, the wide range of our bodily interactions with the world enter into the expressivity of music: The lived body is not simply the site but also the source of musical expressivity. In the same way our bodies and minds might respond to an impending August thunderstorm, we might hear the music of Sessions's Third Sonata as "dark and uneasy." Or in the same way we might see the swirling patterns in a lace curtain, we might experience the melodic ornamentation in Beethoven's variations as a "filigree."

The other issue requiring further discussion is the relation between expressive quality and sound structure.[5] Given that cognitive meanings, as defined by Merleau-Ponty and Johnson, arise from the interaction of person and world, musical meaning,

we could refer each of those expressive qualities to the basic level category "sad" and thus achieve agreement at some cognitive level.

This formulation would allow cognitive creativity and difference for individuals due to metaphorical projection and at the same time cognitive identity at the basic level.

[5]Robert Hatten (1990) summarizes the virtues and pitfalls of various approaches to the relation between expressive quality and musical features. It is also interesting to note that in a recent article,

be it expressive or structural, results from an interaction of listener and sound. A perception of musical meaning will bear the traces of not only how the listener cognitively employs her repertoire of concepts, both musical and not, but also of musical features that are cognitively engaged through metaphorical projection. Rather than trace a path from specific features of a musical event to expressive quality in an attempt to understand how expressive meaning is constrained by structure, a more fruitful line of inquiry would start with the expressive quality asserted by a listener and then proceed to an investigation of what features of musical structure are correlated, through metaphorical projection, with the asserted expressive quality. (My project here subscribes to this sort of approach but without a thorough investigation of how differences in tempo and timing correlate with their metaphorical counterparts. More on this later.)

Let us turn now to the second matter needing definition: tempo and timing. Performers of notated music are both recreative and creative artists, not only following explicit directions in the score but also making interpretive decisions about musical features not specified in notation. Their creative role is fulfilled in part by decisions about tempo and timing. Tempo refers to a generalized speed at which musical events occur and timing to the moments of acceleration and deceleration from the generalized tempo.[6]

Prior to the early 19th century, composers either did not indicate a tempo in the score or indicated relative speeds by conventional terminology such as "Allegro," "Adagio," "Presto," and "Andante." In the former case, today's performers must determine tempo, in the latter case, they must determine the precise degree of fastness or slowness.[7] Since the early 19th century composers have had at their disposal metronomes which allow a clocklike indication of speed. Only in the 20th century have composers made consistent and relatively reliable use of such markings. While the metronome markings, which are most often used in conjunction with the relative terminology, allow more precision, their use does not eliminate the performer's job of choosing a "proper" tempo. Composers today will often indicate a tempo range or say that "precise" metronome indications locate an approximate tempo.

While it is possible to determine "the tempo" or a generalized tempo of a passage, this does not mean that speed remains constant. Performers variously speed up or slow down from the generalized tempo. Strict regularity of a tempo is not an ideal for a "musical" performance, subtle variations in speed occurring as a matter of course during most performances. Some variations in speed are indicated in the

---

Joseph Kerman (1992) writes that, "Music can represent nonmusical feelings, ideas, and action . . ." (p. 82), with little or no discussion of how the "representational" relation between sound and the nonmusical entity arises.

[6]The term *pacing* is used by some authors to refer to the (usually slight) accelerations and decelerations that occur in performance. In as much as there is no consistent preference in literature about music for one term over the other, I have chosen (somewhat arbitrarily) to use "timing."

[7]Scholars assume that performers contemporary with early composers worked with them and were instructed in the appropriate tempi for given pieces. It is also likely that appropriate tempi were "handed down" through the teacher–student relation.

score by the composer, while others are understood as part of the performer's creative role. These creative variations are within the domain of timing.

Many of the decisions performers make over tempo and timing arise from the intent to project a particular expressive character. How a performer makes a piece "take its time" is essential to music as a temporal art: The piece as an entity sustains the interest of listeners because of the possibility of performances which achieve their distinctness, in expressive and other terms, through the variability of tempo and timing. Expressive character is not, however, simply determined by or correlated with tempo and timing. In traditionally notated scores, some musical features are strictly determined while others are not. Thus, on one hand, players have no significant control over some musical features (in particular pitch combinations and proportional relations of rhythmic durations), and on the other, they must make interpretive decisions about articulation, the "coloring" of sound, and dynamics in addition to those over tempo and timing. In as much as all these musical features of a musical performance have a bearing on musical expressivity, it is ultimately impossible to single out tempo and timing as the source of expressive differentiation.

It is nonetheless possible to make meaningful statements about the contribution of tempo and timing to specific expressive characters. That task is greatly enhanced by comparing different performances of the same work and by the investigative strategy of beginning with analytic description of expressive character and then moving toward a correlation of expressive quality with tempo and timing.

The remainder of the paper exemplifies with specific examples the general theoretical position articulated in the preceding discussion. It presents analytic observations of different performances for the three pieces. The analyses report on the expressive qualities of performances as determined by myself after a long period of rigorous analytic scrutiny. They offer at times some explanation for how musical features engage their metaphorical correlate, but at other times expressive qualities are simply stated, implicitly inviting readers to compare their own expressive determinations with mine. The goal is not to encourage disagreement or to determine *the* expressive character of a performance. Rather, it is to demonstrate, through analytic observations, that musical performances support a range of possible expressive characters and that tempo and timing are strongly correlated with expressive differences from performance to performance. The reporting here of analytic observations of a single listener provides an adequate basis for the project: Differences that readers may observe but that occur within some defined realm of possibility attest to the theoretical underpinnings of the study and affirm the cognitive creativity of listeners.[8]

---

[8]The oral presentation of this paper included sound examples. I have indicated in brackets when these sound examples were played and what music they contained. The reader interested in hearing the sound examples is referred to the discographies on Examples 12.1, 12.2, 12.3, and 12.4.

## Example 12.2

J. S. Bach, Goldberg Variations, BWV 988 (Aria and 30 Variations) Published in 1742; written for Johann Goldberg to play for his patron, Count Keyserling.

Glenn Gould, 1955, CBS MYK 38479
Glenn Gould, 1981, CBS IM 37779
Rosalyn Tureck, Everest 3397
Charles Rosen, Columbia Odyssey 32 36 0020

**No tempo marking by composer**

*times shown in seconds*

| Aria | | Gould 55 ♩ = 66 | Gould 81 ♩ = 33 | Tureck ♩ = 36 | Rosen ♩ = 52 |
|---|---|---|---|---|---|
| A) 1 | | 24 (24) ⎱ | 43 (43) ⎱ | 40 (40) ⎱ | 28 (28) ⎱ |
| 2 | | 28 (52) ⎰ 52 | 48 (91) ⎰ 91 | 45 (85) ⎰ 85 | 27 (55) ⎰ 55 |
| 1 | | no repeat | no repeat | 41 (126) ⎱ | 28 (83) ⎱ |
| 2 | | | | 46 (172) ⎰ 87 | 30 (113) ⎰ 58 |
| B) 1 | | 29 (81) ⎱ | 44 (135) ⎱ | 39 (211) ⎱ | 28 (141) ⎱ |
| 2 | | 30 (111) ⎰ 59 | 49 (184) ⎰ 93 | 46 (257) ⎰ 85 | 28 (169) ⎰ 56 |
| 1 | | no repeat | no repeat | 41 (298) ⎱ | 29 (198) ⎱ |
| 2 | | | | 48 (346) ⎰ 89 | 33 (231) ⎰ 62 |
| Var. I | | ♩ = 120 | ♩ = 84 | ♩ = 80–84 | ♩ = 100 |
| Var. II | | ♩ = 104 | ♩ = 80 | ♩ = 52 | ♩ = 80–84 |
| Var. III | | ♩. = 69 | ♩. = 63 | ♩. = 60 | ♩. = 46–48 |
| Var. XXV | | ♪ = 33 | ♪ = 38 | ♪ = 48 | ♪ = 52 |

## J. S. Bach, "Goldberg" Variations

Bach provided no tempo indications for this work, of which I consider only the Aria and four variations. Example 12.2 shows some "clock time" measurements for four performances: two by Glenn Gould separated by nearly 30 years (1955, 1981); one by Rosalyn Tureck, and another by Charles Rosen.[9] In the four columns, directly under the names identifying the performance, the generalized tempo of the aria, or theme of the variations, is indicated according to a metronome measurement.[10] The Example also shows durations in seconds for the two large halves of the theme labeled with a capital A and B on the far left, and for the shorter sections, labeled 1 and 2. Gould does not repeat sections 1 and 2 of each the A and B halves as indicated in the score. In each performance column the number on the left shows the duration of the specified section, the number in parentheses shows running time

---

[9] I have chosen not to consider both piano and harpsichord performances since timbre affects expressive character, and I have chosen piano instead of harpsichord, the instrument most likely used for the first performance, because of the rich history surrounding the particular performances considered.

[10] Metronome measurements indicate the number of specified durational units per minute. So, for instance, the tempo of Gould 55 means there are 66 quarter notes, a notationally relevant unit for this piece, per minute.

of the passage. The subscripted number on the right shows the duration of each presentation of the 1+2 section pairs.

From the metronome indications, we may note that Gould 81 and Tureck are the slowest and have correspondingly longer durations. Although Gould 81 is slower than Tureck according to what I hear as the baseline tempo, Tureck gives the impression of being lethargic, while Gould 81 is slow but still has forward momentum.

[Listening: Gould 81 and Tureck, measures 1–16, first time]

In Tureck's performance the music is torpid, virtually inert. It is the forced motion that follows extreme tragedy. The slowness of Gould 81, on the other hand, pushes forward with a poignant grief aware of itself. The expressive differences between these two very slow performances may be attributed in part to the speed at which embellishing sounds are played—Tureck plays them slower, Gould faster. In this case, the speed of ornaments does not affect the clock-time tempo but they do influence expressive character.

Gould's earlier performance, from 1955, has a tempo twice as fast as the 1988 performance. The quicker version is elegant and dancing, inviting us to participate.

[Listening: Gould 55, measures 1–16, first time]

The Rosen performance, whose tempo is neither so slow or so fast as the others, is stately and somewhat aloof in its deliberateness. From the durations of parts on the left of the Rosen column, we may note the least variation in the lengths of parts; in this case, the clock-time observation is surely correlated with the sense of deliberateness.

[Listening: Rosen, measures 1–16, first time]

Example 12.2 also shows the generalized tempi for the first third and 25th variations, the latter the longest and one of the few variations in a minor key. From these indications we may note that proportional relations between the tempi of aria and succeeding variations are not constant. For instance, the proportion between the tempo of the aria and that of the first variation in both Gould 55 and Rosen is nearly 1:2, while that proportion in Gould 81 is 2:5 and in Tureck 4:9.[11]

In Gould 55 the doubling of speed contributes to the stark contrast between it and the preceding aria.

---

[11]In the performances of Bach, and also of the following performances of the Beethoven and Sessions sonatas, the observation of proportional relations between movements other than 1:2, 2:3, or 3:4 and of differing relations from performance to performance highlight differences between my approach and that of David Epstein (1979) and others. Epstein claims that "Organic unity through continuous pulse is a psychological aspect of musical time . . ." (Epstein, 1979, 78). Such unity is achieved most commonly by tempo relations characterized by the proportions of 1:2:4:8 but also by others such as 2:3 or 3:4. He does not allow for more complex proportions and does not consider actual performances.

Epstein approaches the question of a "proper" tempo from the perspective of the composer and of the "best" or most aesthetically pleasing tempo. While his approach is prescriptive, mine is descriptive. I take the performances of musicians valued for their musical "interpretations" as indicative of intersubjectively established musical ideals and base my exploration of expressive difference on their recorded performances.

Readers interested in pursuing the topic of proportional tempo relations may consult Epstein's (1979) book for a detailed bibliography.

[Listening: Gould 55, End of Aria to Beginning of Var. 1]

This variation rushes along as if carefree and light-hearted. The succession from aria to first variation in Gould 81 has a different expressive effect. Following the elegant and satisfied slowness of the aria, the first variation presents a poised dance with a hint of raucous playfulness.[12]

[Listening: Gould 81, End of Aria to Beginning of Var. 1]

### Beethoven, Piano Sonata, Op. 109, Finale: Theme and Variations

Beethoven did not supply metronome indications for this sonata, although he did for the earlier Op. 106 work. As indicated in Example 12.3 now, the theme of Beethoven's variations, like that of Bach's, has two longer sections labeled A and B and shorter constitutive parts labeled 1 and 2. Repeats are observed by all performers: Andrew Rangell, Richard Goode, and Charles Rosen.

Of the three performances, Rangell's is the slowest. Its expressive character is melancholy, almost dejected, and reflectively drawn into itself.

[Listening: Rangell, measures 1–8, first time]

Rosen's somewhat faster but still slow tempo projects a restful determination and a reflective poise.

[Listening: Rosen, measures 1–8, first time]

The fastest tempo of Goode's performance has an ease and fluid movement that in its contentment is more hopeful.

[Listening: Goode, measures 1–8, first time]

The proportional relations between tempi of theme and variations differ from performance to performance in the Beethoven as they did in the Bach. Between theme and the first variation the proportion in Rangell is 1:2; in Goode the smallest tempo change occurs: a 9:10 proportion; and in Rosen a 4:5 proportion.

In Variation 1 Rangell's generalized tempo is the fastest, but his timing is the most variable. This variability works within the context of a conventionalized waltz pattern in the piano's lowest notes—this is what is sometimes called the "um-pah-pah" pattern.

[Listening: Rangell, end of theme into Var. 1, 1-ca. m. 8]

In the Goode the smallest proportional change is coupled with more regular timing.

[Listening: Goode, end of theme into Var. 1, 1-ca. m. 8]

---

[12]Example 12.2 indicates tempi for variations II, III, and XXV which for present purposes cannot be discussed. The interested reader is invited to investigate expressive differences between performances. In particular, note the proportional differences of tempi between the aria and Variation XXV: Gould 55, 2:1; Gould 81 and Rosen, 1:1; Tureck, 3:4. In the slow tempo of Gould 55 and further through its relation to the aria, an ardent sense of heartache emerges. The somewhat faster tempo of Rosen's Variation XXV, which maintains the tempo of the aria, projects a loquacious pleading.

## Example 12.3

Ludwig van Beethoven, Piano Sonata #30, Op. 109, Finale: Theme and Variations. Composed 1820.

Andrew Rangell, The Late Piano Sonatas, vol. II, Dorian 90158 (recorded 1991)
Richard Goode, The Late Sonatas, Elektra/Nonesuch 9 79211-2 (recorded 1986, 1988)
Charles Rosen, The Late Sonatas, Columbia M3X 30938 (recording date unknown)

*times shown in seconds*

*Gesangvoll, mit innigster Empfindung. Andante molto cantabile ed espressivo.*

|  | Rangell $\quad$ ♩ = 36 | | Goode $\quad$ ♩ = 52 | | Rosen $\quad$ ♩ = 40–42 | |
|---|---|---|---|---|---|---|
| A) 1 | 20 | (20) ⎫ | 14 | (14) ⎫ | 18 | (18) ⎫ |
| 2 | 20 | (40) ⎬ 40 | 18 | (32) ⎬ 32 | 20 | (38) ⎬ 38 |
| 1 | 19 | (59) ⎫ | 15 | (47) ⎫ | 18 | (56) ⎫ |
| 2 | 20 | (79) ⎬ 39 | 18 | (65) ⎬ 33 | 19 | (75) ⎬ 37 |
| B) 1 | 19 | (98) ⎫ | 15 | (80) ⎫ | 17 | (92) ⎫ |
| 2 | 24 | (122) ⎬ 43 | 18 | (98) ⎬ 33 | 18 | (110) ⎬ 35 |
| 1 | 19 | (141) ⎫ | 15 | (113) ⎫ | 19 | (129) ⎫ |
| 2 | 26 | (167) ⎬ 45 | 18 | (131) ⎬ 33 | 19 | (148) ⎬ 38 |
| Var. I | 122 | (289) | 126 | (258) | 136 | (284) |
|  | ♩ = 72 | | ♩ = 56–58 | | ♩ = 50–52 | |
| Var. II | 123 | (412) | 95 | (353) | 92 | (376) |
|  | ♩ = 54 | | ♩ = 63 | | ♩ = 63 | |
| Var. III | 27 | (439) | 24 | (377) | 28 | (404) |
|  | ♩ = 152 | | ♩ = 160 | | ♩ = 144 | |
| Var. IV | 202 | (641) | 153 | (530) | 155 | (559) |
|  | ♩. = 38 | | ♩. = 46 | | ♩. = 44 | |
| Var. V | 56 | (697) | 52 | (582) | 54 | (613) |
| Allegro | ♩ = 84 | | ♩ = 80 | | ♩ = 92 | |
| Var. VI | 133 | (830) | 125 | (707) | 136 | (749) |
|  | ♩ = 52 | | ♩ = 52 | | ♩ = 42–44 | |
| Theme | 91 | (921) | 78 | (785) | 81 | (830) |
|  | ♩ = 36 | | ♩ = 52 | | ♩ = 38–49 | |

The doubling of tempo and great variability of timing in the Rangell contribute to a bittersweet sense, as if one were remembering a past, pleasant event whose bright potential had spoiled. In the Goode, however, the general mood of contentment from the theme continues in the variation, but it is tinged with the sense of an ironic dance and premonitions of ecstasy.

In the score, Beethoven indicates that the fourth variation should be "somewhat slower than the theme." Example 12.3 indicates that only Goode plays the variation slower; his version is fanciful and seems quite groundless.

[Listening: Goode, Var. IV, 1-ca. m. 8]

Rangell's tempo for Variation IV is slower than Goode's. The slower tempo combined with a deliberateness to the timing projects sentimentality.

[Listening: Rangell, Var. IV, 1-ca. m. 8]

## Example 12.4

Roger Sessions's Third Piano Sonata, Third Movement. Composed 1964–65.

> Robert Helps, Acoustic Research 0654 086
> Rebecca La Brecque, Opus One 56/57

*times shown in minutes and seconds*

> *Adagio e misterioso-Sostenuto ($\downarrow$ = 54)*

|  | Helps |  | La Brecque |  |
|---|---|---|---|---|
| Section I 2'8" | (2'8") |  | 2'26" | (2'26") |
| m. 1 | $\downarrow$ = 42 |  | $\downarrow$ = 30 |  |
| m. 8 | $\downarrow$ = 36 |  | $\downarrow$ = 44 |  |
| m. 20 | $\downarrow$ = 40 |  | $\downarrow$ = 33 |  |

> *un poco piu correne ($\downarrow$ = 58)*

| Section II | 2'51" (4'59") |  | 2'29" | (4'55") |
|---|---|---|---|---|
| m. 24 | $\downarrow$ = 42 |  | $\downarrow$ = 56 |  |
| m. 53 | $\downarrow$ = 42 |  | $\downarrow$ = 52 |  |

> *a tempo ($\downarrow$ = 54)*

| Section III | 3'28" (8'27") |  | 4' | (8'55") |
|---|---|---|---|---|
| m. 60 | $\downarrow$ = 42 |  | $\downarrow$ = 46 |  |
| m. 76 | $\downarrow$ = 40 |  | $\downarrow$ = 44 |  |
| m. 83 | $\downarrow$ = 52 |  | $\downarrow$ = 33 |  |
| m. 90 | $\downarrow$ = 42 |  | $\downarrow$ = 38 |  |

## Roger Sessions, Third Piano Sonata, First Movement

The tempi and durations for two performances of Sessions's Third Sonata, First Movement are shown in Example 12.4: Robert Helps and Rebecca LaBrecque. In addition to showing Sessions's relative and metronome tempo markings, the Example also indicates total duration of the three sections in minutes and seconds; the first number measures the duration of the section and the second in parentheses measures running clock time. Metronome measurements at significant musical moments are additionally listed for each section.

Neither Helps nor La Brecque follow Sessions's metronome indication for the first or third sections, and only La Brecque plays the specified tempo for the second section. Helps maintains roughly the same tempo during the entire movement, while La Brecque's tempi quicken dramatically at the beginning of section II and gradually slow through sections II and III. As a consequence, La Brecque's sections I and II are substantially longer than Helps's, but her section II is slightly shorter.

La Brecque's more variable tempi may be correlated with a more volatile overall character her performance projects. The opening music is indeed mysterious, as Sessions's instructions suggest, and the sudden speeding up at measure 8, when the piano plays reiterated chords, gives a sense of expressive diversity, of sudden mood changes.

[Listening: La Brecque, mm. 1–12]

Helps starts the piece at a faster tempo, giving the music a more present and singing quality, which seems more enigmatic than mysterious, but he slows down at the reiterated chords of measure 8, employing a strategy of timing opposite that of La Brecque. Helps's slowing of more active music creates an expressive continuity during the sonata's opening music.

[Listening: Helps, mm. 1–12]

The tempo Helps chooses for the second section sustains the strategy of expressive continuity. The tempo of quarter equals 42, virtually the same tempo as in the first measures of the sonata, has at the beginning of the section a lilting and gentle character that remains enigmatic.

[Listening: Helps, mm. 23–27]

La Brecque's faster tempo dispels the sense of mystery of the first section; its propulsive force is direct and decisive.

[Listening: LaBrecque, mm. 23–27]

The preceding examples exemplify some correlations between *possible* expressive characters and differences of tempo and timing in recorded performances. The wide range of expressive terminology demonstrates how the expressive character of individual performances can be captured through the cognitive constructs of metaphorical projection. This conception of musical expressivity and *its* expression requires not only the abandonment of the attempt to locate a one-to-one relation between particular performance practices and particular expressive characters but also its invalidation. The conception demands instead the delineation of expressive character by means of cognitive constructs which allow different terms to apply significantly to the same or similar musical events.[13]

In *Being and Time*, Martin Heidegger suggests that the human perception of time is the perception of significance, or what we might think of as meaning. His primary concern was for the intrinsic datability of events as future, present, past. Expanding on his general observation, I would add that the generalized tempo and timing of events are also a feature of temporal meaning. Music, as a performed art, plays on the variability of tempo and timing and their role in expressive significance.

## References

Dennett, D. (1991), *Consciousness Explained*. Boston: Little, Brown.

Epstein, D. (1979), *Beyond Orpheus: Studies in Musical Structure*. Cambridge, MA: MIT Press.

Hanslick, E. (1854), *The Beautiful in Music*, tr. G. Cohen. New York: The Liberal Arts Press, 1957.

Hatten, R. (1990), The proper role of metaphor in the theory of musical expressive meaning. Paper presented at the Oakland, California meeting of the Society for Music Theory.

---

[13]I do not mean to imply that a particular performance practice relates causally to several expressive characters but rather that the particular expressive character any individual reports as "being expressed by" the music is conditioned by that person's way of apprehending it. Such apprehension is constrained by intersubjectively established rules of appropriateness.

Heidegger, M. (1926), *Being and Time*, tr. J. Macquarrie & E. Robinson. New York: Harper & Row, 1962.

Johnson, M. (1987), *The Body in the Mind: The Bodily Basis of Meaning, Imagination, and Reason.* Chicago & London: University of Chicago Press.

Kerman, J. (1992), Representing a relationship: Notes on a Beethoven concerto. *Representations,* 30:80–101.

Kivy, P. (1989), *Sound and Sentiment* (includes the complete text of *The Corded Shell* [1979]). Philadelphia: Temple University Press.

Lakoff, G. (1987), *Women, Fire, and Other Dangerous Things: What Categories Reveal About the Mind.* Chicago & London: University of Chicago Press.

Lazarus, R. (1991), *Emotion and Adaptation.* New York: Oxford University Press.

Merleau-Ponty, M. (1962), *Phenomenology of Perception*, tr. C. Smith. London & Henley: Routledge & Kegan Paul, 1978.

Schwartz, J., Kirson, D., & O'Connor, C. (1987), Emotion knowledge. *J. Personal. & Soc. Psychol.,* 52:1061–1086.

Scruton, R. (1983), *The Aesthetic Understanding.* London & New York: Methuen.

Sousa, R. de (1987), *The Rationality of Emotion.* Cambridge & London: MIT Press.

Stravinsky, I. (1942), *Poetics of Music in the Form of Six Lessons*, tr. A. Knodel & I. Dahl. New York: Vintage Books, 1947.

Tovey, D. F. (1931), *A Companion to Beethoven's Pianoforte Sonatas.* New York: AMS Press, 1976.

Varela, F., Thompson, E., & Rosch, E. (1991), *The Embodied Mind: Cognitive Science and Human Experience.* Cambridge, MA, & London: MIT Press.

13

# *Ma*: Time and Timing in the Traditional Arts of Japan

*Lewis Rowell*

*Abstract*   Foreigners cannot help but notice the remarkably precise timing that pervades every aspect of Japanese life, especially the timing of intervals—from the intervals in conversation and the intervals between everyday gestures, to intervals in sumo wrestling, the tea ceremony, various theatrical genres, music, and the visual arts. The keyword for this subtle sense of timing is *ma*, defined in Iwanami's *Dictionary of Ancient Terms* as "the natural pause or interval between two or more phenomena occurring continuously." It is thus not surprising that architecture, dance, drama, the fine arts, and music have been defined collectively as the arts of *ma*. The concept denotes a uniquely Japanese strategy for correlating the dimensions of space and time and anticipates in a striking manner the modern Western concept of space-time. In this remarkable cultural concept of time as a succession of discrete intervals within a continuum, intuitively sensed and precisely timed, Japanese civilization has cultivated the arts of the "in-between" and forged distinctive modes of timing for daily behavior and experience. This article will examine the role of *ma* in the traditional arts of Japan and its meaning in Japanese culture.

This article is about time as designed by culture; the keyword is *ma*, an extremely common word in Japanese, with a wide range of meanings. Its most basic meaning is *interval*—an interval either of space or time. In this article I shall examine the role of intervals in the traditional arts of Japan, their position, their timing, and

This article is based upon a series of interviews with Japanese performing artists and other specialists, in addition to the references cited. Of the many people who have supplied friendly advice, interpretations, and other items of information, I am particularly grateful to Junko Berberich, James R. Brandon, Samuel L. Leiter, and Richard B. Pilgrim. I would also like to express my gratitude to Jeffrey Gillespie for expert research assistance. All Japanese words appearing in this article have been romanized and rendered without diacritics (e.g., *no* rather than *nō* or noh; kyu rather than *kyū*), except of course for quotations and references. Japanese names appear with given name first and surname last (e.g., Motokiyo Zeami rather than Zeami Motokiyo).

most of all, their cultural meanings. It will become clear that *ma* represents a qualitative, rather than a quantitative, concept of time, and that its measures are largely irrelevant. *Ma* is hierarchical and operates on many levels, in the arts and in daily life.

Anything that occupies or bridges an interval is *ma*: Chopsticks are *ma*, in the sense that they bridge the gap between the bowl and the mouth. *Ma* is both static and mobile: the state of being a bridge and the motion across that bridge. *Ma* is also a quintessentially Japanese version of space-time, one in which the two dimensions have been conceived not as unbounded linear continuua but as correlates that depend on each other for their existence. In other words, as Arata Isozaki (1979) explains, in traditional Japanese thought, time "was not abstracted as a regulated, homogeneous flow, but rather was believed to exist only in relation to movements or spaces" (p. 13).

What has this distinctive concept of time to do with life? I shall return to this question in the concluding section of this article. But, without anticipating the details of the argument I shall there propose, it will be useful to suggest some preliminary connections. This article will emphasize the role of *ma* as a link between two worlds—the phenomenal world of art and the supporting world of traditional Japanese culture. We shall therefore be examining time and timing on two different but related levels: (1) the intervals of art, their spans, gaps, accents, sequences, patterns, and their dynamic properties; and (2) the cultural meanings of these intervals: intuitions of endurance, passage, impermanence, loss, inception, and finality, and the emotions that accompany these intuitions in various human contexts.

*Ma* is the emotion that arises in the intervals between actions or events, in the silences, negative spaces, and moments of nonaction. It can be contended that most of life's defining events and processes occur unseen and unheard, from conception to death. The standard Western strategy in art seems to have been to mark these moments with explicit accents, treating them more like points than spans, and thereby compelling recognition from the spectator—on the artist's terms. In contrast, the Japanese strategy is to create conditions that will stimulate the imagination of informed and intuitive spectators, and then grant them the time or space to make their own connections. The arts of Japan seek to suggest, not to make explicit statements. Within a society that seems bound by so many shared cultural associations and symbolic meanings, art actually plays a liberating role by unleashing the imagination and emotional responses of its audiences.

The main introductory point I wish to make is that the concept of time as a succession of qualitative intervals is by now so deeply rooted in Japanese culture that it has become a distinctive mode of seeing and hearing the world, as well as one of the principal features that define what it is to be Japanese. Let us see what understanding we can reach, with the reservation that elusive cultural concepts such as this will always resist foreign understanding.

To demonstrate the range and application of the idea of *ma*, here are twelve concrete examples:

In Act IV of the classic 18th-century Kabuki play entitled *Chushingura* (The Treasury of Loyal Retainers), the noble Enya Hangan has received a message from the shogun ordering him to commit *seppuku*. As Hangan is about to begin his ritual suicide, his mind is preoccupied with the whereabouts of his chief retainer Yuranoske—he is depending upon Yuranoske to finish him off if he should be unable to complete his task honorably. Suddenly Hangan realizes that he has almost stepped upon the ceremonial mat with the wrong foot, his left foot, and he pauses with his left foot upraised for perhaps as long as 20 or 30 seconds, before he corrects himself and resumes the correct etiquette for the ritual. This prolonged pause is an example of *ma* (Izumo, Senryū, and Shōraku, 1748).

In another popular play of the Kabuki theater, *Sukeroku, Flower of Edo*, the swaggering hero Sukeroku has been ambushed by a gang of ninjas with long staves, and there follows one of the stylized fight scenes that are as typical of the Kabuki theater as the obligatory final chase scene in a Hollywood film. At the climax of the fight, Sukeroku mounts the shoulders of a group of his opponents and assumes a tense pose with his eyes crossed. This dramatic moment of stasis, which is accentuated by a loud stroke from the wooden clappers played by a stage assistant, is called a *mie* and is a second example of *ma* (Brandon, Malm, and Shively, 1978, pp. 84–86; Leiter, 1967). Japanese audiences almost invariably react to a climactic *mie* by calling out the actor's name, and their timing is led (as in some Western opera houses) by a paid clacque, who are experts in *ma*.

In the tea ceremony (*chanoyu*), the formal entry of the guests is an interesting example of *ma*. Once they have entered the external gate, they follow the curving "dewy pathway" into the tea house, walking in a rhythm that is dictated by the stepping stones which have been placed in the footsteps of the tea master who originally planned and decorated the house. In this case, both the intervals between the stepping stones and the pathway itself are regarded as *ma*.

In the traditional art of flower arranging (*ikebana*), a typical disposition of three pieces in the nonlinear pattern known as *ten-chi-jin* (heaven, earth, and man) is arranged so as to lead the viewer's eye through the nonaxial space in a dynamic, asymmetric series of intervals. These intervals represent still another example of *ma*.

In a wrestling ring, two giant sumo wrestlers warily circle each other and mutually determine the moment of *ma* when they first clash.

In a small apartment in a Tokyo suburb, the grandmother of the family removes a copy of *TV Guide* that a careless teenager has left lying in the *tokonoma*, a sparsely decorated alcove that represents the spiritual center of the family dwelling—a fixture in virtually every Japanese home and one that is perhaps more meaningful to the older generations. In this spatial manifestation, *ma* is a part of everyday Japanese life.

And, to take a more mundane example, in one of the immensely popular monster movies of the sixties, Godzilla pauses after having taken a huge bite from the top of the tallest skyscraper in Tokyo. This somewhat indigestible example demonstrates that *ma* can be present in modern pop culture as well as the traditional arts.

In one of the ceiling spaces of the main lobby of Kisho Kurokawa's office for the Fukuoka Bank in Fukuoka City, there is an asymmetric opening that offers an unexpected glimpse of the sky. This sudden penetration of the world of nature into

Japanese urban architecture is yet another example of *ma*. *Ma* is not merely an interval—it is an interval of revelation.

In the classical *no* theater, a deranged old woman is brooding upon her unhappy past and is suddenly possessed by the ghost of her former love. She becomes transformed into him and is compelled to act out his bitter memories, all of this in extreme slow motion. This glacial pace, with each word and action prolonged, is typical of *ma*, and the process of possession and transformation is similarly conceived as *ma*—a moment of revelation that pierces through the conventional world of time and space previously depicted on stage.

A typical scene in many Kabuki plays is the *michiyuki*, a travel interlude, often in the form of a lovers' suicide journey in which a pair of "star-crossed" lovers take their last walk together, pausing here and there to savor a particular view, the scent of flowers, or a glimpse of the moon seen dimly through the clouds, before they throw themselves into the river. This type of lyrical interlude, often played on the *hanamichi* (a ramp running from a side entrance to the stage, allowing entrances and exits to be made through the audience), is a poignant example of *ma*. *Ma* is anything that bridges an interval, and the *hanamichi* itself is a manifestation of *ma*. In this case, the journey is a bridge between life and death, and the various pauses during that journey are isolated moments of profound meaning and emotion that fill up and overflow the gaps within the dramatic narrative. *Ma* is nested within *ma*, just as intervals are nested within larger intervals.

I have two more examples. In the well-known rock garden at Ryoanji, a Zen Buddhist monastery in the suburbs of Kyoto, 15 stones have been placed in an unpredictably asymmetric arrangement in a bed of raked gravel, like islands in a small archipelago. The design of the garden, which will strike most visitors as a severe reduction of nature to a few of its most basic elements and a deliberate attempt to suggest that "less is more," corresponds to the drastically distilled *so* style of calligraphy—the so-called "grass" style in which the figures serve mainly to focus attention upon the blank background, the negative space of the surrounding ground (see following section for an example). Both the disposition of the garden and this style of calligraphy are associated with *ma*.

In the final example, once again from the ancient *no* theater, we hear the calls (*kakegoe*) of the three drummers, measuring the sustained intervals of *ma*—the rests that separate the main beats and events of the music. In the West we generally understand musical rhythm in terms of the binary alternation of arsis and thesis, upbeat and downbeat, a concept that has led us to associate attack, accent, and weight with the beginning of a metric unit. In Japan, on the other hand, a unit of musical time begins with the silence of preparation and concludes with the reactive silence that follows a sounding event. It is not at all assumed that the rhythmic flow is governed at all times by a set of duple proportions. Such a proportional structure is usually to be found somewhere in the musical accompaniment, but the melody or chant "rides" the underlying temporal grid in a constantly expanding and contracting rhythmic current. The intervals between attacks, whether silent or marked by the sustained calls of the drummers, are our final example of *ma*.

Let us see what these examples have in common.

**The Grayness of Time**

In this section we shall test the power of a good metaphor. I draw here on various writings of the architect Kisho Kurokawa (Kurokawa [1983] is only one among many), who has described Japanese culture as a "culture of grays." Readers are warned not to jump to the wrong conclusion! Kurokawa's reference has nothing to do with the feelings of dullness, dreariness, boredom, and neutrality that we in the West instinctively associate with the color gray. Quite the contrary. Gray in this context refers to the traditional India ink drawings in which gray is the intermediate stage between the explicitness of black and the nothingness of the surrounding background.

So let us think of gray as the color of everything that lies in between, in time as in space: the color of understated emotion and the meaning that awaits inference, the ambiguous color of a sensitive and active mind, the color of asymmetric arrangement, shifting perspectives, and fluid experience. Gray is the color and texture of impermanence, twilight, clouds, mist, moonlight, and sound on the verge of fading away into silence. Gray is the color of poetry; prose and everything explicit is black. *Wabi* is the Japanese word that signifies this cultural preference for a subdued range of neutral color, the color of rats and of ashes, and the color that helps to flatten space into the typical two-dimensional perspective (Kurokawa, 1983, p. 38). Another favorite Japanese example is the short rainy season that occurs between spring and full summer, an example of *ma* that brings together many of its most characteristic poetic associations.

The above figure illustrates the character of *ma* in both Chinese and Japanese, drawn in the traditional three styles. At the left is the formal or *shin* style, drawn in block letters. In the center is the same character drawn in the cursive, semiformal style (*gyo*). At the right, almost unrecognizable, is the dramatically abbreviated informal or "grass" style (*so*) mentioned in the previous section. It is this third style that suggests so many of the cultural implications of *ma*.

The character as a whole combines the symbols for gateway (the doorposts on either side) and for the moon (the central figure). Today the central part of the character has been replaced by the symbol for the sun, which in itself tells us something about the contrast between the old and the new Japan. The root metaphor revealed in these calligraphic symbols is the lovely image of moonlight shining through a gate, or glimpsed through a break in the clouds. What it means, with respect to our cognitive and emotional life, is a profound moment of immediate, unmediated experience that manages to penetrate through the barriers set up by our language, our habits of thought, and all the other cultural filters that guide and, at the same time, impede our perception of the world.

Perhaps the most important starting point from which to understand the phenomenon of *ma* in Japanese art and experience is the recognition that these moments of pregnant ''in-betweenness'' are molded by traditional Japanese cultural preferences for such things as ambiguity, asymmetry, austere colors, discontinuity, incompleteness, natural patterns and shapes, open spaces, silence, simplicity, transience, and transitions. These are the distinctive qualities of *ma*. Most of these preferences seem to have arisen after Japan's cultural isolation from mainland Asia and the consequent development of a distinctive language, agricultural experience, and adaptation to Japan's special world of nature. Today the above preferences are so deeply rooted that they virtually define what it is to be Japanese, and they help to explain why the Japanese are instinctively repelled by blundering foreigners whose insensitive movements and timings are perceived as cultural violations (Di Mare, 1990). I know of no other civilization in which precise interactive timing has become such a pervasive feature of daily life.

I will mention one other facet of the Japanese experience of rhythm, which contributes to the list of cultural preferences embodied in the idea of *ma*. From early times, the phenomenon of expanding and contracting rhythm, revealing and responding to surges of emotion, has been regarded as a characteristic of ''high'' art; in contrast, the mechanical sequence of regular beats and undifferentiated intervals has been regarded as ''low.'' *Ma* clearly has elitist overtones.

## The Influence of Asian Religions

The concept of *ma* is rooted in the great religions of East and South Asia. No doubt some recent writers may have overinterpreted the contribution of oriental religions to the role of *ma* in the arts, giving the false impression that the traditional arts of Japan should be regarded as religious rituals. They are surely rituals, no doubt about that, and their timing and many other conventions have been influenced by, and can be traced back to, certain religious ceremonies—in everything from drinking tea and arranging flowers to committing suicide! But the intent of art, in Japan as in the West, is to evoke aesthetic savor—not a vision of the divine, a change of heart or behavior, or any of the other aims of religion.

We find important background for the idea of *ma* in the doctrines of Taoism in ancient China, in the scriptures and practices of the eastern branches of Buddhism, and also, perhaps especially, in Shinto, Japan's indigenous religion. In all of these traditions, subject, of course, to individual variations, the emphasis upon art as a spiritual exercise, a *way*, and also upon emptiness, nothingness, and the avoidance of explicit content, attachment to specific form, and ego consciousness, could not fail to leave their marks upon the arts practiced under their guidance. Some brief passages of testimony will make the connections clearer.

"Form is emptiness, emptiness form" is the message of the well-known Buddhist "heart" sutra, and the following oft-quoted passage from Lao Tzu proclaims much the same doctrine—that nothingness becomes form when surrounded by somethingness, and the form that is thereby created becomes a receptacle for content and meaning:

> Thirty spokes
> share one hub.
> Adapt the nothing [wu] therein to the
> purpose in hand, and you will have the use of
> the cart. Knead clay in order to make a
> vessel. Adapt the nothing therein to the
> purpose in hand, and you will have the use of
> the vessel. Cut out doors and windows in order
> to make a room. Adapt the nothing therein to
> the purpose in hand, and you will have the use
> of the room.
> Thus what we gain is Something [yu], yet it is
> by virtue of Nothing [wu] that this can be put
> to use
>     Lao Tzu, *Tao te ching* (as cited in Pilgrim [1986a, p. 264]).

In the cryptic intellectual exercises of Zen Buddhism, *ma* is the moment of enlightenment when knowledge is attained in a flash of intuition. This emphasis upon immediate, unmediated perception adds yet another valuable perspective on the idea of *ma*: Artistic meaning is not something to be spelled out explicitly, nor is it to be grasped by logical reasoning. It is to be inferred; it requires a leap of mind, surrender of self-consciousness, and an effort to get beyond the physical and mental forms that mediate our experience of the world. One Zen maxim offers the following advice to the would-be artist: "Paint bamboo. Devote yourself only to painting bamboo, until you become bamboo yourself. Then forget you are bamboo." And another famous Zen sutra warns us, "Beware of pointing at the moon, lest your fingers be mistaken for the moon." The fingers remain as an obstacle to our full understanding of the moon: Instead of coming into direct contact with our world, we are looking at ourselves and watching ourselves look at the world.

But, still more than the teachings of Taoism and Buddhism, the physical surroundings and devotional practices of Shinto have been particularly influential in shaping the cult of intervals in the traditional arts of Japan. I will mention three specific influences. First, the Shinto shrine, which in early times was little more than a sacred space set apart from the surrounding profane space by a rope wound around four vertical poles driven into the ground, became an obvious paradigm for both the spatial aspect of *ma* and the design of early theater stages. And second, the basic assumption of Shinto—that the formless gods, the *kami*, do not abide in a shrine, but instead enter when summoned by the worshipper and then depart; and that the worshipper's role is therefore to enter, wait attentively, and be sensitive to the transient presence of the *kami*—has instructed generations of artistic spectators in the same attitudes and practices, thereby producing the same set of expectations for the reception of art. The message here is that art requires us to search actively for meaning in the unfilled times and spaces, that revelation is never definite nor explicit, and that art is always suggesting passage. The sensitive perception of art is always colored by an awareness of its impermanence. And as a third important influence of Shinto, the assumption that the world is pervaded by formless, spiritual energy underlies every aspect of the Japanese artistic experience: from the unwavering spiritual energy that Zeami enjoins the *no* actor to maintain during the moments of nonaction (see below), to the spectator's entrainment in the same current of continuous energy as his eyes follow the dynamic disposition of space, his ears fill in the intervals between sounds, and his mind is engaged in a continuous quest for meaning. Cultural assumptions such as these are self-fulfilling: If we believe that certain meanings are embedded in the arts, we will surely find them.

## Literature Review

The present wave of interest in the concept of *ma* and its role in the arts began to develop in the early 1960s, chiefly because of the efforts of a group of young Japanese architects, among them Teiji Itoh and Arata Isozaki. These artists, most of whom were still in their twenties, were the spearhead of an understandable reaction against the postwar surge of admiration for European Modernism and everything American, both of which dominated the rebuilding boom of the late 1940s and 1950s. Isozaki and his colleagues argued eloquently for an eclectic style of architecture that adopted what they saw as the best features of contemporary international design, but which emphasized elements that were in harmony with traditional Japanese preferences and values. Nitschke (1966) is the best and most comprehensive English-language account of these years. The young builders and city planners were not suggesting anything as simple as a return to their roots, but instead an attempt to discover how their roots could coexist harmoniously with what the rest of the world had to offer. This sounds like a sensible solution and a natural reaction against what must have seemed to many an imported and unsympathetic set of foreign values.

Prior to the 1960s, artists and performers were, of course, aware of the function of *ma* in their various activities, but they seem not to have theorized much about it. This should not surprise anyone. Performance in most Asian traditions (and here I include the "performance" of the several visual arts) is still taught primarily by saying "Do it this way!"—by demonstration and imitation, trial and error, in face-to-face instruction. The authority of an oral tradition does not require a body of written theory, and it is easy to understand how such an elusive matter as pausing and timing would resist efforts to quantify it and reduce it to nothing more than a matter of timing. Also, the breadth of the idea of *ma* must have proved an obstacle: The word is so common and used in so many everyday expressions that it seems at times as broad a concept as the English word *rhythm*.

The new consciousness of *ma* that Isozaki and his circle helped to produce arose precisely because artists and critics were beginning to see and hear it as a phenomenon that transcended any single art or artistic dimension. *Ma* was not simply an interesting behavioral characteristic of music, painting, or poetry, but a unique consciousness of spatial and temporal intervals that pervaded not only the traditional arts but all areas of Japanese experience. It is an interesting demonstration of what can happen in a culture when an idea begins to catch on.

The achievements and ideas of the young architects began to receive international recognition in the 1970s, and a turning point came when Arata Isozaki was invited to design an exhibition on *ma*, first in Paris and then at the Cooper-Hewitt Museum in New York City, in 1979. The New York exhibition was enormously popular, and the spectacular catalog that was produced to document the exhibition (Isozaki, 1979) has done more than any other publication to stimulate Western interest in the various manifestations of *ma* and their cultural background. Isozaki's analysis of *ma* under the following nine categories has been widely cited and remains the most definitive formulation of the many aspects of this complex idea:

> *himorogi*, the establishment of a holy place in space and time
> *hashi*, the bridging of spatial and temporal intervals
> *yami*, the force of *ma* as sustained by absolute darkness
> *suki*, *ma* as a structural unit for living space
> *utsuroi*, a precise moment for movement
> *utsushimi*, the distinctive lifestyle associated with certain living spaces
> *sabi*, intervals of time and space that are permeated with an awareness of the
>     transfiguration that precedes extinction
> *susabi*, an alignment of signs in traditional cultural patterns
> *michiyuki*, the coordinated rhythm of movement from place to place, regulated by
>     breath and physical movements [pp. 12–17].

A detailed analysis of these categories would take us too far afield for the purpose of the present article and would largely recapitulate other published commentaries (including Hall [1983]). Readers may find it instructive to match the above categories against the 12 examples of *ma* presented in the introductory section of this article. In particular, *hashi, utsuroi, sabi*, and *michiyuki* are well represented. *Yami, utsushimi*, and *susabi* are not as relevant to the present discussion.

A large number of Western studies on *ma* appeared in the 1980s, although I detect signs that this latest wave of interest has about run its course. Many of these are quite superficial. A welcome exception is the set of articles by Richard Pilgrim, who is a professor of religion at Syracuse University (Pilgrim, 1984, 1986a,b, 1989). So there is now a substantial body of literature on the idea of *ma*. It contains quite a bit of overlap, and many of the authors tend to recycle the same set of examples and quotations (a tendency to which the present article is not completely immune).

Some of the research has not yet passed beyond the stage of cultural voyeurism, but it is a sufficient base from which to make more specific investigations.

## *MA* in Performance

Let us take a closer look at the role of *ma* in the temporal arts. In this section I take up a number of issues pertaining to *ma* in performance, a broad category within which I include such things as acting, dancing, singing, and instrumental music. Because so many of the most venerable and treasured genres of the performing arts of Japan are composite genres, this grouping is more appropriate than treating dance, music, and the theater in separate compartments. I will focus on the timing, function, and meaning of performance intervals, and also on how these intervals are received by spectators and listeners.

To think of performance as a chain of intervals, articulated by actions, is foreign to the Western concept and experience of art, and it requires something of a figure-ground reversal before we can come closer to understanding it. Certainly Western audiences relish pauses, dramatic interruptions, and pregnant silences, but more often as single highlights—not a series of linked events to ponder and respond to with emotional affect. Nor are we accustomed to conceive of performance as a series of individual units—poses, actions, single words, and single musical tones—prefixed and suffixed by moments of meaningful tension and concentration. I shall try to explain the special relationship between performance units and their framing intervals, drawing upon the treatises of Motokiyo Zeami, who laid the intellectual foundations of the *no* theater in the late 14th and early 15th centuries. In a broader sense, Zeami's views on the theater have influenced all aspects of performance in the traditional arts of Japan.

For Zeami, as it is for today's artists and their audiences, the temporal structure of the performing arts is grounded in a universal principle: the principle of *jo-ha-kyu*, three keywords that resist easy translation because of their several levels of meaning. An examination of their functions, meanings, and cultural associations will lead us to a fuller appreciation of *ma* in performance (see Table 13.1).

We begin with the categories at the top of the table. There is, of course, no dispute over the proper order of *jo, ha*, and *kyu*; they represent an obvious instance of the familiar beginning-middle-end paradigm. But what type of beginning, middle, and end? With respect to their proper functions, we find a mixed assortment of descriptors in the literature. I do not interpret this as evidence for conflicting performance traditions, but merely as an understandable consequence of the difficulties that arise when we attempt to translate complex clusters of artistic phenomena into the realm of ideas.

William P. Malm (1959), a musicologist, describes the respective functions of *jo, ha*, and *kyu* as follows: "Jo means the introduction, ha is the breaking apart or exposition, and kyu is the rushing to the finish or the denouement," observing further that "the theory . . . has the tenacity of the theory of question and answer,

**Table 13.1**
**Table of Correspondences for *jo*, *ha*, and *kyu***

| | *jo* | *ha* | *kyu* |
|---|---|---|---|
| Order | First | Intermediate | Last |
| Function | Beginning, preparation, introduction | Exposition, development, "breaking, ruining" | Denouement, "rushing to conclusion," release |
| Tempo | Slow | Faster | Rapid (or slowing) |
| In sound | Silence | Impact | Reactive silence |
| In breathing | Intake of breath | Outflow of breath | Resulting sound |
| In action | Thought | Breath | Action |
| In dimensions | Spatial | Disorder | Temporal |
| In worlds | Heaven | Earth | Man |
| In style | Formal (*shin*) | Semiformal (*gyo*) | Informal (*so*) |
| Location in the theater | Mirror room | Bridge | Stage |
| Stage space | Upstage | Center stage | Downstage |
| Subject matter | Gods | Humans, lunatics | Demons |
| Dramatic plot | Simple, smooth, natural | Many dynamic changes | Exuberant spectacle |

arsis and thesis, in Western music'' (p. 102). Kunio Komparu (1983), a theater critic, presents a more comprehensive scheme:

> *Jo* means beginning, as in beginning-middle-end. It refers to position and thus is a spatial element. *Ha* means break or ruin. It suggests destruction of an existing state and thus is a disordering element. *Kyū* means fast, as in *kankyū* (tempo) or slow-medium-fast. It refers to speed and thus is a temporal element [p. 25].

Zeami described *jo* as a beginning and *kyu* as an ending, but avoided supplying any single descriptor for *ha*. In his 1428 treatise on *Finding Gems and Gaining the Flower*, he refers to *jo-ha-kyu* as the process that determines the proper order of everything in both the animate and inanimate universes:

> Thinking over the matter carefully, it may be said that all things in the universe, good and bad, large and small, with life and without, all partake of the process of *jo, ha,* and *kyū*. From the chirp of the birds to the buzzing of the insects, all sing according to an appointed order, and this order consists of *jo, ha,* and *kyū*. . . . Grasses and trees alike are wet with rain and dew in this rhythm of *jo, ha,* and *kyū*, just as flowers and fruits appear at the proper season. . . . The pattern of *jo, ha,* and *kyū* is visible even in one gesture in a dance, or in the echo of one step [pp. 137–138].

Fulfillment in art, according to Zeami, can come only when this natural principle is observed, along with the tendency toward complementary balance expressed in the ancient Chinese *yin-yang* theory—which he held to be another of the intellectual foundations of the performing arts (pp. 19–20).

As we can see from these explanations, *jo, ha,* and *kyu* have important implications for at least three aspects of the temporal arts: (1) the order of events; (2) the tempo of performance, which includes both the overall speed and the timing of individual intervals; and (3) most important of all, the function of parts within wholes—by which I mean the *qualities* of temporal intervals. Some amplification will be useful.

*Jo* is generally slow, ceremonial, asymmetric in structure, lacks a strongly defined pulse, and has a high ratio of intervals to actions, that is, the background tends to dominate the figures. *Jo* is regarded as auspicious, as are all beginnings in Japanese thought.

With regard to the second component, *ha*, the expression "breaking or ruining" seems to mean simply breaking the mood and pace established by the preceding *jo* section (Zeami, p. 216). *Ha* is usually characterized by greater regularity, a prominent pulse, a faster tempo, more repetition, larger and more relaxed motions, and a clearer sense of process and progression. Because Western spectators often perceive the opening *jo* as irregular and disordered (by their standards), and *ha* as settling into a regular meter, it comes as a surprise to learn that for the Japanese *jo* is experienced as a stable state, and *ha* as relative disorder and instability. Critics tend to interpret these intuitions as a consequence of the traditional Japanese preferences for asymmetry, odd numbers, and fluctuating tempos, all of which are featured in the rhythms of *jo*. *Kyu*, the final element in this aesthetic triad, often takes the form of a rapid and powerful conclusion, especially in theatrical genres; but it can also be seen and heard as a dissolution, trailing off until it leads naturally to a pregnant pause in which the end of the performance (or performance unit) is savored in silence. It is perhaps in this sense that *kyu* is sometimes translated as "release."

It should by now be clear that *jo*, *ha*, and *kyu* are seen as principles that govern, or ought to govern, virtually any kind of performance. In Okuda Shōzō's classic treatise on the tea ceremony (1920), the master instructs the guests in the proper manner for opening the *fusuma* (the sliding entrance to the tea house):

> In opening and closing the fusuma in the formal (*shin*) manner, there are three stages of breath: *jo*, or "prelude"; *ha*, or "development"; and *kyū*, or "climax." *Jo* refers to the first, slow opening of the fusuma a small part of the way.... One should announce one's intention to enter to those within. *Ha* refers to the sliding open of the door. One should position one's hand at the lower portion of the door, so that one is in direct contact with its center of gravity, and open it far enough to allow passage. *Kyū* refers to the final movement, in which—in contrast to the large and strong movements of the *ha* stage—one lightly and gently finishes the opening, transmitting the sense that the action has been brought to proper conclusion. In the *jo* stage, the spirit of harmony manifests itself, and the *kyū* stage harbors the spirit of tranquility [p. 16].

The three phases are also correlated with performance space, as shown in Table 13.1. A performance of *no* begins in the mirror room, unseen by the audience, where the actor slowly transforms himself into his character, before his formal entrance along the bridge and his first step onto the stage proper. On the stage itself, the region of *jo* is the upstage region—the space of greatest formality and the area in which most entrances are made; the region of *ha* is the center stage; and *kyu* is associated with the downstage area. In terms of performance space, this sequence is regarded not only as a progression toward the audience, but also as a progression from greater to lesser formality.

One final understanding of the role of *jo, ha*, and *kyu* in performance is necessary before we can assess their relationships to the idea of *ma. Jo, ha*, and *kyu* are components of a nested hierarchy that incorporates and integrates all levels of performance in time (Zeami, pp. 83–87, 137–140, 148–162). That is, *jo* is not only the first play of an entire day, the first act of a play, and the first section of an act, but the first part of anything—down to the individual word, gesture, step, or musical tone, which Zeami conceived with *jo* as the opening silence of preparation, *ha* as the sound or action, and *kyu* as the silence that follows (p. 205). In this connection I find it interesting that Japanese music displays a strong preference for sharply impacted sounds (as in music for the shamisen and koto) and for performance techniques (as on the shakuhachi and other types of flutes) that maximize the noise elements when a sound is attacked. These techniques virtually insure that the attack is followed by a moment of reflective silence, thus emphasizing the cultural message that art—like life itself—is always in a state of flux, changing in pace, passing from order to disorder and from stability to instability, never for long persisting in anything that suggests a steady state.

Let us return to *ma* to draw the necessary connections. Alert readers will have noted many of the characteristic features of *ma* in the preceding pages, and they will have concluded rightly that the matter goes far beyond any simple correlation between intervals of *ma* and any of the members of the aesthetic triad. *Ma* has more in common with *jo* and *kyu* for reasons that will soon be apparent—less so with *ha* (because of the more explicit and less suggestive nature of *ha*). But *ma* can occur within, be coextensive with, or be situated between any of these three components—on any level of structure.

It is true that there is a natural relationship to *jo*: *Ma* is associated with breathing, and especially with the intake of breath as preparation for a sound or movement. But the function of breath is often ambiguous, as a glance at Table 13.1 will confirm. Sustained intervals of *ma* are often embedded within one of the three components: In the Kabuki play *Chushingura*, a lovers' journey (*michiyuki*) in the middle of the *ha* portion of the play is an important example of *ma*, on the highest architectonic level of plot structure. And finally, as further evidence for the flexibility of these aesthetic correlations, performance units are often deliberately ambiguous and suggestive of multiple interpretations: Final *kyu* may interlock with *jo* of the next unit, and the audience's realization of this structural elision occurs in a revelatory moment of *ma*.

This brings us, once again, to the issue of beginnings and endings, which readers of *The Study of Time* series will recall as the theme of the fourth conference of the International Society for the Study of Time (1979). In the concluding section I shall refer to Pilarcik (1981), one of the outstanding essays in the volume of papers selected from that conference. When the 14th-century Buddhist monk Kenkō wrote that ''In all things, it is the beginnings and ends that are interesting'' (*Essays in Idleness,* fragment 137), he expressed the traditional Japanese view that *jo* and *kyu* are the auspicious phases in all natural and formal processes, the moments that are the most suggestive, meaningful, and full of aesthetic savor. There is no contradiction here with our analysis of *ma*: The intervals of *ma are* the beginnings and the endings—the preparations for, retrospections upon, and transitions between performance actions. In some notes set down by one of Zeami's students, the master is quoted as saying, "Generally speaking, the moment of silence before the person speaks should constitute the *jo*, the word itself constitutes the *ha*, and the moment after the actor's voice stops constitutes the *kyū*" (p. 205).

It is significant that Zeami chose the metaphor of *hana* (flower) to symbolize the highest achievement of the actor's art. By this he did not mean what we in the West might think—performance in "full bloom!" Instead, what he meant was that each phase of each unit of performance follows spontaneously, inevitably, like the development of a flower from seed, to blossom, to fallen petals (p. 52), and is savored by the audience in what Zeami describes as a "moment of profound exchange" between actor and audience, when the eyes and ears of the spectators are suddenly opened in an interval of heightened consciousness and emotional awareness (p. 158).

There are important implications here for the role of the spectator. Western audiences are not usually caught up in a conscious process of savoring each artistic moment by reflecting on what has preceded and what is likely to occur next. Certainly we feel expectation and react (consciously or unconsciously) as individual links in our train of implications are realized or frustrated, but composers and directors seldom provide us with enough time for these reflections. Nor are we in the habit of looking beyond a climax in order that we may savor what might happen next. Western fairy tales usually end with the phrase "and they all lived happily ever after," and Western audiences bring the same expectations with them to the theater, accepting whatever final state has been achieved—be it comic or tragic. (Steven Sondheim's musical *Into the Woods* is a vivid counterdemonstration.) The Japanese relish for *ma* requires a continuous and active deconstruction on the part of the spectator, but not deconstruction in the sense of conscious, intellectual effort. What Zeami sought from his audience was "an intensity of pure feeling that goes beyond the workings of the mind" (p. 91), which is once again consistent with the aims of Zen.

The performer accomplishes his task, not by the careful calculation of intervals and the explicit indication of their contents, but by understatement, concentration, and sustaining what Zeami describes as an unwavering "inner tension" in moments of nonaction (p. 97). One of Zeami's favorite maxims was, "What is felt in the heart is ten; what appears in movement seven," and he maintained that any motion must always be more restrained than the emotion by which it is motivated (p. 75). As he wrote in the major treatise *A Mirror Held to the Flower,* "What the actor does *not* do is interesting," referring to this accomplishment as "the actor's greatest, most secret skill" (p. 96). And he added that the actor's intent to do nothing must never become obvious to the audience, because it would then become an action.

Doing nothing is the most effective technique of the older actor, one who has passed 50 (p. 106).

In the following important passage, Zeami advocates that the performer maintain connections between each successive performance unit by entering into a state of "mindlessness" that conceals his intent even from himself:

> Sometimes spectators of the noh say, "the moments of 'no-action' are the most enjoyable." This is an art which the actor keeps secret. Dancing and singing, movements, and the different types of miming are all acts performed by the body. Moments of "no-action" occur in between (*hima*). When we examine why such moments without action are enjoyable, we find that it is due to the underlying spiritual strength of the actor which unremittingly holds the attention. He does not relax the tension when the dancing or singing come to an end or at intervals between the dialogue and different types of miming. [Not abandoning this mind/heart in the various intervals] he maintains an unwavering inner strength. This feeling of inner strength will faintly reveal itself and bring enjoyment. However, it is undesirable for the actor to permit this inner strength to become obvious to the audience. If it is obvious, it becomes an act and is no longer "no-action." The actions before and after an interval (hima) of "no-action" must be linked by entering the state of mindlessness in which one conceals even from oneself one's intent [as cited in Pilgrim, 1986b, p. 36].

Making connections is what *ma* is about (*hima* is a common synonym for *ma*), and I interpret this as the quintessential Japanese strategy for resolving the contradiction between (1) time as atomistic and (2) time as continuous. The result has been a concept of performance as a chain of units linked by a current of continuous physical, mental, and spiritual energy. Perhaps this state of "mindlessness" in performance is what Zen masters mean when they advise a painter of bamboo first to paint until he becomes bamboo, then to forget he is bamboo! Strength in performance comes from the ability to internalize, from which the outward signs of performance constitute, as Zeami was fond of putting it, "a manner with an owner." The task of the spectator is to use *ma* creatively so as to detect the owner behind the manner, and to savor the emotion that drives both the actions and the nonactions—to look past the skin and flesh and penetrate to the bone (pp. 69–71).

I close this section with some comments on Japanese attitudes toward rehearsing—or rather, *not* rehearsing. From these we gain some final insights on *ma* in performance. In the older genres such as the *no* theater or even the tea ceremony, performances are not conceived as events to be polished, standardized, and then performed over and over again without variation. In Japanese musical genres, there is no conductor. Players are expected to draw upon their base of expertise, decisions affecting the timing of performance pass from person to person during an event, spontaneity is the goal, and a free style of informed and responsive teamwork is usually the best way to proceed toward this goal. Forget about Arthur Koestler's insensitive characterization of the Japanese as robots in his dreadful book *The Lotus and the Robot* (1960), in which he sought to demolish both Indian and Japanese culture.

There is a Japanese saying that each ceremony occurs only once in a lifetime: The tea, the weather, and the company will always be different. A *no* actor may play a particular role only once in his lifetime, and he will never again perform the same play with precisely the same company. They rehearse very little. On the other hand, Kabuki plays were and still are performed in 25-day runs, so by the end of a run I suspect there will not be much variation between one performance and the next. But that is one of the exceptions. What all of this means is that intervals of *ma* in performance are almost always the spontaneous decisions of an individual: His performance partners are constantly reacting to these decisions (like members of a fine chamber music ensemble), and the lead changes hands constantly. There are obviously boundaries beyond which a performer will not and cannot go, but these will already have been taught during his long period of apprenticeship.

## Four Questions

In this final section I shall pose and suggest tentative answers to four questions, questions with implications that go considerably beyond the role, context, and meaning of *ma* in the arts of Japan. Each of these questions is related, in one way or another, to the general theme of time and life—the life of culture, cross-cultural aesthetics, the life of society, political life, and the phenomenon of life imitating art.

First, what will happen, what *is* happening, to the traditional aesthetic sensitivity for the meanings of temporal and spatial intervals in the bright, new world of Panasonic, Mitsubishi, and Hitachi? Is the new Japan in danger of losing its artistic soul and its ability to respond to its own artistic heritage? Many of my Japanese friends are appalled, and rightly so, at the kitsch that pervades today's popular culture, and reactions such as this are by no means limited to Japan.

Remember though that there is another side to Japanese culture, one that has not been emphasized in the preceding pages. I am referring to the counteraesthetic that for at least the last three centuries has coexisted, perhaps not very harmoniously, with the older "gray" culture. This is the counteraesthetic that is responsible for the hard-edge, garish glitz of Nikko with its profusion of red lacquer and exaggerated proportions. The clash of values is clearly illustrated in the contrasts between the baroque splendors of Nikko and the austerity of the great shrine at Ise, between Tokyo and Kyoto, between the Kabuki and *no* theaters.

My guess is that Isozaki and his group saw clearly what was going on and took a productive line of approach: Instead of running the risk of hardening the lines between the old and the new, they argued for an inclusive approach to the arts, one that does not resist but rather encourages change, so long as that change continues to be informed by traditional Japanese preferences and sensibilities. Older genres such as the tea ceremony and the *no* theater will remain in their present form, and will probably continue to be appreciated more by the old than by the young. But the art that is being produced today is more likely to maintain some connections with its own past than if it had continued to imitate Euro-American culture.

Second, does the equivalent of *ma* exist in the Western arts? Unquestionably, it does. Here I will give a few examples from my own field, music, and try to point out both the similarities and the differences.

In Western music of the last four centuries, it is a fairly simple task to identify some well-known examples of silence (both expected and unexpected) that invite and compel a high level of listener interpretation. I am thinking here of the famous passage at the end of Handel's "Hallelujah" chorus from *The Messiah*, when the continuous series of Hallelujahs is suddenly interrupted and a long, unmeasured silence follows—until the chorus and orchestra continue with the final Hallelujah, now in a very slow tempo. I doubt if any listener will fail to take some meaning from this dramatic and pregnant interval of unexpected silence, whether it is used merely to register surprise, to speculate on what is to happen next, or to invoke some extramusical imagery of angelic lauds. But this is, of course, an isolated and extraordinary exception.

The late John Cage's celebrated minimalist piece entitled 4'33" (of silence) is a lesson in *ma*: The "music" in the piece consists of [relative] silence, silent actions on stage, growing awareness of the ambient sounds in the hall, and the listeners' consciousness of their own respiration, mingled with their responses to the situation—amusement, exasperation, fascination, boredom, impatience, and the like. Because Cage argued eloquently against meaning and symbolism in art, I am certain that he would not have endorsed this interpretation, but it is there nonetheless. Japanese aesthetics have been an explicit and acknowledged influence upon Western minimalism as well as the minimalist movement in Japan, as exemplified in the work of the playwright/director Ota Shogo and his *Tenkei Gekijo* (Theater of Transformation). Shogo's style is marked by silence and extreme slow motion, as in "The Water Station" (*Mizu no Eki*), in which the basic tempo of the play is expressed in a ratio of 2 meters: 5 minutes (Shōgo, 1990, p. 151).

I will mention a few other examples of *ma* in Western music, without interpretation: (1) the end of the fourth movement of Joseph Haydn's Quartet Op. 33, no. 2 ("The Joke"), in which unexpected silences are used repeatedly to play with the listener's train of expectation; (2) the phenomenon known (in German) as *Generalpause* or (in Italian) a *gran pausa*; (3) the *fermata*, an unmeasured silence marked by the sign ⌢ ; (4) the many temporal anomalies, interruptions, distentions, and unexpected silences in the late music of Beethoven, especially the quartets of Op. 130, 131, and 132; and (5) the silent, timed opening of the curtain at the beginning of Act III of Alban Berg's opera *Wozzeck* and its correspondence to the fall of the curtain at the end of Act II, a subtle touch that uninstructed spectators are unlikely to notice as they return from their intermission but which helps them reenter the opera. Any musicologist could multiply these examples many times over.

We have noted some similarities to the Japanese experience of *ma*, but the differences are more significant. Although pregnant pauses and meaningful intervals of silence are at times important components in Western aesthetic experience, they generally occur as single, isolated events—not pervading features of the art work. Nor are they among the expectations that Western audiences bring with them into the theater or concert hall. They carry a much different cargo of cultural baggage—certainly not the cultural assumptions we have noted in the case of Japan: (1) that any stable state achieved is about to destabilize, and (2) that any type of motion is likely to be altered. It is surely true that most Westerners have a smaller supply of patience than the Japanese, and have a limited tolerance for such things as silence, slow motion, and stasis. I can suggest, however, an excellent Western analogue to the concept of *ma* as *michiyuki* (the coordinated rhythm of movement from place to place, as in the sequence of 53 views along the Tokaido Road from Kyoto to Tokyo, or the lovers' suicide journey): It is the devotional pilgrimage known as the "Stations of the Cross" in Roman Catholic practice, a series of 14 meditations on scenes from Christ's passion journey.

Third, what are the social and political implications of *ma*?[1] Here some interesting issues arise, issues that provide a vivid demonstration of cross-cultural attitudes toward time and timing. In contrast to modern performative culture in the West, Japanese culture is highly democratic: Virtually every one of the characteristic properties of *ma* (and especially the frequent pauses, silences, slow pace, and spans of stasis) encourages spectators and listeners to exercise their own powers of interpretation, to make their own connections and respond as individuals. *Ma* gives them the time to do this and the opportunity to reenter the narrative world of art without having been passed by.

In contrast, Western performative culture today has developed into a virtual cult of control and domination, where the audience is compelled to reach the conclusions intended by the director. Nowhere is this more evident than in the obligatory chase scenes in Hollywood action films (e.g., *Foul Play*) with their shots constantly cutting from one scene or perspective to another. A mass Western audience is a group to control and manipulate, and artists have learned that this can be accomplished by the denial of time—a tactic that effectively minimizes individual reactions. *Ma* provides an unstructured or minimally structured time or space within which the performers and audience have the freedom to create.

Despite our professions of belief in democracy, Western public rituals often seem to demonstrate a compulsion to divide society into two groups: the holders of power (the so-called movers and shakers) and the passive, powerless receivers of culture. As an example, my former colleague James Brandon suggested professional football as seen on television, in which both the game itself and the format in which it is packaged and presented to the spectators and the television audience are totally manipulated by the owners, managers, advertisers, and TV executives. If you seek to know our aims, look closely at our games!

This line of thought led me to a rereading of Yasunari Kawabata's novel *The Master of Go*, which was first brought to my attention a number of years ago by Pilarcik (1981). The novel is a journalist's account of an epic championship match between the elderly Master and his young challenger, a contest that develops into nothing less than a confrontation between the old and the new Japan. The Master,

---

[1]Most of the ideas and intepretations in this and the following two paragraphs were suggested by James R. Brandon in a personal communication (March 10, 1992).

whose standing allows him to set his own rules and who plays the game with aesthetic satisfaction as his final goal, squares off against his opponent, who expects a "level playing field" and whose objective is to win. The game itself is a series of nested examples of *ma* on several levels—from the individual moves (which include bridges, blocking maneuvers, promotions, and captures) to the conduct of the entire match, with 40 hours allotted for each player, spanning 6 months, with the players and observers in relative isolation from the outside world, moving from one scenic rural inn to another with the match resuming every fifth day.

In this contrast between the world of "power" football and the leisurely, semistructured world of the match of *go*, perhaps we can grasp something of the cultural role of intervals in the East and West. Japan's traditional "gray" culture of inbetweenness and the special qualities with which temporal and spatial intervals have been invested are, for many, a welcome change from the compulsive time- and space-filling we encounter in much of Western art and experience.

Wonderful things happen when intervals are left unfilled: Instead of shopping malls, we have cities; instead of food halls and "grazing," we have meals; instead of prose, we have poetry.

And finally, what does the idea and experience of *ma* have to do with time and life? I have already ventured to suggest some preliminary connections, and I have pointed out some evidence along the way. In these brief closing paragraphs, I shall ask "why?" What is the reason for this predisposition toward meaningful, qualitatively distinctive intervals of time and space? Is it biological? And, if so, in what sense?

I am not qualified to answer these questions within the scientific contexts of biology, microbiology, physiology, zoology, or any of the fields of inquiry that contribute to our understanding of the genesis and evolution of life. As a layman, I find the proposition dubious. Most of the sources cited in this article regard *ma* as culturally conditioned, but, of course, these things have to start somewhere.

I have come across two explanations that I consider naive. First, that the absence of regular beats and spans of time, along with the consequent rhythmic anomalies, are a legacy from Japan's agricultural past—in contrast to the strong, regular beats and repetitive rhythms of hunting, nomadic, and oceanic peoples (Anon., 1981, p. 54). And second, that adaptation to Japan's rugged terrain, quite different from the wide open spaces of much of mainland Asia, led to a revised set of cultural preferences—among them preferences for the asymmetric disposition and demarcation of space, which presumably then transferred to preferences for the disposition of time (Nitschke, 1966, pp. 128–151).

A more credible explanation is based on the distinctive structure of the Japanese language, with its ambiguities of gender and number and its characteristic gaps between linguistic units and the fixed ending particles, both of which require that the listeners infer the syntax that is missing (Pilgrim, 1986a, p. 260; Takehiko, 1978, pp. 13–32). In the famous haikus of Matsuo Basho (1644–1694), we will never know whether the poet had in mind one or many crows sitting on the branch, one or more frogs jumping into the pond:

Old pond!
Frog jumps in;
Sound of water.

Perhaps, the explanation continues, the Japanese have become experts in inference, in "reading between the lines," and have come to regard it as an essential component in art (Anon., 1981, pp. 55–56; Pilgrim, 1986b, pp. 39–41). The religious background of the idea of *ma*, as outlined in an earlier section, must also be a part of any explanation.

After considering these speculations, I wonder if the Japanese relish for *ma* is not another case of life imitating art—an interesting and widespread phenomenon that has not been unknown in the West, and from which more than one moral has been drawn with respect to the influence of art upon society. If there is anything to this line of interpretation, then we may perhaps locate the idea of *ma* in a taste for stylization in the timing and disposition of intervals, which can be traced from ritual, to the visual and temporal arts, and from there spreading into other areas of human experience, including the intuitions and interactive timing of daily life.

## References

Anon. (1981), "Ma": Space full of meaning in Japanese culture. *The East*, 17/7–8:53–57.

Brandon, J. R., Malm, W. P., & Shively, D. H. (1978), *Studies in Kabuki: Its Acting, Music, and Historical Context*. Honolulu: University Press of Hawaii.

Di Mare, L. (1990), Ma and Japan. *Southern Commun. J.*, 55:319–328.

Hall, E. T. (1983), *The Dance of Life: The Other Dimension of Time*. Garden City, NY: Anchor Press/Doubleday.

Isozaki, A. (1979), *Ma: Space-Time in Japan*. New York: Cooper-Hewitt Museum.

Izumo, T., Senryū, N., & Shōraku, M. (1748), *Chūshingura* (The Treasury of Loyal Retainers), tr. D. Keene. New York: Columbia University Press, 1971.

Kawabata, Y. (1954), *The Master of Go (Meijin)*, tr. E. G. Seidensticker. New York: Knopf, 1972.

Kenkō [Urabe no Kaneyoshi] (1330?), *Essays in Idleness: The Tsurezuregusa of Kenkō*, tr. D. Keene. New York: Columbia University Press, 1967.

Koestler, A. (1960), *The Lotus and the Robot*. London: Hutchinson.

Komparu, K. (1983), *The Noh Theater: Principles and Perspectives*, tr. J. Corddry & S. Comee. Tokyo: Weatherhill.

Kurokawa, K. (1983), Rikyu gray: An open-ended aesthetic, tr. Alfred Birnbaum. *Chanoyu Quart.*, 36:33–51.

Leiter, S. L. (1967), The frozen moment: A Kabuki technique. *Drama Survey*, 6/1:74–80.

———— (1979), *The Art of Kabuki: Famous Plays in Performance*. Berkeley & Los Angeles: University of California Press.

Malm, W. P. (1959), *Japanese Music and Musical Instruments*. Tokyo: Tuttle.

Nitschke, G. (1966), "Ma": The Japanese sense of "place" in old and new architecture and planning. *Architect. Design*, 36:116–156.

Pilarcik, M. A. (1981), Beginnings and endings: Hesse and Kawabata. In: *The Study of Time*, Vol. 4, ed. J. T. Fraser, N. Lawrence, & D. Park. New York: Springer.

Pilgrim, R. B. (1984), Foundations for a religio-aesthetic tradition in Japan. In: *Art, Creativity, and the Sacred*, ed. D. Apostolos-Cappadona. New York: Crossroads.

———— (1986a), Intervals (*ma*) in space and time: Foundations for a religio-aesthetic paradigm in Japan. *Hist. Religions*, 25/3:255–277.

———— (1986b), *Ma*: A cultural paradigm. *Chanoyu Quart.*, 46:32–53.

———— (1989), The Japanese noh drama in ritual perspective. *The Eastern Buddhist*, 22/1:54–70.

Sen, Soshitsu, XV (1986), *Ma*: A "usefully useless" thing. *Chanoyu Quart.*, 46:5–6.

Shibayama, Z. (1987), One-word gates: *ro, kan, jaku, kan. Chanoyu Quart.*, 50:7–22.

Shōgo, Ō. (1990), The water station (*mizu no eki*), tr. M. Boyd. *Asian Theatre J.*, 7:150–183.

Shōzō, O. (1920), The taste of tea: Excerpts from the *Chami*, tr. D. Hirota. *Chanoyu Quart.*, 64:7–29, 1991.

Takehiko, K. (1978), *Ma no Nihon Bunka* (the Japanese culture of *Ma*). Tokyo: Kodansha.

Zeami, M. (1363–1443), *On the Art of the Nō Drama: The Major Treatises of Zeami*, tr. J. T. Rimer & Y. Masakazu. Princeton, NJ: Princeton University Press, 1984.

# III.
# Defining Order in Time and Life

## Reflections

In this section anthropological, sociological, and historical perspectives are sketched in the context of relationships between time and life. Various models of perceptions of temporality provide a means for understanding the events and complex interactions of life. Social synthesis relies on commonly acknowledged and accepted temporal structures to which the individual conforms. Some would say that society fashions and molds the shape of individual lives in terms of its temporalities.

> Keeping time,
> Keeping the rhythm in their dancing
> As in their living in the living seasons
> The time of the seasons and the constellations
> the time of milking and the time of harvest
> The time of the coupling of man and woman
> And that of beasts. Feet rising and falling.
> Eating and drinking. Dung and death.
>
> [Eliot, 1943, "East Coker," p. 24].

The temporal frameworks defining the lives of individuals, their social structures, and the natural world interact, intersect, and envelope one another in the course of human existence. Cycles and repetitive occurrences define life in terms of regenerative and ecological processes. Linearities define the limits of birth and death, giving meaning to the events in between. Ideas of progress and decline, ethics and purpose emerge from the directional quality attributed to time. Clocks and chronologies differentiate and order individual lives in terms of commonly accepted standards. But the still points, dreams and myths of human life also define a sense of who we are and how we measure our lives against the days, nights, and passing years.

> Home is where one starts from. As we grow older
> The world becomes stranger, the patterns more complicated
> Of dead and living. Not the intense moment
> Isolated, with no before and after,
> But a lifetime burning in every moment
> And not the lifetime of one man only
> But of old stones that cannot be deciphered.
>
> [Eliot, 1943, "East Coker," p. 31].

## Reference

Eliot, T. S. (1943), *Four Quartets*. New York: Harcourt, Brace & World, 1971.

14

# Time: A Tripartite Sociotemporal Model

*J. M. Halpern, T. L. Christie*

*Abstract*  This paper outlines and illustrates a model of sociotemporality based upon a tripartite paradigm using the concepts of linearity, cyclicity, and liminality as basic sociotemporal primes. The paper reviews the study of time in anthropology, arguing that much sociocultural anthropology has been concerned with studies focused ''on time''—either based upon the analysis of time reckoning systems and concepts of time, or with the use of time as a parametric device to measure social change. The paper suggests that societies and social interaction must be situated ''in time''—as processes located upon and determined by the temporal landscape. The application of the model utilizes ethnographic data on Canadian Inuit and South Slav peasants.

An Inuit hunting trip is a journey in time as well as space. The 18,000 Inuit of the Canadian Arctic live today in some 46 communities scattered across the northern Canadian mainland and the islands of the Arctic Archipelago. As recently as 30 years ago, Inuit groups still followed a seasonally migratory cycle of movement between the winter sea ice and the summer hunting grounds and fishing lakes and streams. Contemporary Inuit are sedentary, working at service occupations in their communities, such as commercial trapping, and depend on Canadian government transfer payments (old age pensions, family allowance, welfare, and hunter/trapper subsidies). Their lives are regulated by the technical, chronometric constraints of the Euro-Canadian institutions around which Arctic community life is structured: the schools, churches, stores, government offices, and maintenance facilities which link the settlement to the Canadian state.

Hunting for seals, whales, caribou, and birds, however, remains economically important in Inuit life. ''Land food'' is ''real'' food and seal and fox pelts (despite depressed prices and widespread boycotts on wild furs) valuable sources of cash. Just as importantly, the land symbolically represents the Inuit ties to their past and their peoplehood—''Nunamiut,'' people of the land, is a common self-designation.

Moreover, trips onto the land which must nowadays be wedged into weekends or the summer school holidays, represent a return to a traditional, temporal order which is in stark contrast to the regimen of school and work in the town.

Inuktitut is still spoken by the vast majority of Inuit and is the language of 80 to 90 percent of the homes. The official languages of the Arctic are English and French, however, and Inuit children are schooled in these languages from about grade 4 onward. With the southern languages came southern technology and ideas. Inuktitut has had to enlarge its vocabulary to accommodate talk about government, education, business, and medicine. With the state came "Canadian time"—chronometric timekeeping, dependent on clocks, watches, schedules, and time zones. But to the student of the contemporary Canadian Arctic, it is clear that the temporal stress and distortion of Inuit life cannot be simply explained by the problems of learning to use watches, be on time for work, be at the airstrip for a scheduled flight. Temporal difficulties do not arise because Inuktitut lacked words of the days of the week, the hours of the day, minute, second, clock, or watch. The problems lie in contrastive "temporal morphologies"—in the "shapes" which events, activities, and institutional structures take on the temporal landscape. Beyond time reckoning and the conceptualization of time itself, there lies a "cultural ecology" of temporality—patterns of adaptation to certain key temporal realities: the inevitability of change, and the perception of recurrence. The model outlined in subsequent sections of this paper represents an attempt to grapple with the cultural ecology of time (Halpern, 1967b; Damas, 1984; Canadian Dept. of Indian Affairs, 1990; Halpern and Christie, 1990; Christie and Halpern, 1990).

**The Triangle of Time: A Tripartite Model**

The model sketched in this paper focuses upon temporality, the time dimension of human thought and social organization. We suggest that there is heuristic potential in conceiving of temporality in terms of three dominant dimensions: linearity, cyclicity, and liminality (stasis). The first two of these are, of course, very familiar in discussions of time. Linearity is Fraser's fletched and pointed arrow (1992); it embodies the personal and social sense of duration: autobiographical, familial, communal, and historical. We identify it with *change*. Metaphorizing linearity or linear time as line or arrow, however, masks the fact that its "motion," its "trajectory" is perceived as a series of identified alterations in the world around us: The minute hand on the clock moves on a point or two, the sun changes its position, the moon its phases. Duration itself, with its implication of changeless continuity, is identifiable only because of the contrast between that which alters and that which endures. "Generations pass while some trees stand, and old families last not three oaks," Browne reminds us (quoted in Carter, 1958, pp. 45–46). The cycle, or circle, that other familiar geometric temporal metaphor, we identify with recurrence and repetition. One's day-by-day experiences provide both novelty *and* recurrence; our lives are quintessentially linear in their inevitable progress toward their ends, yet we see recurrent stages in them and we continue, Giddens (1979, p. 217) points out, to speak of "the life cycle." Society itself exists because of the persistent recurrence of familiar behavioral patterns regulated by custom, tradition, and ritual regulation. Bourdieu's (1990, pp. 106–107) use of the term *habitus* to refer to social convention is a useful reminder that the predictability of social life is a reflection of the habitually recurrent behavioral patterns that custom ingrains in us. Our third temporal modality is less familiar. We use the term *liminality* to suggest a form of temporality distinct from either linear change or cyclical recurrence. It may be thought of as the absence of "real time" or "real event" experiences. On the subjective level, liminality is experienced in psychologically altered states such as trance and epiphanous visionary events, whether induced ritually or by means of substances such as psychedelic drugs. It conforms to the "flow" experience of artistic creativity and the sense of transport which may accompany religious ritual or occasions of extreme joy, sorrow, shock, or passion. On the conceptual level, we see liminality as the quality of sacred things and events. Monuments and sacred places may be imbued with a timeless aura which lifts them out of the flow of ordinary temporality. "Time stands still" in cathedrals, cemeteries, and climactic battlefields. In fact, we physically and socially circumscribe such settings so that the very proxemics of their placement (behind fences, in parks or parkettes, on hills and knolls) encodes semiotically their special, separate status.

Temporal modalities do not exist in isolation from one another, they are inter-active and enmeshed. The cycle is not really a circle, but a helix. This spring season reiterates the solar, meteorological, and biological changes that took place last spring and the springs before that, but it is also *another* spring in the lives of individuals who are a year older. Even for peoples without chronometric and calen-drical reckoning, aging and other marks of unidirectional change are ever present. In Michael Young's (1988) apt and homely illustration: "The child is bowling the hoop along the road; each individual cycle can be unrolled, at least inside the mind, and its period measured. A mathematical snail can be made to crawl along a line on the inside of the moving hoop" (p. 12). Liminal periods implicate within the social and conceptual flow and flux of linearity and cyclicity. Sleep and sacred times punctuate normal life events (even for the nonreligious, "time outs"—Robert Grudin [1982] has called them "nests in time" [pp. 90–91]—interrupt the flow of things quotidian). On the conceptual level, the liminal may be "dreamtime" or mythtime, time before time, when another order of life prevailed. In historiographic societies, "charter" events may be lifted from the flow of historical chronology to be enshrined: Paul Revere's ride, Wolfe scaling the ramparts at Quebec. Liminal conceptual time is not necessarily past time. It may be (indeed, is, in most religious systems) a coexistent "other" time: The supernatural world exists just on the other side of death, the gods and spirits are present during ritual. There is a future dimension as well: There will be a "New Heaven and a New Earth," the buffalo will return, and the white people will depart; the earth will end in fire or ice: Whether sacred or scientific, belief systems are seldom eschatologically agnostic.

In the remaining sections of this paper we will attempt to situate our sociotemp-oral model in its relation to anthropological and sociological thought, explicate its features more fully, and offer some applications to some ethnographic case mate-rial—in particular, the Inuit of the Canadian Arctic and Serb peasant peoples.

## "On Time": Time in Anthropological Theory

The study of time in sociocultural anthropology has a long and extensive history. In general, anthropologists have tended to write "on time"—treating time either as a datum, an aspect of cognitive culture, or as a parametric quantity (i.e., Western chronometries) against which to measure cultural evolution or change.

As a trait of cognitive culture, time and temporal concepts are implicitly or explicitly compared with those of developed or "Western" societies. The most cited here (and certainly the most lucid and seminal) is Evans-Pritchard's account of Nuer time reckoning and temporal concepts (1940). Earlier, Hallowell, indepen-dent of Evans-Pritchard and certainly uninfluenced by him, had presented a very similar picture of time reckoning and temporal perspectives for the Saulteaux (Ojibwa) of Manitoba (1937). More recently, Bohanan (1953) and Beidelman (1963) reported on Tiv and Kaguru time reckoning respectively, as did Ohnuki-Tierney (1973) for the Ainu. Numerous studies of the calendrical systems of ancient civiliza-tions have been provided by historians, archaeologists, sinologists, and Eqyptolo-gists. Among recent summaries of this data are those by Boorstin (1985), Whitrow (1989), and Aveni (1989). The mesoamericanists Thompson (1954) and Soustelle (1961) have written in detail on Mayan and Aztec calendrics. Parametrically, time and time scales have been of major importance and the source of fundamental conflicts in anthropology since its emergence from evolutionary social philosophy.

The central explanatory mode of the period was development—evolutionary change over time. From the work of Bachofen, Maine, Morgan, Spencer, Tylor, and Frazer, anthropology inherited a quest for origins and patterns of development parallel to that of Darwin's organic evolutionary model and the schemes of prehis-tory of Thomsen and Lubbock. Prevalent into the first decade of this century, unilineal evolutionary theory faded as empirical field research replaced armchair speculation and as theoretical interest shifted to the cultural and social study of societies yet extant. Significantly different from each other in their interpretation of sociocultural function, A. R. Radcliffe-Brown and Bronislaw Malinowski shared a common focus on the synchronic study of extant societies and a disdain for "pseudohistory" or "conjectural history." Their Africanist and Oceanist students maintained their theoretical focus into the 1960s (Firth, 1957; Kuper, 1983). In North America the principal countervailing influence was provided by Franz Boas. Here, the problems of "salvage ethnography" and a focus on material culture (influenced certainly by the "salvage" aspect of museum collecting) eclipsed unili-neal evolutionary speculation in diffusionist, cultural holistic, and historical particu-laristic studies. Like Radcliffe-Brown and Malinowski, Boas was a clan patriarch and his intellectual grandchildren are still to be found in anthropology departments and museums in the United States (Lowie, 1937; Harris, 1968).

Evolutionary interests revived on a somewhat more empirical and modest scale between the 1930s and 1950s under the influence of Leslie White (1959), Julian Steward (1955), and Marshall Sahlins and Elman Service (1960); as a consequence of the work of Marvin Harris and his fellow cultural materialists, the biocultural evolutionary perspective yet persists in some academies.

Another approach to the problem of synchronicity and the desire for a temporal component in anthropological studies led to the development of community studies and restudies. Redfield's Folk-Urban Continuum and the investigation of change, especially studies of Latin American peasant communities by Redfield, Oscar Lewis, and others, as well as acculturation studies of North American Indian peoples, were all inspired by the desire to "bring time in" to the analysis of social and cultural life (Redfield, 1930; Lewis, 1951; Halpern, 1967a; Halpern and Kerewsky-Halpern, 1972). Michael Young (1988) has noted a similar sense of frustration among sociologists. He quotes Anthony Giddens (1979): " 'sociologists have been content to leave the succession of events in time to historians, some of whom, as their part of the bargain, have been prepared to relinquish the structural properties of social systems to the sociologists' " (p. 18).

In anthropology, critiques of more modest studies of diachronic change have been as vigorous as those directed against longer term evolutionary analysis. Such studies all fall within modern times, and the forces of modernization are alike at work in all peasant and tribal societies; therefore we are not looking at individual patterns of change of "accounting for sociocultural change" (to use the title of an "inventory" of such factors [Kushner and Gibson, 1962]); "before" and "after" "snapshots" of a society do not tell us precisely *why* it changes as it does, but merely provide historical documentation of the changes that have occurred.

## "In Time"—Another Perspective

The model sketched here is by no means original, nor even particularly new. More than 30 years ago Raymond Firth wrestled with the often cited contradiction in structural–functional analysis: the antinomy of change on the one hand, and functionally integrated structure on the other. Firth (1963) suggests that the answer lies in what he calls "social organization": "the systematic ordering of social relations by acts of choice and decision. . . . Here is room for variation . . . Time enters here" (p. 40).

More recently, sociologists, perhaps in the first instance Anthony Giddens (1979), have argued for a much more effective integration of time in social research. Giddens points out that the "synchronicity" of sociological (and, of course, anthropological) functionalism is illusory: (1) social activities take place over time (as does the research work which describes and analyzes them); (2) "stability" in social structure is not stasis (an idea which is certainly at least partly attributable to the concrete metaphorical word *structure* itself). Society maintains itself through dynamic processes of recurrent interaction and recurrent social habitual repetition.

"Social reproduction" and "structuration," Giddens' terms for societal and institutional maintenance, both imply process rather than entity—society is "inherently recursive" (p. 217). In Michael Young's image: "Society is less like a structure made of stone than like a bubble tent kept standing by constant puffing" (1988, p. 41). We no longer have Firth's "choice and decision" as some sort of collateral feature of an otherwise perduring "structure": Choice, decision, repetition, and change *are* the constitutive qualities of social interaction. We have opted for Zerubavel's term "sociotemporality" to refer to this processual, temporal perspective (1981). In a construct not dissimilar to, but much simpler than, Fraser's hierarchically nested "canonical" temporal forms, Zerubavel suggests that time regulation can be seen as rooted in three "temporal orders." The physicotemporal and biotemporal orders are those of the nonhuman world, which they obviously influence. Like Fraser's "nootemporal" sphere (and perhaps comprehended by it, since the "no-sphere" would incorporate human thought about time as well as all else in its umwelt [Fraser, 1992, p. 164]) the sociotemporal order regulates human social life. Zerubavel (1992) sees it as a wholly socially constructed artifact, arbitrary and rooted in social convention. Other theorists would probably disagree, certainly Young (1988), who argues that societal rhythms are at root biological imperatives:

> Day after day, we get up in the morning as though it were the most natural thing in the world for us to do as diurnal animals, which indeed it is; we go to work and return home and eventually, like the dormouse and the pottos, we curl up in the same warm, dark place as before to sleep . . . [p. 41].

In similar fashion, the necessities of work and rest and change and sex, filtered through the contraints of social custom, circumscribe human temporal behavior. It is, however, on a rather smaller scale that Zerubavel sees sociotemporality in operation. He suggests that there are four forms of sociotemporal regulation: "rigid temporal structures" (the sequential patterning of social behavior); "fixed durations" (patterns determining the appropriate length of activities); "standard temporal locations" (the "whens" of social life, and uniform rates of recurrence). As his descriptive accounts indicate, Zerubavel's temporal constraints provide admirably illuminating insights on the structuring of life in urban institutional settings: hospitals, homes, workplaces, religious orders, and schools. Even slight cross-cultural comparisons indicate, however, that such regulators are rooted in modern Western society. Two of them, at least, are clearly the products of chronometric regulation: "duration" and "temporal location" are certainly dependent on the ordering of life by arbitrary clock time.

The model presented here is at once simpler, more diffuse, and abstract. Linearity and cyclicity are broadly defined characteristics of social phenomena. They include concepts and cognitive orientations. They may help to explain constraining customs and rules, but they are not regulatory in quite the immediate sense that Zerubavel's temporal constraints are. The model is not based in chronometric regulation, and, indeed, seeks to dispense with all considerations of "time" itself—either as an abstract entity, or as the periodicities of life—days, months, years, and so on.

In the following sections of the paper we examine in somewhat greater detail the tripart elements of the model.


## Linearity

Linear time is frequently associated with novelty, directionality, and change, as well as with duration. "While stretching from past to present and into the future, it admits the unexpected or non routine; innovation, even progress is in the cards" (Payne, 1990, p. 22). It has been frequently argued that since the perception of change is the essence of an awareness of duration, the linear is the intuitive, the primary temporal dimension, inherent in the autobiographical perception of our life course and that of others and other things. Certainly it is the temporal aspect most compatible with concepts of evolutionary change, progress, and, in contemporary physical science and cosmology, with Eddington's "arrow" of cosmic unidirectional movement. But the perspective here is forward-focused, turned toward what is to come, toward some perceived "telos," even if only vaguely perceived as different from the present. Such a perspective is often attributed to contemporary time perspectives as contrasted with non-urban-industrial "others." Serbian peasant pastoralists migrating northward into the central *Šumadja* region of Yugoslavia during the 19th century, however, transformed the oak woodlands into open and increasingly fragmented fields, changing themselves in the process into settled agriculturalists whose population expansion brought about irreversible ecological change and created an increasing premium on arable land. In the minds of the migrants they were pioneers, their irrevocable environmental changes "progress" (Halpern and Kerewsky-Halpern, 1972). Long ago the American sociologist W. F. Ogburn (1950) incorporated similar thought patterns into his "cultural lag" theory, arguing that concepts of conservation with respect to American forests had developed as late as they did because of the persistent (and to Ogburn's mind, dysfunctional) pioneering spirit which saw the deforestation of the land as "progress" and the forests themselves as a virtually inexhaustible resource.

On the other hand, retrospective linearity has a stable, perduring quality. The "structural time" of Evans-Pritchard's account of the Nuer is the backward reach of linked lives, and overlapping generations to those who first entered Nuerland (1940, p. 107). The lineage is a quintessentially linear sociotemporal form. On the other hand, the Inuit are a lineageless people who can rarely trace their ancestry more than a few generations back. Older Padlei Inuit people, however, will place their own births or those of their parents or grandparents at Padlei itself—"the source" (i.e., the source of the Maguse River in central Keewatin). If they have not been taught alternate (European scientific) views, they will also claim continued residence for their people in the Keewatin area since the time of the *tuneraat*—the giant people who left their tent rings on the land and placed the large erratic boulders that dot the landscape.

Lévi-Strauss echoes Eliade's notion of nonhistoric, "archaic man," who, with "antihistorical" intent, abolishes "concrete time." For Lévi-Strauss the distinction lies between "cold" and "hot" societies:

> [T]he former seeking by the institutions they give themselves to annul the possible effects of historic factors on their equilibrium and continuity . . . the latter resolutely internalizing the historical process and making it the moving power of their development [p. 233].

Recent studies in "semiotic history" (Harkin, 1989, pp. 1–2) suggest that narrativization of historic events is a culture-bound form of discourse and, as Ricoeur (1983) has put it, " 'the most complete sense in which narrated events are true is in the merger between narrated events and a bounded community, producing a historical consciousness' " (Ricoeur, quoted in Harkin, 1989). Moreover, the "signs" of historical discourse need to include those meaningful symbols of the past which when "read" constitute the texts from which historic meaning is drawn: the battlefield of Kosovo, ancient *inukshuks* (anthropomorphic stone markers) in the Arctic, the Lincoln and Washington memorials.

We suggest, then, that while linearity may be identified with irreversible change, progress, innovation, its retrospective facet is linked with rootedness and enduring stability that inhere in the backward reach of kinship and other social institutions and the perduring emblems of a known and linked past.

## Cyclicity

> Cyclical movement with its constant lively repetition of what has been is the nearest response that life provides to the human aspiration for permanence [Young, 1988, p. 18].

The South Slav *zadruga* perpetuates itself by renewal each generation (Halpern and Wagner, 1984). In fact, as Fortes has pointed out, the domestic group is everywhere a cyclical process in time, passing through phases of formation, expansion, dispersal, and renewal across the married lives of its founders (1969). For the Inuit, name souls, embodying the essence of a deceased relative, replaces that person in the child that bears their name (Balikci, 1970, pp. 199–200).

With proper treatment the animals may cyclically return as well. The Tsimshian place the bones of eaten fish in the river so that they can return to the sea, don salmon garb, and again swim back to the people's weirs and nets. Placing the bones of cooked animals in the stove achieves somewhat the same end for the Cree, as the soul of the animal leaves the house with the smoke and returns to the forest to be born in another moose or beaver.

Leach has argued that the cyclical represents an "artificial" temporal concept, that only change and duration can be intuitively known (1966). Following a similar argument advanced by Eliade, he suggests that the source of concepts of cyclicity derives from religious belief and represents efforts to avoid the terminality that on-going flow suggests. But cycles would seem as manifest in human life as linearities: sunrise, noon, and sunset repeat themselves day by day, as do the mensual phases of the moon and the seasonal solar round of the sun.

The "ecological" time of the Nuer, the Tiv, the Cree, and the Inuit is cyclical. The Inuit (understandably) use the word *ugyug* for both "year" and "winter" and consider that together with a summer it constitutes a year (the transitional seasons are brief and inconsequential—"almost summer" and "almost winter").

Everywhere in the world, daily, lunar, seasonal, and sidereal reiteration is built into systems that regulate life. Clock time, while artificial and arbitrary, is itself cyclical in its hourly and daily flow—even 24-hour digital clocks return to 1:00 A.M. once in each daily cycle (Captains James Kirk and Jean-Luc Picard with their perpetually accumulating "star dates" have yet to be born). Canadian Inuit have been obliged to enlarge their vocabularies in order to deal with the technological, social, and political complexities that whites have brought into the Arctic. Nowhere is this more apparent than in the temporal vocabulary which must incorporate a plethora of chronometric terms absent from traditional Inuktitut speech—days of the week, terms for the hours of the day, and so on. The Inuit deal with this in two ways: They borrow the phonological words from the European language of influence (English, French, German), or rather, its equivalent in their own phonemic system (*sabat*—"seven"), or they create or adapt an existing word or phrase. These latter neologisms are the most interesting because they reflect the Inuit conception of, or response to the Euro-Canadian introduction (Saturday, Sunday, Monday: the day before taboo day, taboo day, the day after taboo day). The word for *hour* in one Labrador dialect is *kaivallagusig*—"it turns back upon itself"—the minute hand of the clock has returned to its original position (Dorais, 1980).

## Liminality

We have borrowed the term which Victor Turner (1969) so insightfully borrowed from Arnold Van Gennep to refer to the third element in our sociotemporal triangle. Elaborating imaginatively on Van Gennep's "in-between" stage in *rites de passage*, Turner has demonstrated that the antistructure, the "betwixt and betweenness" of the middle stage in passage rituals has wide applicability in numerous ritual situa-tions. Our suggestion is that there is a temporal reality at the psychological and institutionalized social level which corresponds neither to the reiterative return of cyclical events nor the onward sense of lineal events. The perception of liminality is characteristic of certain beliefs and is part of the ritual atmosphere in some social settings.

Van Gennep was preceded by Durkheim with his distinction between "sacred" and "profane" times. As Leach has pointed out, Durkheim's "sacred time" is not merely different from the quotidian round of ordinary time, it is typically its reverse (Leach, 1966, p. 134). Similarly, the antistructural and "communitas" features of

Turner's liminality may reverse roles and eliminate structural order in favor of a felt quality and bonding. We find it useful to introduce this interstitial temporal feature to take account of those situations, settings, and beliefs which are marked by an absence of the other qualities of sociotemporality—an absence of "time," stoppage of its iterative or progressive quality. Sleep is liminal. It is socially structured (in a bed, at a specific time, with or without a legitimately authorized companion, appareled in some appropriate way). But the event so structured is outside of normal interaction patterns, it is a temporary death in the sense that it involves a removal from life's activities.

Mayan and Aztec intercalary days were not the leftovers of a calendar that erroneously stopped $4\frac{1}{2}$ days short of the length of the solar year. They represented a socially constructed period which divided off that which has ended from that which was yet to begin. As a liminal period it was marked by apprehension concerning the portents for the future which would be revealed during that time.

Duerer has described the Australian dreamtime as "everywhen" (1985). In a number of studies of mythtime (Hallowell, 1937; Evans-Pritchard, 1940; Carpenter, 1968), it is placed back of time—in an extension of retroactive lineal time—beyond the last lineal ancestor, in the time when what are now totemic eponyms lived and interacted with the First People.

This "time out of time" does not, in fact, reside only in the past but continues to exist as a ritual step beyond everyday life. Boas tells us that it returns to dominate Kwakiutl life and obliterate mundane time each winter season as the village is taken over by supernatural beings who spend the rest of the year in the forest (Boas, 1966). Harner's account of the Jivaro indicates that for them the supernatural world is real existence while the mundane life of everyday is only a vague shadow (1969).

Liminality may be considered as lying in the future in some belief systems. Millennial concepts of the time after the end of the world's existence may place the event 120,000 years in the future, or at the whim of God or the gods.

Funerals and other religious rituals are doubly sacralized in that they take place in liminal time but also represent the human approach to the perpetually liminal place/time of gods and souls.

In an earlier paper we used MacAndrew's and Edgerton's concept of "time out of time" to discuss the phenomenon of Inuit alcohol consumption patterns (1969). Sanctioned or unsanctioned release time from social responsibility as well as from the consciousness of regulatory norms can frequently characterize certain patterns of binge drinking. For some Inuit, the ordered and temporally structured environment of the town or village is an empty place when work is not available and life on the land is not possible. Under such circumstances alcohol may provide access to a different form of liminality (Brody, 1975).

## Conclusion

We began this paper with the observation that the Inuit hunt can be seen as a temporal as well as a spatial journey. Like Turner's liminal pilgrimages to sacred shrines, the intermittent trips onto the land take Inuit families out of the temporal world of the town, onto the land dominated by ecologically cyclical realities, toward the sites and monuments that represent both the linear and liminal past.

If we were to spatialize this three-part model, we would suggest that there may be heuristic value in a concept of sociotemporality pictured as a helix in a box. The cycles of the world, of society, and biology both loop back upon themselves and move along the linear axis of the known past and the predicted future. Round about—the space within the box—lies the liminal, that time which is out of time (Halpern, 1967).

# References

Aveni, A. (1989), *Empires of Time: Calendars, Clocks and Cultures*. New York: Basic Books.

Balikci, A. (1970), *The Netsilik Eskimo*. New York: Natural History Press.

Beidelman, T. O. (1963), Kaguru time reckoning: An aspect of the cosmology of an East African people. *Southwest. J. Anthropol.*, 19:9–20.

Boas, F. (1966), Unpublished field notes. In: *Kwakiutl Ethnolography*, ed. H. Codere. Chicago: University of Chicago Press.

Bohanan, P. (1953), Concepts of time among the Tiv of Nigeria. *Southwest J. Anthropol.*, 9:252–262.

Boorstin, D. J. (1985), *The Discoverers*. New York: Vintage/Random House.

Bourdieu, P. (1990), *The Logic of Practice*. Stanford, CA: Stanford University Press.

Brody, H. (1975), *The People's Land: Whites and the Eastern Arctic*. Harmondsworth, U.K.: Penguin.

Canadian Department of Indian Affairs and Northern Development (1990), *Canada's North: The Reference Manual*. Ottawa: Canadian Communication Group.

Carpenter, E. (1968), The timeless present in the mythology of the Aivilik Eskimos. In: *Eskimo of the Canadian Arctic*, ed. V. F. Valentine & F. G. Vallee. Toronto, Canada: McLelland & Stewart.

Carter, J., ed. (1958), *Erne Buriell and the Garden of Cypress*. Cambridge, U.K.: Cambridge University Press.

Christie, L., & Halpern, J. M. (1990), Temporal constructs and Inuit mental health. *Soc. Sci. & Med.*, 30:739–749.

Damas, D., ed. (1984), *Handbook of North American Indians*, Vol. 5. Washington, DC: Smithsonian Institution.

Dorais, L. J. (1980), *Lexique analytique du Labrador Inuit*. Laval, Canada: Laval University Press.

Duerer, H. P. (1985), *Dreamtime*, tr. F. Goodman. London: Blackwell.

Evans-Pritchard, E. E. (1940), *The Nuer*. Oxford: Oxford University Press.

Firth, R., ed. (1957), *Man and Culture: An Evaluation of the Work of Bronislaw Malinowski*. London: Routledge & Kegan Paul.

——— (1963), *Elements of Social Organization*. Boston: Beacon Press.

Fortes, M. (1969), Introduction. In: *The Developmental Cycle in Domestic Groups* (Cambridge Papers in Social Anthropology, No. 1), ed. J. Goody. Cambridge.

Fraser, J. T. (1992), Human temporality in a nowless universe. *Time & Soc.*, 1:159–173.

Giddens, A. (1979), *Central Problems in Social Theory*. London: Macmillan.

Grudin, R. (1982), *Time and the Art of Living*. New York: Ticknor & Fields.

Hallowell, A. I. (1937), Temporal orientation in western civilization and in a preliterate society. *Amer. Anthropol.*, 39:647–670.

Halpern, J. M. (1967a), *A Serbian Village*. New York: Harper & Row.

——— (1967b), Inuit aspects of mental health: An approach utilizing linear and cyclical time concepts. *Colleg. Anthropologicum*, 1:213–226.

—————— Christie, L. (1990), Temporal constructs and "Administrative Determinism": A case study from the Canadian Arctic. *Anthropologica,* 32:147–165.

—————— Kerewsky-Halpern, B. (1972), *A Serbian Village in Historical Perspective.* New York: Holt, Rinehart & Winston.

—————— Wagner, R. (1984), A microstudy of social process: The historical demography of a Serbian village community (1775–1975). Paper presented at the Fifth Congress of Southeast European Studies, September.

Harkin, M. (1989), New directions: Symbolic ethnohistory. *Semiot. Spectrum,* 11:1–2.

Harner, M. (1973), *The Jivaro.* Garden City, NY: Doubleday/Anchor.

Harris, M. (1968), *The Rise of Anthropological Theory.* New York: Thomas Y. Crowell.

Kuper, A. (1983), *Anthropologists and Anthropology. The Modern British School,* rev. ed. London: Routledge & Kegan Paul.

Kushner, G., & Gibson, M. (1962), *What Accounts for Sociocultural Change? A Propositional Inventory.* Chapel Hill: University of North Carolina, Institute for Research in Social Science.

Leach, E. (1966), Two essays concerning the symbolic representation of time. In: *Rethinking Anthropology.* New York: Athlone Press.

Levi-Strauss, C. (1966), *The Savage Mind.* London: Weidenfeld & Nicolson.

Lewis, O. (1951), *Life in a Mexican Village: Tepotzlan Restudied.* Urbana: University of Illinois Press.

Lowie, R. (1937), *The History of Ethnological Theory.* New York: Holt, Rinehart & Winston.

MacAndrew, C., & Edgerton, R. B. (1969), *Drunken Comportment.* London: Nelson.

Ohnuki-Tierney, E. (1973), Sakhalin Ainu time reckoning. *Man,* 8:285–299.

Ogburn, W. F. (1950), *Social Change with Respect to Culture and Original Nature.* Gloucester, MA: Peter Smith.

Payne, R. (1990), The law and the land: Tensions of time in Israel. Paper presented at the Conference on Time and Space, Sir Wilfred Grenfell College, Cornerbrook, Newfoundland, May.

Redfield, R. (1930), *Tepotzlan, A Mexican Village.* Chicago: University of Chicago Press.

Ricoeur, P. (1983), *Temps et Récit.* Paris: Editions du Seuil, Tome I.

Sahlins, M., & Service, E. (1960), *Evolution and Culture.* Ann Arbor: University of Michigan Press.

Soustelle, J. (1961), *Daily Life of the Aztecs on the Eve of the Spanish Conquest.* Palo Alto, CA: Stanford University Press.

Steward, J. (1955), *The Theory of Culture Change: The Methodology of Multilinear Evolution.* Urbana: University of Illinois Press.

Thompson, J. E. S. (1954), *The Rise and Fall of Maya Civilization.* Norman: University of Oklahoma Press.

Turner, V. (1969), *The Ritual Process.* Chicago: Aldine.

White, L. (1959), *The Evolution of Culture: The Development of Civilization to the Fall of Rome.* New York: McGraw-Hill.

Whitrow, G. J. (1989), *Time in History.* New York: Oxford University Press.

Young, M. (1988), *The Metronomic Society.* Cambridge, MA: Harvard University Press.

Zerubavel, E. (1981), *Hidden Rhythms: Schedules and Calendars in Social Life.* Chicago: University of Chicago Press.

15

# A Lawyer's Perspective: Re-Presenting Human Life to Society

*Mark H. Aultman*

> When in the Course of human events, it becomes necessary for one people to dissolve the political bands which have connected them with another, and to assume among the Powers of the earth, the separate and equal station to which the Laws of Nature and of Nature's God entitle them, a decent respect for the opinions of mankind requires that they should declare the causes which impel them to the separation.
>
> Thomas Jefferson—*The Declaration of Independence*

> Young life is breathed
> On the glass;
> The world that was not
> Comes to pass.
>
> James Joyce—*Ecce Puer*

*Abstract*  Social time is not, and should not be thought of as, independent and autonomous. Though one may speak of human life and consciousness as "socializing" (i.e., as individual and group lives and minds increasingly coming into contact with other lives and minds), social time should not be viewed as hierarchically above the times of human life and mind, but rather as interacting with them. Otherwise there are insurmountable problems for democratic theory and politics. This is especially true if the powerful control social ideology and imagery, and either claim the right to redefine the past, or in fact redefine the past without public knowledge, in order to sustain social structure and imagery at the expense of individual or group memory and experience.

This paper considers the relationship between the time of human life and mind and the time of society. It argues that there is no separate, autonomous social time, irrespective of the fact that social time may be a legitimate object of study apart from other levels or kinds of time, and irrespective of the fact that human consciousness may legitimately be said to be ''socializing'' in the global village sense of instantaneous communication creating an awareness of worldwide events.[1] In particular the paper seeks to warn of the danger inherent in viewing social time as hierarchically superior to, or above, other levels of time, and seeks to have social time viewed as an interactive level of time that depends upon the time of human life and mind.[2]

I will first consider the conflict between individual and social memory, and the problem created where the social is viewed as hierarchically superior. Next I will consider the advantages of viewing the different temporal levels as interactive, with the past as a real constituent structure that binds the present, but nonetheless leaves it open to the future. I will argue that an essential task of the human mind is to ''re-present'' human life to social institutions. Then I will consider the tendency to reify social time, arguing that it is the human mind, sometimes working through, and sometimes against, social institutions, that maintains vitality and flexibility in society.

**Individual Versus Social Memory**

In the spring of 1989, while European communist systems were undergoing convulsive change, Chinese students were killed on a bloody night that has come to be symbolized by the name *Tiananmen Square*. For many in China, though, the event never happened. Official government reports minimized the carnage, or did not report it at all. For many the dissidents became nonpersons.

In his novel *1984* George Orwell described a fictional social process, official and formal, by which the past was similarly erased. Winston worked in the Ministry of Truth, and his job was to rewrite history to make it conform to the current party line. Mementoes of people whose existence did not conform were destroyed or altered. People who remembered incorrectly disappeared, or were tortured until they no longer remembered.

For society such conflicts are an aspect of social power—one group or institution maintaining its view of reality over others. Lawyers who represent individual

---

[1]This should not be taken to mean that human consciousness is the sole agent of socialization, nor, on the other hand, as a description of a merely passive process in which human consciousness *is* socialized. The truth is somewhere in between.

[2]This paper accepts J. T. Fraser's distinction between levels, or at least kinds, of time that are ontologically as well as epistemologically distinct. The paper does not consider alternative theories that posit only a single kind of time (e.g., Whitrow, 1980, p. 374). Hierarchical concepts such as ''levels,'' and related issues of defining a ''place'' in a structural system are, I believe, epistemological tendencies that result from thinking spatially when considering temporality. Nonetheless those tendencies are useful, and many nonhierarchical theories that avoid this problem (but provide less definitional clarity) are subject to the same criticism as hierarchical systems to the extent they view social evolution as a temporal end rather than a hierarchical summit.

clients in matters initiated by social institutions (as in the criminal process) often see the tension between the rationalizations of social institutions and the interests of individuals. Cases can be constructed based upon circumstantial evidence or mistaken testimony. A defense attorney often must reconstruct the past as it happened from the point of view of the client, that is, must "re-present" the person to a social institution that does not know, and may not want to know, the truth of what happened.

A defendant, for example, may have an alibi defense—he may have been somewhere else in the presence of witnesses when a crime was committed. The prosecution may believe that a social interest in having society seen as stable, in having people feel secure, and in having people believe crime is generally punished, overrides lingering doubts as to the actual guilt of this particular individual. It may even rationalize its actions on the basis that if the defendant did not commit the particular crime, he committed crimes like it. A criminal case may thus become a kind of social fiction whereby individual guilt or innocence becomes secondary to maintaining an image of society as a stable force capable of punishing what it defines as wrongdoing.

The problem with this view, if it were to become generally accepted as a dominant policy, is that it would leave innocent individuals, that is, eventually all of us, at the whim of what would be, from our point of view, random state power. Theories of time, as we shall see, have a bearing on this problem.

## The Problem of a Hierarchical Theory of Time

J. T. Fraser's hierarchical theory of time (Fraser, 1975, 1992) posits separate biotemporal, nootemporal, and sociotemporal levels of time. Time in this theory is a nested hierarchy of qualitatively different temporalities, correlates of different structural and functional complexities ranging from an atemporal level of chaos at one extreme, through a biotemporal level where life gives direction to time, a nootemporal level which extends and modifies the biotemporal horizons through conscious choice, to a level, the sociotemporal, now being formed. In this theory there are functions and structures that fall between two stable integrative levels. Across these intermediate levels laws and regularities—the generalized languages—of nature change. Higher order languages are unintelligible at lower levels—the principles peculiar to biology, for example, are inexpressible in terms of the principles of quantum physics. Languages of higher levels subsume lower ones, but not vice versa.

This hierarchical system has broad explanatory power, but it creates problems when applied in social, and especially political, theory. If the sociotemporal is viewed as a separate and higher integrative level above both the biotemporal and the nootemporal, and if in principle its language and laws may be, and if in fact they tend to be, unintelligible to lower levels, then the theory of time is incompatible with democratic theory to the extent that democracy requires that citizens understand and give informed consent to government.[3]

Some commentators have tried to work around this problem. Michael Young and Tom Schuller (1988, p. 15) suggest that, if we use the word *hierarchy* to describe temporal structures, we should not understand it in an authoritarian sense, but must recognize structures at all levels as quasi-autonomous wholes retaining a level of self-sufficiency within the overall system. But if the sociotemporal is a stable level which integrates levels below it in a hierarchy, then on principle, threats to the existence of the whole not only may, but should, be repulsed irrespective of the quasi-autonomous nature of its parts. Tiananmen Square, from this view, not only was justifiable, but is what should have happened.

The problem is not limited to actual threats to an overall system. In Nazi Germany Jews killed in concentration camps were not a threat to the society, although official ideology made them out to be. If in principle generalized languages and laws at the social level may be unintelligible to people (the constituents of society) at the lower level, official social ideology, even though wrong, may go uncorrected and be used as a justification to do harm to both individuals and to society.

Theory and practice reinforce one another at the level of, and at all levels above or following, nootemporality. Once one accepts the principle that the social is autonomous and hierarchically above the constituents of society, those who control society will tend to use the argument to justify, in the name of stability or social order, actions which, despite ideological justification, are not in the interests of society or its constituent members.

## Toward an Interactive Theory of Sociotemporality

Most of the problems outlined can be resolved within the terms of the hierarchical theory of time by recognizing that the sociotemporal does not yet, and need not, exist in finally stable form; is something we participate in creating; and is something which, to some extent, can "go different ways"; that is, is open to the future. A distinction must be drawn, however, between openness to the future and the openness of the past.

Barbara Adam, in *Time and Social Theory* (1990, p. 143), describes a process whereby the past is known as reconstituted in the present, that is, as mediated by

---

[3]There is a difference between laws of nature and laws of society (and a further distinction between laws as they exist in legal systems and of laws of sociology). To the extent the objection is made that laws of nature (gravity, for example) are not subject to political debate (cannot "go different ways"), we are not, as I understand Fraser's distinctions, talking of sociotemporality, but rather of different levels of temporality where laws of probability are not overridden by the nootemporal evolution which makes possible conscious choice. One can, for example, given certain assumptions, measure and predict rush hour traffic at an intersection. But in the sociotemporal *umwelt*, a conscious decision can be made to change the timing of the traffic light to alter that traffic flow.

intervening experience and events. She suggests, at least in passing, that this may make the past revokable and hypothetical, like the future. But while it is true for individuals that memory of past events is colored by present perception, this does not make the past hypothetical or revocable for them. In cases of incest, for example, adults who were victims often repress the memory in order to maintain present mental images. The past, however, remains unrevoked, and its effects may linger below the surface to emerge later in full light.

We may ask if the situation is any different for society as a whole. There is a sense in which the past is more hypothetical and revocable for societies than it is for individuals. In the case of China, as in the case of the incest victim, the past can be repressed in order to maintain images of legitimized power. At some point individuals who remember events will die, though social histories will live on. The history of a social group can be told leaving out some of the events which formed it. Unpleasant events or institutions, such as Tiananmen Square or slavery, can be ignored. If, however, those events or institutions are still maintained and justified as a significant part of the social structure, their effects will remain—violent suppression of ideas and the forced labor of one human being for another create different *kinds* of societies than those without them.

An essential task of the human mind, or of the nootemporal level of time, I submit, is to ''re-present'' facts, the realities of human life, to social institutions that might not want to hear or see them. The openness of society to the *future* does not mean that the past, constituent members, or other temporal levels are similarly open, hypothetical, or revocable. To the contrary, the other temporal levels bind the social in much the same way the past of individual memory binds the human mind—as a constituent structure that is needed to ''make sense of'' the whole even while it may not, ultimately, limit its freedom of action.

The previously mentioned example of a criminal case involving an alibi illustrates the significance of a commonly viewed past. In legal theory both the prosecution and the defense proceed as if they are attempting to establish past facts that actually happened. It is this assumption that provides rational structure to the process. If the prosecution were to assert the right to proceed despite knowing the truth of the defendant's alibi, or if the defendant were to assert the right to present false evidence of an alibi, the assumption of a common reality based upon actual past facts would be replaced by a process of competing fictions which would override the shared view of a common or real past.

In particular cases one side or the other might find justification for such fictions (a defendant, for example, might believe it justifiable to fabricate an alibi to overcome falsified evidence being presented by the prosecution). In the long run, however, the effect of a general acceptance of this malleability of the past would be to give to the powerful and to those who manipulate social ideas and images the ability to define what constitutes reality. On this view Tiananmen Square maybe did not happen, and those who state that the Holocaust did not occur may, if they become powerful enough, turn out to be right.

Using the language of political theory, it may be said that humans have ''rights'' which limit, and which bind, the society. My argument, however, is not limited to political theory—artists or journalists can do the same thing; that is, they may remind a collective audience of facts or experiences in human life that are real and assert some kind of claim on the social. It is not the assertion of an abstract right that is determinative—it is the concrete re-presentation of the reality of human life that can evoke both images and feelings of common humanity. In the Rodney King/Los Angeles police brutality case, for example, it was the televised image of another human being being beaten by police that provoked outrage. And it was the failure or inability of the prosecution to present Rodney King's point of view that resulted in the acquittals in the first trial.

## Demythologizing Social Time

Sam Macey suggests in a recent book (1989) that social time has come to be defined by increasing rationalization, or efficiency of measurement, resulting in increasing efficiency of industrial production. The Japanese, for example, increase efficiency at their automobile plants by requiring that parts be delivered at carefully timed stages in the production process, rather than, as American companies have tended to do, stockpiling them until needed (1989, p. 214). Better organization is in fact better "timing."

This does not mean, however, that there is one social time. There can be many social times, corresponding to different organizational social structures. Still, as human social evolution becomes more complex and interdependent, these different social times interact, and tend to become, through organizational interaction, subsumed under one time, an awareness of which is sustained by ever more instantaneous communication.

Even in this view, though, social time is not separate from human or biological time, but rather is an extension of it, both depending upon and subsuming it. When, for example, businesses organize work, they must take into account the human need to sleep. As organizational considerations come to dominate, the human need may be ignored—attempts may be made, for example, to have flight controllers work with too little rest. This will, however, eventually create problems for the effective functioning of the organization. And there is, as Giddens points out (1987, p. 163), a dialectical or countervailing tendency by which organizational subordinates reassert their positions.

Barbara Adam suggests (1990, p. 157; 1992, p. 182) that the view of social time created by modern technology and organization is artificial. The need for a single point of view to coordinate activities causes us to reify the view of time fostered by these activities. We tend to adopt metaphors of time created by dominant technologies—from Newtonian views of clockwork time, through entropic views popular in the heyday of steam engine technology, to a view of time as instantaneous information exchange fostered by communications technology. Marshall McLuhan (1964), taking his cue from Harold Innis' book *The Bias of Communication* (1951), argued much the same thing in the 1960s, as summarized by the title of his 1964 book *Understanding Media—The Extensions of Man*.

Another way of saying this is that as a society we think and communicate through technologies of communication, and that these technologies influence us in ways we do not always recognize. They condition our views of basic cultural values, such as neutrality and objectivity of legal process (Aultman, 1969). Recognizing the bias, however, does not by itself change the social effect of a technology. We are, for good or for ill, in a global village where instantaneous communication creates at the local level an awareness of far-off events such that the distinction between the local and global is more difficult to maintain.

The suggestion that there is a separate autonomous "social time" is not a good lens through which to view the modern world. Rather, there are competing social and other times being drawn together and coordinated in a social process that is also a mental process. Another way of saying this is that social time is still nootemporal—rationalizing and stabilizing social structures is a mental as well as social process, and does not happen automatically. Because traffic can be organized to permit more traffic does not mean doing so is in the social interest; and because concentration camps can be organized to kill people more efficiently does not mean they should be. There must be rational choice, and moreover, for this to happen at the social level there must be some way to overcome the tendency of social institutions to foster decisions that, though they may perpetuate the interests of particular organizations or groups, are irrational from the point of view of society and its constituent members.

Theories of social criticism, as institutionalized in the first amendment to the U.S. constitution, and theories of organized protest, as utilized by Gandhi, Martin Luther King, or Greenpeace, expressly recognize not only the legitimacy, but also the utility for the overall social structure, of criticizing, and selectively refusing to follow, socially promulgated norms. Communication from the bottom up, in opposition to dominant social images, helps maintain, in the long run, legitimate social structures. Structures with only automatic responses (as totalitarian societies tend to be) lose the ability to adapt, and eventually lose legitimacy.

This suggests again that it is the human mind (or nootemporality), free and aware of options, that renders social structures capable of adapting. This further suggests that the social should not be viewed as hierarchically above the nootemporal, so as to require deference. The human mind exists in both the biotemporal realm and in a social world where individuals and groups act in conflict and cooperation. The human mind does not always integrate, at least from the point of view of claimed social imperatives. Its task can be, and sometimes must be, to maintain the vitality of human life, in face of the sometimes irrational rationalizations of the sociotemporal.

## References

Adam, B. E. (1990), *Time And Social Theory*. Philadelphia: Temple University Press.
——— (1992), Modern times: The technology connection and its implications for social theory. *Time & Soc.*, 1:175–192.

Aultman, M. H. (1969), Law, communication, and social change—A hypothesis. *Fordham Law Rev.*, 38:63–72.

Fraser, J. T. (1975), *Of Time, Passion, and Knowledge*, 2nd ed. Princeton, NJ: Princeton University Press, 1990.

——— (1992), Human temporality in a nowless universe. *Time & Soc.*, 1:159–173.

Giddens, A. (1987), *Social Theory and Modern Sociology*. Stanford, CA: Stanford University Press.

Innis, H. (1951), *The Bias of Communication*. Toronto: University of Toronto Press.

Macey, S. L. (1989), *The Dynamics of Progress: Time, Method, and Measure*. Athens: University of Georgia Press.

McLuhan, M. (1964), *Understanding Media—The Extensions of Man*. New York: McGraw-Hill.

Orwell, G. (1949), *1984*. New York: New American Library.

Whitrow, G. J. (1980), *The Natural Philosophy of Time*, 2nd ed. Oxford: Clarendon Press.

Young, M., & Schuller, T., eds. (1988), *The Rhythms of Society*. New York: Routledge.

16

# The Technology–Ecology Connection and Its Conceptual Representation

*Barbara Adam*

*Abstract*   Today we live in a world where not merely the natural environment but also cultural life is globally connected. It is a world where actions in one part of the globe affect the lives of people, animals, and plants in faraway places and different historical periods. Social scientists therefore need to extend their conceptual frameworks and their areas of concern. They need to embrace a basic under-standing of the processes of life, of the connecting webs of relations, and of contemporary metaphors. The focus on time is an ideal aid to such a redirection of social science emphasis: Time permeates every aspect of our being from the eotemporal to the sociotemporal sphere. The paper focuses on the artefactual times of our technological environment and relates these to the temporal ongoings of living ecological systems. It proposes that such an exploration will help us the better to understand some of the environ-mental crises brought about by the scientific, industrial way of life.

Concern with life is not high on the social science agenda. Indeed, the social sciences have a long history of defining their subject matter in distinction to the world studied by biologists and physicists. This particular division of labor has not only helped to establish social sciences such as sociology and anthropology as independent academic disciplines but has also provided a wealth of knowledge about the cultural sphere, its socioeconomic structures and political processes, its educational and religious traditions. Social time (Adam, 1990, chaps. 4 and 5; Bergmann, 1992) in turn has been researched in relation to these human social phenomena. It too has been clearly delimited against the subject matter of the natural sciences, from physical and biological times (Adam, 1990, chaps. 2 and 3). In this quest for the distinctiveness of human culture social scientists have elimi-nated from their analyses the biological and inorganic aspects of our existence. As a basis of both knowledge and action, this stratifying approach is becoming increasingly problematic.

Today we live in a world where not merely the natural environment but also the cultural life is globally connected, where the earth is embraced in a network of information, political allegiances, and economic relations. It is a world where actions in one part of the globe affect the lives of people, animals, and plants in faraway places and distant times. This means that the fragmenting, isolating division of labor, so important for the development of the industrial way of life, is no longer adequate to comprehend this new era of connectedness, relations, and implication. Social scientists therefore need to extend their conceptual frameworks and their areas of concern. They need to embrace a basic understanding of the processes of life, of those connecting webs of relations. Moreover, they need to recognize the mediating role of metaphors and the potential of contemporary technology for understanding environmental and cultural processes. The necessity for such a shift in approach is most clearly apparent in analyses relating to the present environmental crisis: Industrial societies have created life-threatening conditions that have no regard for boundaries and affect on most life on earth. Global warming, acid rain, and ozone depletion are cases in point. They demonstrate the need for this new stress on relations and connections which inevitably extend beyond traditional, disciplinary boundaries. The focus on time is an ideal aid to such a redirection of social science emphasis: Time permeates every aspect of our being from the eotemporal to the sociotemporal sphere (Fraser, 1992, pp. 159–173), from atomic particles via the rhythmic organization of life to time reckoning systems based on water, shadows, calendars, and clocks.

## Reality Conceptualized: Images and Metaphors

Since their inception the social sciences have been tied to the project of the Enlightenment thinkers; they have been associated with the development of objective science, with the control of nature, and with the rational organization of every sphere of life (Giddens, 1976; Harvey, 1989). Sociology in particular was concerned to establish itself as an objective science. Its aim, to some extent at least, was to offer a social science equivalent to the natural sciences' control over nature. Like the natural sciences, it focused on the rational elements of its subject matter and emphasized the rational features of its theories and methods. Its development and its fate therefore are linked to industrialization, to the rise of science as the dominant cosmology, and to the commodification and rationalization of every sphere of social life (Marx, 1857, 1867; Weber, 1904–1905; Giddens, 1981, 1990).

Given this history, it is not surprising that analyses of social time emphasize the organization of Western life to the metronomic beat of the clock (Moore, 1963; Zerubavel, 1981; Young, 1988). It is almost to be expected that these studies of social time stress the control over time and the rational standardization of time across the globe (Kern, 1983). The standardization of time at the turn of the century, the commodification of time in all areas of social exchange, and the separation of time and space through technological developments in communication and transport, all these form central components of analyses of social time. Time theories, in other words, complement the social sciences' explicit concern with objectivity, rationality, and the scientific study of Western, post-Enlightenment society (Adam, 1988, 1990, 1992). Like the social sciences in which they are embedded, time theories have been tied to metaphors premised on dominant technologies, in particular to clocks and heat engines, the technologies of the 18th and 19th centuries. The image of the clockwork relates to predetermined motion and mechanical interaction; that of the heat engine to a dynamic system that exchanges energy in a one-way direction. When we explore these images in more detail, we find that they entail different kinds of time.

The time of the clock ticks away evenly and objectively. It is linked to abstract motion and to distance traveled in space. The clock marks time by dissociation, by abstracting it from human events, and assigning it a number value. Emphasis on the clockwork highlights mechanical relationships: cogs and springs interacting to form an integrated whole. It accentuates parts. The smooth running of the whole depends on accurate timing, tempo, sequence, duration, and periodicity. Breakdowns can usually be traced back to the source by logical pathways. If one fully understands the working of the clock, one can isolate the cause of the breakdown and, in principle at least, repair it by replacing the appropriate parts (Prigogine and Stengers, 1984, pp. 27–68; Adam, 1990, pp. 48–55).

If we turn to the image of the heat engine we find that it not only retains all the functional characteristics of the clock but also contains some new features which are no longer reducible to those of the mechanical clock. The heat engine and the clock share the functional principles of machine time: Both rely for their smooth functioning on accurate timing, tempo, sequence, duration, periodicity, and the spatial measure of time. In contrast to the clock, however, heat engines use and dissipate energy.[1] They consume nonrenewable resources and dissipate heat to the environment. Theirs is a system that is running down and as such it also changes the surrounding environment in the direction of a slow heat death. These new features are intimately tied to the different tracking of time: The primary time of the heat engine is that of unidirectional, irreversible change (Prigogine and Stengers, 1984, pp. 257–290; Adam, 1990, pp. 61–65).

Clocks and heat engines are metaphors for an inanimate nature, a nature in motion, locked in the predetermined pathways of energy exchange in the direction of decay. They are expressions of a world of parts: simple, demystified, measurable, and predictable. They represent a reality that can be taken apart and reassembled both physically and conceptually. Furthermore, emphasis on these metaphors puts humans in the role of machine operator, even that of maker. As Sheldrake (1990, p. 3) points out, "the idea of nature as a mechanical, inanimate system is in some ways more comforting, it gives a sense that we are in control."

Today, those conceptualizations are found wanting. They are considered inappropriate for understanding the processes of life and ecological relations on the one

---

[1]Clocks too need energy input—they need to be wound—but their design is based on mechanical principles and a theory of motion where the need for external energy input is explained on the basis of friction.

hand and for sociocultural globalization on the other. Moreover, the belief in control is slowly recognized as a misplaced faith, and there is a dawning realization that those traditional metaphors stand in the way of our coming to grips with the contemporary industrially induced environmental crisis. A brief comparison between the temporalities of living organization and machines will bring to the fore some distinctions and similarities which help us the better to understand the reasons for the dissatisfaction.

## Life and Machine Time: Similarities and Differences

Organic life is self-organizing, self-healing, and self-regenerating. Machines, in contrast, are always dependent on designers, makers, and attendants. On the basis of that difference alone, life eludes mechanistic design and organization. In addition to this fundamental distinction, however, there are some important similarities between the temporalities of living beings and machines: Both combine variance and invariance, stability and change. Both are marked by an arrow of time. Both are temporally organized and entail in their processes the time aspects of succession, duration, timing, tempo, and periodicity. Both entail a multitude of time spans. Those shared characteristics, however, take on a different meaning in artefacts and living systems.

First, machines are designed for precision. In convertional machines, therefore, variance beyond a certain tolerance means failure; it means they are going wrong. For living systems, in contrast, variance is the source of life, stability, continuity, and evolution while invariance spells death. Without genetic variance, in other words, there would be no evolution. For living systems invariance means the cessation of growth, development, and regeneration. This centrality of variance is nowhere more pertinent than in human cultural life where it constitutes a defining feature. The sources of cultural change, in other words, are cultural universals such as language, external representation, and storage of knowledge, tools for the transcendence of physiological limitations and modes of production, all of which are by definition variable.

Second, while the machines' arrow of time points in the direction of increasing entropy and decay, in living beings decay and the march toward heat death are opposed by the development of individuals and the evolution of species, their growth and aging. In living processes, therefore, the use and depletion of time are counterbalanced by evolution, generation, and replenishment: Decay is compensated by repair through healing and by ''superrepair'' through speciation.

Third, all organisms from single cells to ecosystems display interdependent rhythmic behavior. They are synchronized in an all-embracing web of complex interconnections. Linear sequences take place but these form part of a wider network of cycles as well as finely tuned and coordinated temporal relations where ultimately everything connects to everything else. The structure of an ecological system is temporal and its parts resonate with the whole and vice versa. While it is possible to talk of the rhythm of a machine, this rhythmicity is qualitatively different from the orchestration of internal and external rhythmic exchanges of living systems. In machines, time is always running down and out. It is not generated in interaction. With machines there is no rhythmic embeddedness, no transformation, regeneration, or creation of time. Machines are not designed as an integral part of the give-and-take of ecological exchange. Their waste energy is not designed to be another being's basis of existence. In their design machines are abstracted from the life-generating creativity of organic being. They are creations apart; externalized, abstracted, and bounded. Theirs is a finite time: isolated, encased, and fixed. It is a time of product rather than process, a time that can be quantified and measured.

Finally, natural and artefactual processes entail a multitude of time spans. Rhythms and cycles which are of relevance at the biological level of human existence extend over far greater periods than those conventionally used by social scientists. Biological time scales range from the atomic and microscopic to the evolutionary and even cosmic, from nanoseconds to millenia. Human perception and thinking, in contrast, operate within the much more restricted time span of seconds to decades and centuries. The traditional social science response has been twofold: to bracket the biological time scales and to focus within the human range on one aspect at the expense of others, to study microinteractions or macrosystems, detailed daily rounds or life-cycle events. With the creation of microchip technology and nuclear power, however, neither the narrow time scale of seconds to centuries nor dualistic choices are sufficient for the explication of contemporary human existence. Artefactual time scales join those of living processes in their extension beyond the human capacity of perception and consciousness: imperceptibly fast and incomprehensibly long periods have become an integral part of contemporary life. Thus, where our biology fails the new technologies may succeed in persuading social scientists of the need for expanded time spans and for grasping together what traditional analyses have kept apart.

Taken together, these similarities and differences mean that the ecological give-and-take of living beings cannot be understood through the metaphors of clocks and heat engines. It cannot be grasped without evolution, development, the generation of time, expanding as well as multiple time spans and, most importantly, without rhythmic embeddedness and the constant energy exchange with the environment. Collectively they demonstrate the inappropriateness of the classical technological metaphors not only for the conceptualization of life but also for a globally networked world, and they show the inadequacy of traditional assumptions for comprehending a cultural world marked not by certainty and predictability but indeterminacy and processes beyond human control.

## The Globally Networked World and Its Representation

This loss of human control is demonstrated on a daily basis through such contemporary phenomena as environmental problems, the stockmarket, and political unrest.

Scientists, economists, and politicians are no longer the designers, makers, and controllers of the systems in their charge: The scientific control of nature is out of control, and so are the economic control of markets and the political control of war and peace. Hailed scientific developments from CFC's to nuclear power, technological developments from steam engines to jet planes, and the use of nonrenewable energy for domestic and industrial purposes, have brought devastation to our earth, water, and air. All have unintended and uncontrollable effects. "Yesterday's" benign inventions are "today's" sources of cancers, global warming, ozone depletion, and the dying of lakes and forests. Those expressions of environmental crises force the recognition that the nature–culture intersection is no longer separable. True wilderness, that is, nature unaffected by science and technology, seems no longer to exist. "Nature" has been acculturated and in that form it feeds back into cultural processes and permeates contemporary existence. Moreover, outcomes of those complex interpenetrations are not necessarily tied to the place of action. In other words, the impact of scientific technology and the industrial way of life is global: Local action has national, international, and global outcomes which have progressed through the global network of nonlinear, ecological relations. Such effects are not reducible to the processes of clocks and heat engines and they elude economic solutions and scientific control. The longevity of nuclear matter, for example, means that we have nothing equal to its life span of existence, no material with which to make safe its radioactive waste. Consequently, we have to take responsibility today for hazards that may outlast us by millenia, and we need to take account of the potential consequences of actions which can only be known with the benefit of hindsight. In such situations, the past is no longer a reliable guide to the future.

Time lags and latency periods create further problems for amelioration and control: They make the links between causes and effects invisible. Not surprisingly, nuclear accidents fall outside the capacity of systems of insurance (May, 1989, p. 13; Beck, 1991; Renner, 1991, p. 146). In combination with the relativity of place, invisibility creates an illusion of safety and security. This has dramatic effects on knowledge: When the connections between input and output are no longer direct and traceable, those relations cease to be amenable to scientific certainty and verification, quantification and control. Scientific "proof" consequently becomes a misplaced goal. Clocks and heat engines, the key metaphors for a world of bounded nation states, dominated by Newtonian and thermodynamic science, are no longer appropriate for a social world in which knowledge, information, politics, business, finance, transport, and environmental problems are globally networked.

Electronic communication is a key factor in this contemporary globalization. It enmeshes our earth in a network of sounds and images. It is not subject to the constraints of time and space in the same way as were the earlier technologies: It is in principle available, storable, and retrievable instantaneously all over the globe (McLuhan, 1964; Rifkin, 1987; Poster, 1990). Since the limit to the speed of that information transfer is the speed of light, those communications are practically simultaneous and facilitate the transcendence of space. The age-old link between time and space is unsettled (Adam, 1992, p. 185). Electronic information is dispersed across time and space. Decentered and able to be infinitely multiplied it no longer has a fixed location: "it is everywhere and nowhere, always and never" (Poster, 1990, p. 85). Like the processes of life it is both material and immaterial. Moreover, as Poster (1990) points out further, "the new language structures refer back upon themselves, subverting referentiality and thereby acting upon the subject and constituting it in new and disorienting ways" (p. 17). Analogous to living processes this brings with it a loss of certainty and predictability. It facilitates nonlinear connections where the initial cause becomes disconnected from and irrelevant for the outcome and vastly increases the fragility of social networks: Effects in one part of the world reverberate through the global network and insignificant local happenings may have dramatic impacts at the national and global level. Unpredictability and uncertainty, the production of the unknown rather than the known, are therefore integral aspects of the networked connections of electronic information. Predictability, of course, has not disappeared but unpredictability has taken a prime position. A further irreducibly new feature of technological artefacts relates to the loss of identity. Machines, be they mechanical or thermodynamic, are temporally organized *within* each machine without being rhythmically embedded in their environments. They may be coupled to other machines but this machine interdependence does not alter any of the machines' individual functional identities. Computers, in contrast, are so profoundly interconnected that their identities are no longer sacrosanct. Just like living interactive systems, they are able to pass on complex information, change each other in the process, even get "infected." In their mutuality they are subject even to epidemics (Poster, 1990, pp. 1–20).

With electronic communication, artefactual and living systems have moved closer together. Both display features which cannot be grasped through the metaphors of the previous centuries. A fundamentally connected, open universe, an emerging reality of which we can have no foreknowledge, a reality that implicates operators in their subject matter, is consequently better served by the metaphor of electronic communication. It allows for generative and productive processes and is thus much more closely allied to living being and ecological processes than the metaphors of the technologies of the 18th and 19th centuries. This means that the metaphors of electronic communication technologies offer the potential for more appropriate intervention in the environmental, economic, and political crises of the late 20th century.

## Conclusion

If our contemporary reality exhibits global features and is marked by historically unprecedented characteristics, and if it is distinguished by a multitude of times that coexist in an embedded and mutually implicating way, then I feel that these features must be allowed to permeate social science assumptions, theories, and methods. With the aid of appropriate metaphors those characteristics can inform our understanding in the same way as the key features of the Enlightenment had penetrated the work of the pioneers of the natural and human sciences and their successors.

Such a shift in conceptualization will be essential if we are to respond more appropriately to the environmental crisis of the late 20th century.

# References

Adam, B. (1988), Social versus natural time, a traditional distinction reexamined. In: *The Rhythms of Society*, ed. M. Young & T. Schuller. London & New York: Routledge, pp. 198–226.

—— (1990), *Time and Social Theory*. Philadelphia: Temple University Press.

—— (1992), Modern times: The technology connection and its implications for social theory. *Time & Soc.*, 1:175–192.

Beck, U. (1992), From industrial society to risk society: Questions of survival, social structure and ecological enlightenment. *Theory, Cult. & Soc.*, 9:97–123.

Bergmann, W. (1992), The problem of time in sociology: An overview of the literature on the state of theory and research on the "Sociology of Time," 1900–82. *Time & Soc.*, 1:81–134.

Fraser, J. T. (1992), Human temporality in a nowless universe. *Time & Soc.*, 1:159–174.

Giddens, A. (1976), *New Rules of Sociological Method*. London: Hutchinson.

—— (1981), *A Contemporary Critique of Historical Materialism*. London: Macmillan.

—— (1990), *The Consequences of Modernity*. Cambridge, U.K.: Polity Press.

Harvey, D. (1989), *The Condition of Postmodernity*. Oxford: Blackwell.

Kern, S. (1983), *The Culture of Time and Space 1880–1919*. London: Weidenfeld & Nicolson.

Marx, K. (1857), *Grundrisse*. Harmondsworth, U.K.: Penguin, 1973.

—— (1867), *Capital*, Vol. 1. Harmondsworth, U.K.: Penguin, 1976.

May, J. (1989), *The Greenpeace Book of the Nuclear Age*. London: Victor Gollancz.

McLuhan, M. (1964), *Understanding Media*. London: Routledge, Kegan & Paul.

Moore, W. E. (1963), *Man, Time and Society*. New York: John Wiley.

Poster, M. (1990), *The Mode of Information*. Cambridge, U.K.: Polity Press.

Prigogine, I., & Stengers, I. (1984), *Order Out of Chaos*. London: Heinemann.

Renner, M. (1991), Assessing the military's war on the environment. In: *The State of the World 1991*. ed. L. R. Brown. London: Earthscan.

Rifkin, J. (1987), *Time Wars*. New York: Henry Holt.

Sheldrake, R. (1990), *The Rebirth of Nature*. London/Sydney: Century.

Weber, M. (1904–1905), *The Protestant Ethic and the Spirit of Capitalism*. London: Unwin & Hyman, 1989.

Young, M. (1988), *The Metronomic Society*. London: Thames & Hudson.

Zerubavel, E. (1981), *Hidden Rhythms*. Chicago: University of Chicago Press.

# 17

# Some Perspectives on the Idea of Progress as a Problem in the Study of Time: The Cases in China, Japan, and Russia in Comparison with Modern Europe

*Masaki Miyake*

*Abstract*   The concept of time inherent in the idea of progress is that of linear time going upward in the direction of perpetual betterment.

In this paper I discuss aspects of the idea of progress in the light of various notions of historical time. This discussion draws on the works of European, Chinese, and Japanese thinkers: Bernard de Fontenelle, K'ang Yu-wei, and Fukuzawa Yukichi. The discussion then turns to the critical attitude that Westernization is an enforced modernization in the non-Western world, examining the parallels between Sōseki in Japan and Chaadayev in Russia.

In this article I follow the Japanese and Chinese practice of writing the family name first.

## The Emergence of the Idea of Progress

It is to the first half of the 17th century that we should look for an understanding of the origins of the assumptions inherent in the idea of progress. As S. L. Macey (1986) points out, "the meaning of the word *progress* began to change during the first half of the seventeenth century" and he says that "the earlier meaning associated with a journey through space" is still found as late as 1678 in John Bunyan's *The Pilgrim's Progress from This World to That Which Is to Come* (Macey, p. 98).

Macey also points out the importance of the "Quarrel of the Ancients and Moderns" for the history of the idea of progress (p. 94). This quarrel began with Alessandro Tassoni in Italy, but its main protagonists were the 17th-century French authors Charles Perrault and Bernard de Fontenelle for the moderns and Nicolas Boileau-Despréaux for the ancients. The quarrel continued in England as late as the early 18th century with Jonathan Swift's *Battle of Books* (1704), which ridiculed the moderns. The American sociologist Robert Nisbet (1980) summarizes the quarrel as being "waged around this question: which are superior, the literary, philosophical, and scientific works of classical Greece and Rome, or, instead, the works of the modern world; that is, the sixteenth and seventeenth centuries?" (Nisbet, p. 151).

Throughout the quarrel it was Fontenelle's writings which contributed the most to the theory of the idea of progress (Pollard, 1971, pp. 40–41). Nisbet sees particular significance in Fontenelle's assertion that "men will never degenerate, and there will be no end to the growth and development of human wisdom" (Nisbet, 1980, p. 155; quoting Fontenelle, 1688, p. 426). J. B. Bury (1955), in his classic work on the idea of progress, says that "Fontenelle asserts implicitly the certainty of progress when he declares that the discoveries and improvements of the modern age would have been made by the ancients if they exchanged places with the moderns; for this amounts to saying that science will progress and knowledge increase independently of particular individuals" (Bury, 1955, pp. 109–110). It was on the basis of these assumptions, according to Bury, that Fontenelle became the first to formulate the idea of the progress of knowledge as a complete doctrine (Bury, 1955, p. 110; cf. Miyake, 1992, pp. 6–8).

If we accept Bury's analysis, we may assume that the great discoveries made by scientists like Newton were a decisive factor for the formation of the idea of progress developed by the moderns, among others by Fontenelle. This factor came to be called the Scientific Revolution in the Western world in the 17th century (Macey, 1986, p. 94).

The metaphor used for the concept of time inherent in the idea of progress is that of a linear line going perpetually upward. It is often said that this concept of time derives from the Christian concept of linear time running between the Creation of the world and the Final Judgment. A typical example of an interpretation of this kind can be found in the works of the German philosopher Karl Löwith (1949, pp. 60–61). The German theologian Rudolf Bultmann is also of the opinion that the idea of progress is the secularization of the Christian view of history and says that "the concept of providence is replaced by the idea of progress promoted by science" (Bultmann, 1957, p. 73; cf. Fraser, 1990, pp. 371–373; Berthold, 1991, p. 345).

Works such as the British historian Arnaldo Momigliano's article (1966) suggest that it is too sweeping a generalization to insist that the ancient Greek concept of time was cyclic. Momigliano says that the concept of time cherished by the representative Greek historians such as Herodotus and Thucydides is not cyclic at all. Momigliano demonstrates this assumption by scrutinizing the text of their works (Momigliano, 1966, pp. 11–12). He further insists that Polibius, who is often thought of as showing a typically cyclic view of history, operates, outside the famous chapters on the constitutions of Rome and Carthage, in the rest of his book *World History (Historia Katholike)* as if he did not hold a cyclical view of history at all (Momigliano, 1966, pp. 12–13).

The 8th century B.C. Greek poet Hesiod can be said to have developed something like a concept of deteriorating time. He spoke in his work *Works and Days* of a myth of five races, the races of golden men, silver men, bronze men, the race of heroes, and the race of iron men. In their voluminous work *Utopian Thought in the Western World* (1979, p. 70), Frank E. Manuel and Fritzie P. Manuel say that, of the entire myth of five races, it is often only the utopia of the golden-age which was remembered. The influence of Hesiod's myth of the golden age was persistent and it survived up to the 18th century. The root of Hesiod's four metallic ages can be found in oriental myth, whose offshoot can also be traced in the Book of Daniel (Bultmann, 1957, pp. 25–26; Manuel and Manuel, 1979, p. 71). The four metals composing the colossus in Nebuchadnezzar's dream, gold, silver, bronze, and iron, correspond to the kingdoms of the Babylonians, the Medes, the Persians, and the Greeks; namely, the Seleucids from Alexander the Great to Seleucus IV or Anti-ochus (Bultmann, 1957, pp. 25–26).

In contrast to Hesiod, the Book of Daniel does not seem to show the concept of a deteriorating time, for all these empires are treated as equally corrupt. The Book of Daniel is important because of its eschatological implication. In Hesiod there is "no anticipation of the last things" (Bultmann, 1957, p. 27).

## The Various Concepts of Historical Time

In my report to the methodological session dealing with the concepts of time in historical writings in Europe and Asia of the 17th International Congress of Histori-cal Sciences, Madrid, August 27, 1990, I attempted to classify the various concepts of time appearing in historical writings in Europe and Asia, and identified eight distinct types: (1) oscillating time, (2) cyclic time, (3) Newtonian linear time, (4) Christian linear time, (5) linear time going upward, (6) linear time going downward, (7) time as a series of points, and (8) spiral time (Miyake, 1990, pp. 134–135; cf. Miyake, 1991, pp. 321–327; Michon, 1986, p. 59, Table 3).

The idea of progress is, as I have suggested above, often taken to be a variation of Christian linear time. This is an issue which must be carefully scrutinized. The idea of progress can be characterized as type 5, that is, time going upward in the direction of eternal betterment. Christian linear time has its beginning, the Creation, and its end, the Final Judgment. The metaphor for this concept of time is a segment of a line, showing both ends. Type 5, the concept of time inherent in the idea of progress, will have as its metaphor a linear line with a rising gradient, but with no end.

The concept of time prevailing in Confucian thought is seen as type 6, that is, time going downward. In the Confucian classics there is frequent reference to the degeneration from the golden age in a remote antiquity to the turbulent and polluted present. The metaphor for this concept of time can be then that of a linear descent.

In the following section I would like to show how this pessimistic concept of time was converted to an optimistic one by analyzing the ideas of the 19th-century Confucian thinker K'ang Yu-wei (1858–1929), who lived in China under the Ch'in Dynasty. His case will show the birth of the idea of progress in a country in the non-Western world, a country where Confucianism, and not Christianity, has been most influential among the intellectuals.

## K'ang Yu-wei's Idea of "One-World" (Ta T'ung)

Among the outstanding intellectuals of 19th-century China, I would like to focus on K'ang Yu-wei. I choose him because his way of thinking has much to do with the concept of time prevailing among those Chinese intellectuals who were deeply influenced by Confucianism. To introduce K'ang to the reader, I will quote a very concise description by Ivan Morris, former Associate Professor of Chinese and Japanese at Columbia University. Morris notes that K'ang was both statesman and philosopher and that he inspired the abortive Hundred Days of Reform in 1898, directed toward modernizing and strengthening the declining Ch'ing Empire. Morris says that K'ang, while advocating Confucianism as a State religion, believed that "technical knowledge from the West was indispensable for a strong China" (Maruyama, 1963, p. 306).

In the section "*Li Yün*" of one of the Chinese classics *Li Chi* (The Records of Rites), a collection of canonical writings compiled presumably in the Former Han dynasty (302 B.C. – 8 A.D.), K'ang found the concept of *Ta T'ung* (One World, or the Age of Great Universality) (K'ang, 1958, p. 27; Tsuchida, 1927, pp. 194–196), and the concept of *Shao K'ang*, which means the Age of Little Peace-and-Happiness (K'ang, 1958, p. 72). It is difficult to translate into English the Chinese word *Ta T'ung*. Laurence G. Thompson shows thirteen different translations such as: The Age of Great Peace (Tsuchida, p. 197), The Great Unity, Grand Union, Grand Course, Cosmopolitan Society, The Great Similarity, etc. (K'ang, 1958, pp. 29–30). Thompson himself chooses the expression "One World" (K'ang, 1958, p. 30). It goes without saying that this expression has nothing to do with "One World Movement." In "*Li Yün*," China is thought to degenerate from the former age to the latter age, and the golden age of *Ta T'ung* is vividly described with a strongly nostalgic tone (K'ang, 1958, pp. 27–29; Tsuchida, 1927, pp. 194–195).

K'ang changed the traditional interpretation of this passage, which was a pessimistic one assuming a degeneration from such an ideal age to the age of little peace and happiness. He thought, quite on the contrary, that the ideal age (*Ta T'ung*) would emerge in the future. With this utopian idea he mixed up the evolutionary theory of three ages: the Age of Disorder, the Age of Increasing Peace-and-Equality, and the Age of Complete Peace-and-Equality (Tsuchida, 1927, pp. 196–198; K'ang, 1958, p. 72). In his main work *Ta T'ung Shu* (The Book of Ta T'ung) he says:

> Even if there be sages who establish the laws, they cannot but determine them according to the circumstances of their times, and the venerableness of customs. The general conditions which are in existence, and the oppressive institutions which have long endured, are accordingly taken as morally right. In this way, what were at first good laws of mutual assistance and protection end by causing suffering

through their excessive oppressiveness and inequality. If this is the case, then we have the very opposite of the original idea of finding happiness and avoiding suffering.

India is like this, and China likewise has not escaped it. Europe and America are rather near to [the Age of ] Increasing Peace and Equality; but in that their women are men's private possessions, they are far from [according with] universal principles, and as to the Way of finding happiness, they have likewise not attained it. The sage-king Confucious, who was of godlike perception, in early [times] took thought [of this problem], and grieved over it. Therefore he set up the law of the Three Governments and the Three Ages: following [the Age of] Disorder, [the world] will change to [the Ages, first] of Increasing Peace-and-Equality, [and finally], of Complete Peace-and-Equality; following the Age of Little Peace-and-Happiness, [the world] will advance to [the Age of] One World [K'ang, 1958, p. 72].

K'ang was without doubt influenced by the Darwinian theory of evolution (K'ang, 1958, p. 50). It is also possible that he was influenced by T. H. Huxley's work *Evolution and Ethics* (1893) translated into Chinese by Yen Fu (1853–1921) (Schwartz, 1964) and Charles Lyell's work *Principles of Geology* (1830—1838), which was also translated into Chinese by D. J. MacGowan and Hua Heng-fang in 1872 (Sakade, 1976, pp. 50–51; cf. Bedini, 1975, p. 476). But K'ang remained within the tradition of Confucianism. The concept of time inherent in the Confucian view of history has been, up to the days when K'ang tried to change this concept, that of retrogression. For Confucious and Mencius, two founders of Confucianism, the ideal rulers emerged in the early stage of Chinese history. The golden age was for them at the beginning of Chinese history. These ideal rulers were the legendary Chinese kings Yao, Shun, Yu, and the founders of the Chou Dynasty Wu-wang (King Wu) and his younger brother Chou Kung Tan (ca. 11th century B.C.).

It is interesting that K'ang made a dramatic conversion from the concept of time indicated above as type 6 into that indicated as type 5. In the 19th century, and early 20th century, reference to Confucian thought was indispensable, if one were to preach reform to the Chinese intellectuals, so deeply rooted was it in the intellectual life in China. So K'ang chose to change the interpretation of this thought from a pessimistic view of history into a progressive view of history, strongly imbued with utopianism. As in the case of Condorcet, the idea of progress is often connected with utopianism. K'ang's *Ta T'ung Shu* is an Asian example of this connection. Thus, we can see the emergence of the idea of progress in China as a spiritual drama performed, not within the tradition of Christian eschatological theology, but within the tradition of Confucianism.

## Fukuzawa's Idea of Progress

In Japan, Fukuzawa Yukichi (1834–1901) combined the roles of Fontenelle, Condorcet, and K'ang Yu-wei. To introduce him to the reader, I borrow again a concise description by Ivan Morris. He says that Fukuzawa was a prominent educator and writer "who played a leading role during the Meiji Period (1868–1912) in introducing Western civilization to Japan" and was a "firm believer in a responsible Cabinet system and in securing human rights" (Maruyama, 1963, p. 296). In 1868 he founded a private school, "Keiō Gujuka," which became Keiō University in Tokyo. The British Japanologist Basil H. Chamberlain (1850–1935), who lived in Japan from 1873 to 1911, describes in his work *Things Japanese* that "the number who flocked to learn of him (Fukuzawa)" were "so great and so easily moulded, that it is no exaggeration to call Fukuzawa the intellectual father of more than half the men who now direct the affairs of the country" (Chamberlain, 1904, p. 367).

In his intellectual history of Japan in the Meiji period, the Japanese philosopher Kōsaka Masaaki describes the main characteristics of Fukuzawa's thought as a combination of protest against feudalism and advocacy of modernity. Fukuzawa admired two aspects of the West in particular: rationalism and the spirit of independence and self-respect (Kōsaka, 1958, p. 71). He thus admired Western civilization and insisted that Japan should "make Western civilization its objective" (Kōsaka, 1958, p. 73; cf. Fukuzawa, 1940, p. 229). At the same time, he had some reservations. He often spoke of knowledge and virtue, and by that he meant intellect and morals (Kōsaka, 1958, p. 73). According to him, intellect makes progress, morals do not. His view was different from that of Condorcet, who believed that intellectual and moral progress were parallel (Manuel and Manuel, 1979, p. 503). Fukuzawa says:

> Morals have made no progress from generation to generation. From the beginning of the world to the present their nature has not changed. This is not so with intellect. For every one thing the ancients knew, modern men know a hundred; what the ancients stood in awe of, we belittle; what they thought wonderful, we laugh at. The subjects of intellect increase day by day, its discoveries are manifold . . . Progress in the future is beyond all imagining [English translation: Kōsaka, p. 74; Japanese original: Fukuzawa, 1875, pp. 116–117].

In this distinction between the progress of morals and progress of intellect, Henry Thomas Buckle's influence on Fukuzawa is evident. Fukuzawa eagerly studied both Buckle's work (1857–1861) *History of Civilization in England* (Maruyama, 1986b, p. 4) and François Guillaume Guizot's work (1870) *General History of Civilization in Europe* (Maruyama, 1986a, p. 134). On the other hand, we can see here the intellectual tradition in Japan of trying to combine Eastern morality with Western technology. Sakuma Shōzan (1811–1864), a Samurai of the Matsushiro clan, may be characterized as a prototype of the intellectuals of the "Enlightenment" in Japan, of which Fukuzawa was one of the most representative. Sakuma once wrote a poem in a letter to a friend in which he said: "Eastern morality, Western technology, mutually complete a circular pattern." To clarify the meaning of this poem, he wrote: "Unifying morality and technology—Asia and Europe—is like completing a circle. If one part is missing, the circle remains incomplete" (Kōsaka, 1958, pp. 26–27; Miyake, 1992, pp. 9–13; Minamoto, 1993, pp. 29–30).

For Fukuzawa, the course Japan had to choose was to Westernize the country by learning science, technology, and the political systems of the West. But, as modernization and Westernization continued in Japan, the skepticism regarding its superficiality grew. Both modernization and Westernization were eagerly accepted in Japan as the only way to "enrich and strengthen the country." The situation in Japan in the end of the 19th century was similar to that in Russia during the period after the reign of Peter the Great. Modernization and Westernization were thought of as nothing but the progress of civilization in the two non-Western countries, namely Japan and Russia.

### Sōseki and Chaadayev: The Non-Western World's Skepticism and Pessimism with Regard to Progress and Westernization

One representative intellectual in Japan who tried to clearly formulate the skepticism toward modernization and Westernization was Natsume Sōseki (1867–1916). He sensed in Japan's rapid Westernization the danger of the loss of Japan's cultural identity. To introduce this writer, I will quote from the afterword to an English edition of his last and unfinished novel *Meian* (Light and Darkness). The description is by the translator, V. H. Vigliemo.

> It is no exaggeration to state that Natsume Sōseki [I follow Japanese practice of writing the surname first. Kinnosuke was his given name, but he is always referred to by his sobriquet, Sōseki] is at once the greatest novelist and literary figure of modern Japan. His novels have become modern classics, and a large section of his works has been written into the textbooks of the Japanese higher and middle schools. Moreover his influence on subsequent Japanese literature was, and continues to be, immense [Sōseki, 1985, p. 376].

In spite of his great success as a novelist, Sōseki grew more and more skeptical and pessimistic about the future of Japan, which, in his eyes, was in too great a hurry in its efforts to superficially Westernize itself. Sōseki's most important remarks on this problem are contained in a lecture, which he gave in Wakayama in August 1911 with the title "Gendai Nihon no kaika" (Enlightenment of Modern Japan). In this lecture, Sōseki establishes a distinction between the Enlightenment in the West, which he called "general enlightenment," and the enlightenment in Japan. The former is an *internal (endogenous) development* and the latter is *externally derived (exogenous)*. By the word *kaika* (enlightenment) Sōseki does not mean the "Enlightenment" in 18th-century France, but the progress of the civilization in general as "ningen katsuryoku no hatsugen no keiro" (the process of the manifestation of human energy). Since the Meiji Restoration in 1868, Japan was exposed to the overwhelming impact of Western culture, which made it impossible for Japan to continue its internal enlightenment. Sōseki describes Japan's fate by saying: "It is as if within the two important domains of economizing and expending human energy, which hitherto were on the scale of 20, suddenly, as a consequence of

outside pressure, the scale of economizing and expenditure was forced up to 30''
(Kōsaka, 1958, p. 446).

In Japan, the late professor Yamamoto Shin (1913–1980) of Kanagawa University was a pioneer of the study of the theory of civilizations. He tried to establish
a general theory of civilizations, which would fully explain, among other things,
the history of Japan from antiquity to the present. Comparing the reactions of two
writers, Sōseki in Japan and Peter Chaadayev in Russia, toward the dilemma of
Westernization, Yamamoto tries to elucidate the spiritual reaction of the intellectuals
in the non-Western world toward the Westernization forced upon it by outside
forces. He points out that modernization and Westernization are two different categories. The West modernized itself in the modern age, but cannot Westernize itself.
The Westernization only occurs in non-Western societies, when they try to borrow
the knowledge, technology, institutions, and ideas of the West (Yamamoto, 1974;
1985, chap. 3).

For Fukuzawa, "progress" was identical with Westernization. But he was
brought up in the Asian cultural tradition of Confucianism, and it was no easy task
for him to replace an Asian cultural heritage with Western civilization (Fukuzawa,
1969, introduction, p.xi).

Sōseki was far less optimistic than Fukuzawa on the issues of progress and
Westernization. Sōseki's lecture shows his own mental anguish in this regard.

Yamamoto finds parallels between Sōseki's lecture, which is attracting attention at the moment in Japan, and *"Les lettres philosophiques"* of Peter Yakovlevich
Chaadayev (1794–1856). Chaadayev wrote eight *"Philosophical Letters."* Only
the first letter, dated December 1, 1829, and written in "Necropolis" (the city of
the dead), was published in the Russian journal *"The Telescope,"* 19, in 1836
(Chaadayev, 1836. Togawa, 1962, p. 67; Zenkovsky, 1953, vol. 1, p. 150). The
impact of this letter was enormous. The journal was banned by the Russian Ministry
of Internal Affairs and the editor was exiled from Moscow. As is widely known,
Chaadayev himself was officially declared by Czar Nicholas I to be insane (Katsuda,
1961, p. 47).

According to Yamamoto, three assumptions are necessary in order to perceive
the similarity between Sōseki's lecture and Chaadayev's *"Philosophical Letter No.
1."* First, we must understand that Russia does not belong to the West. Second, we
must understand, as is already suggested above, that Westernization is not equal to
modernization. In regard to the experience of Westernization, the West has nothing
to teach the non-Western world from its own past, though the West has much to
teach the non-Western world about the problem of modernization. Third, the model
used to understand the phenomenon of Westernization should be molded by a
comparative study of Westernization in non-Western countries. On the basis of
these three premises Yamamoto develops a comparative study of Sōseki and Chaadayev. He says that Chaadayev's *"Letter"* is a total and highly critical reflection
on the Westernization of modern Russia. For Chaadayev, modern Russian civilization is superficial and imitative, inconsistent and unaccumulative. For Chaadayev,
there was no inner development, no natural progress in Russia, and all the current
thought in Russia was ready-made in the West. Thus, Chaadayev's diagnosis of

Russia is very similar to the message in Sōseki's lecture, which characterizes Japanese development as "gaihatsu-teki" (exogenous) in contrast to "naihatsu-teki" (endogenous) development of the West, Yamamoto asserts (Yamamoto, 1974; 1985, chap. 3).

Chaadayev was strongly influenced by such French Catholic thinkers as Comte de Maistre, envoy of Savoy to St. Petersburg, Vicomte de Bonald, and François Chateaubriand (Zenkovsky, Vol. 1, p. 152). So Chaadayev showed in the first *"Philosophical Letter"* the belief that Roman Catholicism was the true nucleus of the West and only Roman Catholicism was able to afford to guarantee true progress. He lamented the fact that Russia had adopted on liberation from the Mongollan yoke not Roman Catholicism but the Byzantine Empire's Greek Orthodoxy. He was also highly critical of the idea of progress developed by Voltaire, Diderot, and Condorcet (Katsuda, 1961, p. 60). But later, in his *Apologie d'un fou* (1837), he came to the conclusion that Russia, because it was backward in comparison to the West, was able to proceed faster than the West. Russia was able to select the good from the bad in learning from the accumulated experience of the West. He was not, however, a fanatic "Slavophile," for he still believed that the Westernization ever since Peter the Great had been the right course for Russia to follow (Katsuda, chap. 3; cf. Togawa, 1979, pp. 30–32).

Chaadayev's *"Letter"* shocked Russian intellectuals. Aleksander Herzen said that it was like a "shot ringing out in a dark night" (Zenkovsky, Vol. 1, p. 150). The Russian intellectuals were awakened by it and divided into two parties: the Slavophiles and the Westerners (Nolte, 1991, p. 66). In a similar way, Sōseki's lecture suggests two solutions for Japanese intellectuals. One is equivalent to "Westerners" and the other to that of "Slavophiles." An extreme example of the latter was a symposium in 1942 on "Overcoming the West" published in the journal *Bungaku-kai* (The Literary World) in the same year (Kawakami et al., 1979). It was argued in this symposium that democracy, capitalism, and liberalism, and the idea of progress had to be overcome by Japan, because they were all without exception of Western origin (Hiromatsu, 1980). Neither Sōseki himself, nor indeed Chaadayev, were at all partisan or dogmatic with regard to the two sides.

After Japan's defeat in 1945, the "Westerners" gained ascendancy over Japan's equivalent of the Slavophiles, and the latter were discredited. A fact symbolic for this situation was that French literature was very popular among Japanese intellectuals for a while just after 1945. Marxism was also influential. But the controversy over the "Westerners" and "Slavophiles" is still not totally closed, either in Japan or in Russia. So long as the search for cultural identity in the non-Western world continues, the issue of external and internal development, or, in other words, the issue of exogenous and endogenous development, which Sōseki, Chaadayev, and to some extent Fukuzawa also raised, remains unresolved for the non-Western world. Especially in Russia, where the socialist system based on Marxism has collapsed, the search for cultural identity will now assume importance. The issue "What is Russia?" will be a burning one, and Chaadayev will be once more vividly remembered.

As regards Japan, the issue "What is Japan?" does not seem to be a burning issue now. But, through rapid Westernization, Japanese civilization has lost much of its traditional character. Neither Confucianism, which had been an Orthodoxy during the Tokugawa period (1603–1868), nor things like the tea ceremony, which are often much admired by foreign visitors, are any longer integral aspects of Japanese civilization. The problem of clarifying the nature of Japanese cultural identity, which has long troubled Japanese intellectuals such as Sōseki, is obscured at present by Japan's enormous technological success and economic prosperity. It will inevitably take a long time for this question to be resolved. And its resolution is indeed necessary, if Japan is to find its role in the world. In this respect, Yamamoto's assumption that both Russian and Japanese civilizations are "peripheral" civilizations seems to me to be highly suggestive (Yamamoto, 1974; 1985, chap. 3; Miyake, 1992, p. 8).

## The Plurality of Life's Areas and Plurality of Time

Today, it is not possible for us to have the optimism that Fukuzawa retained all his life. Macey says that during the mid 1960s, "progress—and more specifically material progress through technology—came to be questioned on a scale that had not occurred for some three hundred years" (Macey, 1986, pp. 93, 101). The increasing force of the environmental movement is an example of this questioning. Of course, as Macey points out, there is a paradox "when people want more and more material goods and yet complain bitterly about the inevitable rationalization of their work and their lives, as well as the increasing contamination of the earth" (Macey, 1989, p. xiii).

As to the contamination of the earth, environmental pollution and degradation threaten highly industrialized and less industrialized Western and non-Western countries. It is a global issue. A legacy of the Marxist regime of the Soviet Union has been the danger of nuclear pollution on a vast scale. This is underscored by the meltdown of the nuclear reactor at Chernobyl, of course, and also by such problems as the Russian government's current dumping of nuclear waste into the Japan Sea. It can be useful to consider this coincidence of the global issue of environmental pollution and the demise of Marxism in the light of the idea of progress.

Marxism can be characterized as a mixture of the idea of progress, eschatology, and utopianism. The Marxist "utopia" in Russia, which attracted many intellectuals of the non-Western world, collapsed, and as a result, utopianism in general is now rather discredited. We must ask ourselves whether the idea of progress without a utopia can continue to exist at all. In any case, we are now far from the conviction that either human beings or human society can attain perfection. This conviction of perfectibility was shared by the defenders of the idea of progress, such as Condorcet, K'ang Yu-wei, and also Marx.

We should remember here the plurality of the areas of our life. Progress in some areas, such as technology, will still inevitably continue. In other areas, such as the environment, it will be difficult for us to make great progress so long as we pursue a better material life. Fukuzawa himself says that intellect makes progress, but morals do not.

Suggestive in this respect is the concept of the plurality of time of diverse events and areas at a given historical moment proposed by the American sociologist Siegfried Kracauer. He says that at a given historical moment, "we are confronted with numbers of events which, because of their location in different areas, are simultaneous only in a formal sense" (Kracauer, 1966, p. 68; 1969, chap. 6).

Viewed from this angle, we may say that time underlying some areas of life will pursue the course characterized above as type 5, namely progress, and time inherent in other areas will pursue the course characterized as type 6, namely retrogression. But the Russian examples of nuclear pollution warn us that technology contains in itself the danger of destruction on a large scale and therefore that of retrogression. We do not really know whether technology has the potential of progressing to the point where it can be relied on to control all dangers.

## References

Bedini, S. E. (1975), Oriental concepts of the measure of time. In: *The Study of Time II*, ed. J. T. Fraser & N. Lawrence. New York: Springer.

Berthold, W. (1991), Grundformen der Geschichtsauffassungen unter dem Aspekt der Zeit: Kreislauf, Rückschritt und Fortschritt. In: *Nachdenken über Geschichte: Beiträge aus der Oekumene der Historiker in memoriam Karl Dietrich Erdmann*, ed. H. Boockmann & K. Jürgensen. Münster: Karl Wachholtz.

Buckle, H. T. (1857–1861), *History of Civilization in England*, 2 vols. London: J. W. Parker.

Bultmann, H. (1957), *History and Eschatology*. Edinburgh: Edinburgh University Press.

Bunyan, J. (1678), *Pilgrim's Progress from This World to That Which Is to Come*. New York: Oxford University Press, 1966.

Bury, J. B. (1955), *The Idea of Progress. An Inquiry into Its Growth and Origin*. New York: Dover Publications.

Chaadayev, P. Y. (1836), *Philosophical Letters and Apology of a Madman*, tr. with introduction M. B. Zeldin. Knoxville: University of Tennessee Press, 1969.

Chamberlain, B. H. (1904), *Japanese Things. Being Notes on Various Subjects Connected with Japan* (Originally entitled *Things Japanese* and published in 1904). Rutland, VT, & Tokyo: Charles E. Tuttle, 1971.

Fontenelle, B., de (1688), Digression sur les Anciens et les Modernes. In: *Œuvres Complètes*. Texte revu par A. Niderst, Tome 2. Paris: Fayard, 1991.

Fraser, J. T. (1990), *Of Time, Passion, and Knowledge*, rev. ed. Princeton, NJ: Princeton University Press.

Fukuzawa, Y. (1875), *Bunmei ron no gairyaku* (The Outline of the Theory of Civilization). Tokyo: published by the author. Partly translated and summarized by M. Kōsaka, ed., *Japanese Thought in the Meiji Era* (see below: Kōsaka). New revised edition, Tokyo: Iwanami-shoten, 1962. The quotation is from this edition.

——— (1940), *The Autobiography of Fukuzawa Yukichi*, 3rd rev. ed., tr. E. Kiyooka. Tokyo: Hokuseido Press.

——— (1969), *Fukuzawa Yukichi's An Encouragement of Learning* (Gakumon no susume, originally published by the author in 1872–1876), tr. with an introduction by D. A. Dilworth & U. Hirano. Tokyo: Sophia University (Monumenta Nipponica).

Guizot, F. (1870), *General History of Civilization in Europe*, tr. with occasional notes by C. S. Genry. New York: D. Appeleton.

Hesiod, *Works and Days*, tr. R. Lattimore. Ann Arbor: University of Michigan Press, 1959.

Hiromatsu, W. (1980), *"Kindai no chōkoku" ron. Shōwa shi e no ichi dansō* (Discussions on the Theory of "Overcoming the Modern Age." Fragmentary Thoughts on the Intellectual History of the Shōwa Era). Tokyo: Asahi-shuppan-sha.

Huxley, T. H. (1893), *Evolution and Ethics (The Romanes Lecture, 18 May 1893)*. London & New York: Macmillan.

———— Huxley, J. (1947), *Evolution and Ethics 1893–1943*. London: Pilot Press.

K'ang Yu-wei (1958), *Ta T'ung Shu. The One-World Philosophy of K'ang Yu-wei*, tr. with introduction & notes L. G. Thompson. London: Allen & Unwin.

Katsuda, K. (1961), *Kindai Roshia seiji shisō shi* (History of Political Thought in Modern Russia). Tokyo: Sōbun-sha.

Kawakami et al. (1979), *Kindai no chōkoku* (Overcoming the Modern Age). Tokyo: Huzanbō, 1979. The reprint of the articles and the symposium published in the journal *Bungakukai* (The Literary World) September and October 1942, with a foreword by Matsumoto Ken-ichi and detailed comments by Takeuchi Yoshimi. The symposium was presided by Kawakami Tetsutarō, a literary critic, and published in *Bungakukai* (Tokyo: Bungei-shunjū-sha) in October 1942.

Kōsaka, M., ed. (1958), *Japanese Thought in the Meiji Era, Centenary Culture Council Series, Japanese Culture in the Meiji Era, Vol. IX, Thought*, tr. & adapted by D. Abosch. Tokyo: Pan Pacific Press.

Kracauer, S. (1966), Time and history. In: *History and the Concept of Time: History and Theory, Studies in the Philosophy of History*, Beiheft 6. Middletown, CT.: Wesleyan University Press, 1966.

———— (1969), *History. The Last Things Before the Last*. New York: New York University Press.

Löwith, K. (1949), *Meaning in History. The Theological Implications of the Philosophy of History*. Chicago: University of Chicago Press.

Lyell, C. (1830–1838), *Principles of Geology*, 3 vols. London: John Murray; New York: Johnson Reprint, 1969.

Macey, S. L. (1986), Literary images of progress: The fate of an idea. In: *Time, Science, and Society in China and the West: The Study of Time V*, ed. J. T. Fraser, N. Lawrence, & F. C. Haber. Amherst: University of Massachusetts Press, 1986.

———— (1989), *The Dynamics of Progress. Time, Method, and Measure*. Athens: University of Georgia Press.

Manuel, F. E., & Manuel, F. P. (1979), *Utopian Thought in the Western World*. Oxford: Basil Blackwell.

Maruyama, M. (1963), *Thought and Behaviour in Modern Japanese Politics*, ed. I. Morris. London: Oxford University Press, 1963, Glossary and Biographies I. Morris.

———— (1986a), *"Bunmei ron no gairyaku" o yomu* (Reading *"The Outline of the Theory of Civilization"*), Vol. 1. Tokyo: Iwanami-shoten, 1986.

———— (1986b), *"Bunmei ron no gairyaku" o yomu* (Reading *"The Outline of the Theory of Civilization"*), vol. 2. Tokyo: Iwanami-shoten, 1986.

Michon, J. A. (1986), J. T. Fraser's "Level of Temporality" as cognitive representations. In: *Time, Science, and Society in China and the West: The Study of Time V*, ed. J. T. Fraser, N. Lawrence, & F. C. Faber. Amherst: University of Massachusetts Press, 1986.

Minamoto, R. (1993), Confucian thinkers on the eve of the Meiji Restoration: Sakuma Shōzan and Yokoi Shōnan. In: International Christian University (Mitaka, Tokyo), Publications III-A, *Asian Cultural Studies*, 19.

Miyake, M. (1990), General comments on the concepts of historical time. In: *17th International Congress of Historical Sciences, Grands Thèmes, Méthodologie, Section Chronologique 1, Rapports et abrégés*. Methodology 1, Concepts of Time in Historical Writings in Europe and Asia. Ed. Comité International des Sciences Historiques. Madrid: Comité Español de Ciencias Históricas, 1990.

———— (1991), The concept of time as a problem of the theory of historical knowledge. In: *Nachdenken über Geschichte: Beiträge aus der Oekumene der Historiker in memoriam Karl Dietrich Erdmann*, ed. H. Boockmann & K. Jürgensen. Münster: Karl Wachholtz.

———— (1992), The concept of time as a problem of the theory of civilizations and history. *Bull. Inst. Soc. Sci.* (Meiji University), 15/3.

Momigliano, A. (1966), Time in ancient historiography. In: *History and the Concept of Time: History and Theory, Studies in the Philosophy of History*, Beiheft 6. Middletown, CT.: Wesleyan University Press.

Nisbet, R. (1980), *History of the Idea of Progress*. New York: Basic Books.

Nolte, E. (1991), *Geschichtsdenken im 20. Jahrhundert. Von Max Weber bis Hans Jonas*. Berlin & Frankfurt am Main: Propyläen.

Pollard, S. (1971), *The Idea of Progress. History and Society*. Baltimore: Penguin Books.

Sakade, Y. (1976), *Daidō-sho* (Ta T'ung Shu). Tokyo: Meitoku-shuppan-sha.

Schwartz, B. I. (1964), *In Search of Wealth and Power. Yen Fu and the West*. Cambridge, MA: Harvard University Press.

Sōseki (penname for Natsume, K.) (1911), Gendai Nihon no kaika. In: *Watakushi no kojin shugi* (My Individualism). Tokyo: Kōdan-sha, 1978.

―――― (1985), *Light and Darkness* (Meian). *An Unfinished Novel*, tr. V. H. Viglielmo. London: Pan Books.

Swift, J. (1704), *The Battle of Books*. New York: Oxford University Press, 1958.

Togawa, T. (1962), P. Ia Chaadaev, Philosophical Letters (Translation with Comments) 1. In: *Slavic Stud. J. Slavic Inst. Hokkaido University*, 6.

―――― (1979), Pierre Tsaadaev: Fragments et Pensées Diverses (Inédits), présentés par Tsuguo Togawa. In: *Slavic Stud. J. Slavic Inst. Hokkaido University*, 23.

Tsuchida, K. (1927), *Contemporary Thought of Japan and China*. London: Williams & Norgate.

Yamamoto, S. (1974), 'Ohka' no sotai-teki hansei: Chaadayev to Sōseki (Total Reflections on 'Westernization': Chaadayev and Sōseki). *Bull. Inst. Humanities*, 8, Kanagawa University (Yokohama), in Japanese with English summary. This article is now included in his posthumous work indicated below as chapter 3 "Ohka no hansei" (Reflections on Westernization).

―――― (1985), *Shuhen bunmei ron. Ohka to kokusui* (Theory of Peripheral Civilizations. Westernization and the Indigenous). Tokyo: Tōsui-shobō, 1985.

Zenkovsky, V. V. (1953), *A History of Russian Philosophy*, Vol. 1, tr. G. L. Kline. London: Routledge & Kegan Paul.

# 18

# Time and the Individual in the Contemporary World: The Meaning of Death

*M. H. Oliva-Augusto*

*Abstract*   Personal relations bind members of a society together and allow them to formulate shared ideas about the world, for example as the ways in which the nature of time is perceived. They also permit them to attribute specific meanings to the various dimensions of their existence such as to death. This paper deals with the meaning of death for members of contemporary society and with the related issue of the meaning of life. I will discuss certain aspects of modern social life, as conceived of in the late 18th and early 19th centuries. Principally, I will deal with the emergence of the individual, the idea of freedom, and changes in the notion of time—features which will then be identified in the contemporary world. Changes in the meaning of life and death as perceived in contemporary experience will become clear by comparing these two moments.

The representations which are shared by members of a society are fundamental to the character of that society, accounting for its true nature, and simultaneously allowing its members to recognize themselves as participants in it. This sociological principle means that the entire complex of shared meanings, of representations about ''their'' society by the individuals who together constitute it, the ways in which they think about it and understand it, is what allows that society to exist with a certain identifiable and recognizable profile. The nature of a given society's participants, however, is derived from those very representations through which society creates people suitable for its needs; in them the process of introducing those born in the society takes root.

The views men and women have of their society are maintained by each individual as well as by the whole. The issue here is how society represents itself

---

English translation by William J. Shelton.

and creates meanings that are peculiar to it. Members of a given society become social beings by incorporating those representations and meanings. The process of socialization, by which that society's members internalize those representations and meanings, allows its members "to become human" in a collectively acceptable and specific manner. At the same time, all of the institutions that are particular to that society also grant concrete expressions to these meanings.

Institutions exercise a triple function, therefore: (1) they structure representations of the world in general, without which human beings cannot exist; (2) they assign goals to the actions which members of a given society will develop, indicating what should or should not be done; and (3) they establish the types of affective relations and inclinations which are characteristic of that society (Castoriadis, 1990, p. 125).

The most important of all meanings produced in this manner is that which refers to society itself, its representation of itself as *an entity:* this representation is inextricably linked to *a particular way of wishing itself* to be *this* particular society, and *loving itself* as *this* particular society, different from all others. This is what allows each individual to identify him- or herself with a "we," with a collectivity that, in principle, is indestructible. "A sense that concerns society's self-representation, a sense that can be shared by individuals, a sense that allows them to create a meaning of the world for their personal benefit, a sense of life, and, finally, a sense of their death" (Castoriadis, 1990, pp. 126–127).[1]

During the late 18th and early 19th centuries, two representations were at the basis of how men and women saw society and how society represented itself. The first referred to the belief in the possibility of unlimited progress, guided by human reason. This progress, spurred on by scientific and technological developments, was accompanied by a belief in the possibility of continuous growth in industrial production and the accumulation of wealth. That vision suggested that humanity could emerge from its condition as victim of unknown forces and instead could dominate them. It presupposed a progressive mastery over nature by humans, as well as the abandonment of superstitions which, as in the case of religious beliefs, placed their lives beyond human control. It was taken for granted that this development would allow them to totally dominate nature, making possible, in turn, the gratification of fundamental human needs.

The second representation drew on the belief in human creative capacity, in the possibility that people would grow in freedom, achieving the common good through free participation in business, public affairs, and social processes. This representation generated a particular meaning that referred to individual and social autonomy, to freedom, to the possibility of creating forms of collective freedom, corresponding to a democratic, emancipatory, revolutionary project (Castoriadis, 1990, p. 127). Therefore, on the one hand, stood the belief in progress; while on

---

[1]"Sens qui concerne l'autoreprésentation de la société; sens participable par les individus; sens leur permettant de monnayer pour leur compte personnel un sens du monde, un sens de la vie et, finalement, un sens de leur mort" (Castoriadis, 1990, p. 127).

the other was the belief in humanity and its freedom. We may call these two representations the capitalist meaning and the meaning of individual autonomy.

These representations are mutually antagonistic, leading us in opposite directions. Indeed, the capitalist meaning points toward centralization and disciplining; the meaning of individual autonomy, by contrast, leads to the idea of participatory democracy. Being contemporaneous, however, and coming to be concurrently effective, they reciprocally contaminate each other in the end (Castoriadis, 1990, p. 127).

Thus, the representation which modern society has of itself is derived from these two interrelated meanings. Modern society views itself as being in the time and place of progress and uninterrupted rationalization leading to an enlarged process of production and accumulation. Simultaneously, it presents itself as allowing, more than in previous forms of social interaction, the successful realization of the human being. The sense which results from these meanings is that the convergence of progress, reason, production, and accumulation implicitly makes possible the existence of freer, happier, and more fully realized people.

That representation and the sense it conveyed have, today, suffered setbacks. We must understand how this double and contradictory meaning that emerged with modernity is itself actualized in the contemporary world. Similarly, we must evaluate the extent to which, in today's society, the implementation of the notion of time linked to that representation interferes with the possibility of the individual fully realizing his or her potential as a human being. If, between the moment in which modern society emerged and the present, the predominant representations and notion of time have been altered, the sense of life and perception of death which prevailed in the late 18th century up to the late 20th century, have certainly also undergone significant changes.

## Time, the Individual, and Modernity

We must not forget that human beings are formed by the society into which they are born. The importance of that link is highlighted in social theory which holds that society "forges" its members according to the meanings by which it is characterized, providing itself, and them, with an identity. Only when the notions of progress, reason, production, and accumulation of wealth acquire emphasis, is it possible to conceive of the idea that isolated individuals, independent of their local or family groups, are the ones who construct the world.

The concept of the *individual* develops along with the very process by which the two contradictory meanings emerge in modern society. While this concept presupposes human competence for designing life projects, it also suggests the capability of self-control and self-regulation. It assumes that individual potentials are not hindered by any ties to the past, and that one is capable of creating a personal history, independent of the group to which one belongs. Simultaneously, it indicates the possibilities of "self-made" persons who can project a future. This requires the belief that human fate is not predetermined. Implicit in this idea, are the notions that each person's life is his or her own possession and that human beings will become whatever they make of themselves.

When one speaks of individuality, the possibility of self-reflection, of criticism, of freedom, is also implied. In this sense, the course taken by the individual's life is, partly at least, the result of *choice*. The individual's *destiny is not beyond* him or her; it is not determined externally: it is the *individual's own destiny*, in the full meaning of the term. Consequently, individual realization demands that people leave signs of their passage, signs characterizing the plenitude or the emptiness of their own individual existence.

The historical form of sociability that emerged in the modern world and allowed the concept of the free individual, as well as his or her empirical existence, to come into being, also produced the experience of a new notion of time. At this point, we are dealing with linear time, perceived as a measurable, divisible, homogeneous, uniform, arithmetized flow. This is also progressive time, accumulatory, rationalizing, time conquering nature, experienced in terms of unlimited growth, of ever greater approximation to exact total knowledge (Castoriadis, 1982, p. 244).

This new time makes possible a clear distinction between "before," "now," and "after." This temporality supposes, for the individual as well as for society as a whole, the existence of a past, present, and future. The present appears simultaneously as a moment of passage between past and future and as the point of departure for new experiences. Life surfaces as building space—of people themselves, of society, of the future, of a project—made possible by past experience (Heller, 1982, pp. 141–162).

Future orientation, which tends to prevail, and the absence of bonds with the past, are linked to the manner in which humanity came to face its destiny. The latter is not something derived from the will of the gods; nor is it imposed externally, but, rather, arises from human action itself (Heller, 1982, pp. 141–162). Nevertheless, a personal history can only be constructed within a definite period of time: the life span of each person. This presents boundaries beyond which one cannot venture and within which each will be summoned to trace one's own path and leave one's marks in passing, as indications that one's life was successful.

Changes in the perception of time and consciousness of the end of earthly life is a definitive limit. The notion of the individual emerges along with changes in the notion of time and its experience, and a recognition of the finiteness of life. This convergence consequently involves a profound alteration in the meaning of death. Death ceases to be the moment of passing on to another existence, where compensation—positive or negative—for the life lived will be given. It acquires the sense of an inexorable end. Recognition of this limit sets up an opposition between the idea of eternity, which oriented earlier ages, and the acknowledgment of human mortality. It likewise highlights the notion of time as an irreversible dimension of human existence, as opposed to previous perceptions of time as a cyclical repetition of situations. The prospect of that limit, which points toward the need to live the present moment to its fullest, since it cannot be repeated, also impels people to take maximum advantage of the time available so as to fill it with events and deeds. The time available is used in such a way as to extract as much as possible from it, through one's accomplishments.

Recognizing this process, Max Weber stated that, in the contemporary world, human beings may feel disgusted, worn out, or weary of life, never fulfilled by it (Weber, 1958, p. 140). It has also been said that if death did not exist, most people would be honest, for dishonesty frequently results from lack of time: the fear of losing forever what was not obtained today (Heller, 1987, p. 387). In a certain way, consciousness of the end is what feeds the present. In this sense, one's relationship with death expresses the way in which one's relationship with life is assumed, as well as its meaning.

It is at this point that the representations which modern society constructs with regard to itself—its desire to be, its idea of the human being, and its notion of time—intersect.

## Life and Death Today

After almost two centuries and two world wars, the persistence of misery and hunger, together with the perception that inequality among people continues, made these representations undergo certain transformations. Furthermore, we now perceive that the ways in which people establish their relationships and exploit nature are not unrelated. We also perceive that unlimited domination of nature is impossible since nature is not inexhaustible. There is a limit to its exploitation, beyond which nature begins to revolt: the hole in the ozone layer, the depletion of natural sources of energy, the consequences of indiscriminate destruction of forests, the rise of the earth's temperature, the climatic inversions which we have witnessed, all demonstrate the need to change the ways in which mankind exploits nature.

The still existing perception of the possibility of uninterrupted progress is now accompanied by the recognition that this immense and irrefutable development does not always better people's lives. We can observe a breathtaking scientific and technological development that daily achieves wonders which, only a short time ago, were thought to be unobtainable, and which, in turn, are constantly superseded by new conquests. Nevertheless, it remains clear that, while we can develop the most advanced scientific and technical understandings, the economic and social distances separating different social strata are progressively increasing.

Thus, of the two opposite views, the meaning of individual autonomy and the capitalist meaning, that the representation of modern society sought to reconcile, only the latter is truly present and dominant in contemporary society. Nevertheless, what today's capitalism seems to propose is the indefinite expansion of the rational matrix, emptied of whatever humanistic content that gave it vitality in the past. The exercise of reason is not carried out for the sake of a better human life, but rather for the sake of greater wealth, or progress for the sake of progress. It is appropriate for us to ask whether, in many cases, we are really speaking of an exercise of reason, or of its negation. The very ideology of uninterrupted progress, which guided history and projects into the future, is now being held up to question or, for many, has lost its meaning.

The other world view, which presaged the possibility of an emerging free humanity, capable of autonomously constituting a history that would simultaneously provide for individual happiness and the common good, has been visibly weakened.

Consequently, for many contemporary men and women, the experience of the present moment does not allow them to see themselves as whole beings, or as individuals in the full meaning of the term, but rather causes them to feel like disconnected beings with neither roots nor prospects.

As a result, most people today have lost the feeling of belonging, of participating in a "we." The subjective translation of the meaning of individual autonomy and of the reality that sustains it therefore results in a profound individualism in which each person turns selfishly to his or her desires and expectations, not recognizing a fellow being in the other. The result of this process is none other than the continuous growth of consumption and leisure, which have become ends in themselves, the fragmentation of life into an array of meaningless acts, and the extreme solitude that haunts people, even though they live in society.

This state derives partially from the contemporary experience and meaning of temporality. The processes at work during the emergence of modern society have been carried out to their ultimate consequences, but are now deprived of their transforming potentials. The demands and dominant logic of the social order require a linear perception of time, with emphasis on the quantitative and utilitarian to the detriment of the qualitative. This is fundamentally a progressive time, centered on efficiency, on the need to exhaustively drain the present's potential, but which somehow no longer carries the prospect of global domination of nature, the possibility of total knowledge, the idea of mankind constructing its own destiny.

Human beings, having been atomized, become dominated by an external rhythm, and, instead of regulating their own time, are made into its victims. They no longer see themselves as building their life and their world. Rather, they feel susceptible to threats whose origins they can neither detect, nor can they control their development. With this, they tend to discipline themselves in a complete and uniform manner, in almost all aspects and on almost all occasions. Discipline presents itself as a characteristic of contemporary society's model of self-control. Its model of civilization is represented by the regulation of time typical of it: it is no longer punctual and specific, but penetrates all of human life, without allowing for oscillations. This feature is uniform and inevitable (Elias, 1989a, p. 162).

In addition, another feature characterizes the contemporary world. In the most developed societies, people think of themselves as individual and independent beings, separated from one another by a sort of invisible wall. Since an individual life is seen as isolated from the lives of others, and hermetically separated from the world, it should have meaning in and of itself. But when people are unable to find this type of meaning, human existence seems absurd, and they feel disillusioned. Nevertheless, according to Elias, it is important to remember that the "category of meaning cannot be understood when referring to the human individual or to a universal derived from this notion. The existence of a plurality of beings, who are interdependent in some way and communicating with one another, constitutes what we call meaning'' (Elias, 1989b, p. 68). In other words, ''meaning'' is a social category and the subject corresponding to it is a plurality of human beings. To the extent that people tend to see themselves as individual and independent beings, dissociated from and indifferent to those with whom they live, their life (and their death) is lived as if devoid of any meaning.

At this point, we need to consider how these elements relate to one another. We must emphasize the nexuses that articulate social meanings at work in the contemporary world: possible individuality, the experienced notion of temporality, and the perception of death.

As stated earlier, each historical moment and each society creates its specific type of human being. Considering all the changes that have taken place in the representations that contemporary society and humankind make of themselves, the typical character of our epoch has been presented by various authors as the artificial and passing union of a disperse set of traits that do not quite constitute a clear human profile.

Some refer to the individuality, which is possible in the contemporary world, as a heterogeneous patchwork or as a series of collages (Castoriadis, 1990); others compare it to a video-clip identity (Lipovetsky, 1986). Still others assert something that is almost paradoxical: despite the fact that individualism is progressively being established, the perception that humans have of themselves is filtered through the way they believe others perceive them. It is as if people were using radar in an attempt to grasp the perception that others have of them, molding themselves according to these external expectations (Riesman, 1964). In other words, their yardstick is outside of themselves.

What thus arises is a manifestation of generalized conformism, rather than autonomy. But, the possibility of controlling their own lives, or providing for their own future and that of their children, of leaving enduring marks of their passage through the world, remains ever more distant. Insecurity, the inability to predict tomorrow, prevails in their lives (Horkheimer, 1976, pp. 168–169).

As Foucault has described, in this type of sociability where producing is considered to be so important, an ever more fragmented division of time tends to permit its more effective utilization. Similarly, guaranteeing the quality of the time used has become increasingly significant. The goal of a totally useful time, necessitates efficiency and speed and presents the possibility of a theoretically ever-increasing utilization. What occurs, consequently, is an ever more intense acceleration of time's rhythm (Foucault, 1977, pp. 136–141).

People today perceive that ''time flies.'' Time's velocity has made the endeavor of planning the future obsolete, if not almost impossible. By the same token, the *now*, as well as the need to consume it exhaustively, have come to reign absolutely. ''Making'' time and not ''wasting it'' have become an obsession. People are crushed by the rhythms and programs imposed upon them by all social webs, at the workplace and elsewhere. The need to adequately administer time is internalized, as are all of the most important social rules. It is converted into an imperative. Individuals must adjust their own behavior to the ''time'' established by the group to which they belong (Elias, 1989a, p. 135). Personal temporality, whose rhythm does not accompany the swift pulse of external time, is overpowered and converted into its "colony." Men and women thus become their own internal clocks and the instruments of their own temporal servitude. The pressure to rigidly program time penetrates daily life, both socially and individually (Chesnaux, 1983, p. 40).

On the other hand, the "empire of the ephemeral," the emphasis on the instantaneous which has become dominant, the importance of a "now" devoid of meaning, end up removing the significance of the past while emptying the possibility of a future. The notion of both individual and social history that marked the emergence of these forms of sociability, temporality, and individuality, as well as the very possibility of establishing an identity, are devastated along with the loss of sense that social life presents, with the ever greater fragmentation of the time and with the significance which instantaneousness acquires.

To the best of its ability, each historical epoch elaborates its own mechanisms for facing the problem of death. As has already been said, consciousness of their very finiteness and of the need to "eternalize" themselves through deeds realized during their lifetimes provided modern men and women with their way of confronting death.

In contemporary society, since life has lost its meaning—to the very extent that the sense of one's own history or the very sense of history have disappeared—death is also meaningless. There are various mechanisms that attempt to repel it, as if denying it would somehow keep it away. We are dealing here with the same mechanisms involved in making life "go by": taking a refuge in the immediate, the generation gap, loss of the sense of continuity. In today's world, the individual lives a frenzied race in order to forget that he or she is going to die and that, strictly speaking, all that he or she does has no meaning. Thus, people succumb as individuals, since their sense of belonging is obscured and the experience of their singularity is annulled.

At the same time, and as a consequence, there is ever greater insensitivity concerning the way life is lived and death presented. This is the dominant mode of existence, even though, in isolated points, rituals and behaviors recalling old patterns of sociability remain. In distant regions of Brazil, among mestizos and Indians, or scattered here and there in the cities, in urban shantytowns and suburbs, funeral rites and conceptions of death rather distinct from those now prevailing still exist (Martins, 1983, p. 9). Nevertheless, the latter are insidiously gaining ground. This is reflected in a number of attitudes regarding current social issues: popular efforts to implement the death penalty, the indifference of young murderers to the taking of human life; the way in which children, the "promise of the future," are treated, child abandonment and even murder; lack of respect for the elderly, which ranges from disregarding their experience to denying them a dignified end of life, expressed by how difficult it is for them to receive a reasonable pension after many years of work.

On the other hand, we can perceive in the contemporary world a parallel process. Currently, because of the extent to which society's lack of security has increased, making it ever more difficult for individuals to foresee—and to exert a certain control over—their own long-term future, as was considered possible when modern society emerged, the need for supernatural protection is emerging (Elias, 1989b, p. 15). It is as if a "reenchantment" of the world were taking place as can be seen by the great vitality with which new forms of religion surface (or resurface) and mystic experiences of all sorts proliferate. For Lipovetsky, the resurfacing of spiritualities and esotericisms of all kinds does not contradict the principal logic of our time. Rather, it is a way of enforcing it, "allowing for an individualistic cocktail of realization" (Lipovetsky, 1986, p. 119).

## Conclusion

The argument presented in this paper proposes that human beings have reached a critical point where they have lost a sense of life, a sense of death. Social life has lost its meaning, individuality has become impossible. Is there some way to remake meanings, to project senses again, to reconstruct the promise of free individuals?

Some authors point to redimensioning the present as a possible way out of the morass. That redimensioning demands the rediscovery of the future, a new relationship with tradition,[2] and also with death, as well as a different way in which the individual can confront time.

There are also those who remind us of the need to take the "struggle for time" into the field of politics. That reaction should be present in the workplace—as a struggle for internal organization and control over the length of time worked. It should also be present in private life, through an administration of personal time that makes room for the unexpected, prevents the imprisonment caused by commitment to a schedule, and that also rejects mechanisms of consuming the time, in wasting it (Chesnaux, 1983, pp. 52–53).

Society can allegedly make other meanings emerge if it is capable of helping us recognize our finiteness. Here another way of seeing the world and human mortality is supposed, and the obligation that men and women today have in relation to those who came before them and those who will follow is recognized. Our debts to future generations are like our debts to those of the past, for contemporary men and women would not be what they are were it not for the hundreds of thousands of years of work and effort of their predecessors (Castoriadis, 1990, p. 134).

On the other hand, a new historical vision capable of effectively and lucidly opposing itself to this shapeless and kaleidoscopic world, this bazaar in which we live is inconceivable unless a new and fertile relationship with tradition is established. This does not mean restoring traditional values as such or restoring them because they are traditional, but, rather, recovering a critical attitude capable of

---

[2]This feature must be highlighted in that it implies a reorientation in the way we consider the past. The Enlightenment considered that breaking all links with the past was a sign of progress, which was harshly criticized by conservative thought. The latter view the past as a source of life and knowledge. The demand to use the past as a reference for new experiences, expressed by some authors who certainly cannot be identified as conservatives, deserves more careful analysis.

recognizing values that have been lost (Castoriadis, 1990, p. 135). According to another approach, the past is the only concrete reference available for us to consider the possibility of other forms of social organization, which means that we can look to the past in search of references for another future. In other words, the past can help us confront the present (Chesnaux, 1983, pp. 53–54).

Both of these approaches suppose a linkage between past and future, by way of the present, and both recover the observation which Tocqueville had already made in the 19th century: "Since the past ceased to cast light on the future, the human mind has wandered in darkness."

## References

Castoriadis, C. (1982), *A instituição imaginária da sociedade*. Rio de Janeiro: Paz e Terra. (Available in English as *The Imaginary Institution of Society*. Boston: MIT Press, 1987.)

—— (1990), La crise du processus identificatoire. Toulouse: Ed. Erès, *Connexions 55*, 1:123–135. *Malaise dans l'identification*. Toulouse: Ed. Erès.

Chesnaux, J. (1983), *De la modernité*. Paris: La Découverte/Maspero.

Elias, N. (1989a), *Sobre el tiempo*. Mexico City: Fondo de Cultura Económica. (Available in English as *Time: An Essay*. Cambridge: Blackwell Scientific Publications, 1992.)

—— (1989b), *La soledad de los moribundos*. Mexico City: Fondo de Cultura Económica. (Available in English as *The Loneliness of the Dying*. Cambridge: Blackwell Scientific Publications, 1986.)

Foucault, M. (1977), *Vigiar e Punir: Nascimento da prisão*. Petrópolis: Vozes. (Available in English as *Discipline and Punishment: The Birth of the Prison*. New York: Random House, 1979.)

Heller, A. (1982), *O homem do Renascimento*. Lisbon: Editorial Presenča. (Available in English as *Renaissance Man*. New York: Schocken Books, 1981.)

—— (1987), *Sociología de la vida cotidiana*. Barcelona: Península. (Available in English as *Everyday Life*. New York: Routledge, Chapman & Hall, 1984.)

Horkheimer, M. (1976), *Eclipse da razão*. Rio de Janeiro: Editorial Labor do Brasil. (Available in English as *Eclipse of Reason*. New York: Continuum, 1973.)

Lipovetsky, G. (1986), *La era del vacío*. Barcelona: Anagrama.

Martins, J. S. (1983), *A morte e os mortos na sociedade brasileira*. São Paulo: Hucitec.

Riesman, D. (1964), *La muchedumbre solitaria*. Buenos Aires: Paidos. (Available in English as *The Lonely Crowd*. New Haven: Yale University Press, 1950.)

Weber, M. (1958), "Science as a vocation." In: *From Max Weber: Essays in Sociology*, ed. H. H. Gerth & C. Wright. New York: Oxford University Press.

19

# Anthropomorphic Operators of Time: Chronology, Activity, Language, and Space

*Jens Brockmeier*

*Abstract*  Human temporality has been understood as a kind of objective entity which is (in a Kantian sense) preconditionally "given" to every individual. The question, not only for philosophy but also for psychology and anthropology, is *how* time is "given" and individually recreated. How does the metaphor of time transform the multitude of very different diachronic experiences into a gestalt of human proportions? It is the function of cultural-specific "anthropomorphic operators" (such as chronology and language) to construct these psychosocial time syntheses and to give to the meanings of our world and ourselves the structure of temporal order and coherence.

## Culture and the "Givenness" of Time

Most of us would probably have no difficulty with the idea that time is something which has to be understood as a kind of objective entity: a structure which is "given" to every individual. How often do we complain about our constricting and restraining schedules, forcing us to obey temporal orders which appear to be external, overwhelming, threatening? We are also familiar with the extreme ends of this spectrum of experiences. On the one hand we have known time as the judge, as the patriarch, as the god: personifications of time in the Western world for 4000 years which, as Sam Macey (1987, p. 178) has emphasized, are not easily swept aside. On the other hand we know the idea of the absence of time, the dream, as Rainer Maria Rilke wrote in *The Duino Elegies*, of hours of childhood, "hours when behind the figures there was more/than the mere past, and when what lay before us/was not the future!" (1963, p. 45). But, first of all, we are convinced that the orders of time are in no way individual when we experience time in a moment of Joycean epiphany. The epiphany was for Joyce the sudden revelation of the "whatness" of a phenomenon; as, for example, the moment in which time seems to manifest itself as a universal fabric.

Time as the judge, as the absent present, as epiphanic experience: whenever we start to think about our world, about ourselves and about the time in which we live, we find ourselves already caught by this multifaceted but nonetheless invisible web of the Augustinian time experience. If there is a place to use the term *ubiquitousness*, then it is here. Since all our modalities and modes of knowledge are immersed in the dimension of diachronicity, we never reach any stable epistemic distance regarding either diachronicity or our experience of it. This is the point of that inexpressible experience of something to which we feel intimately close. We are like any social anthropologist, a living part of the field we want to explore, a diachronic moment of the stream of "nows." Wherever we try to grasp it, the "now" is only to be found in the intersubjective and societal coordination of our actions as well as in the astrophysical and philosophical extrapolations of our minds, in our dreams and autobiographical constructions, as well as in love and death.

A very successful intellectual strategy, which we often employ when we are in trouble, is the use of metaphors. But what makes it so difficult in this case is that our countless experiences of the diachronic dimension of our being are veiled by the fact that there is just *one* metaphor through which we usually picture the whole diversity of these experiences. This is the metaphor of time, of its flux and passage. Maybe one could understand the metaphor of time as a kind of rhetorical umbrella which casts its shadow over a wide range of diverse phenomena emerging in a fundamentally heterochronous world. If we start to spell out the nature of these multiple manifestations of diachronicity gathered under this umbrella, we will soon find that they are only connected through a kind of context which Wittgenstein might have called "family resemblance" (Hernadi, 1992, p. 147).

Thus, in coming to terms with the metaphor of time in this way, we must recognize that time is always already presupposed in the human condition and is of course no answer to any question. Rather this *is* the question. For what is the meaning of this being "given" and of our "living in it"? What is beyond the metaphorical picture by which these words try often enough to comfort us? There is a long philosophical tradition which has focused on this question; when I used the Kantian term *given*, I do not claim by any means that this is the last word in the discussion. Rather, I would like to argue that time is an objectively given precondition for our activities as well as for our thoughts. Time implies nothing but a certain point of departure for further inquiries and an invitation to further interpretation. For example, the protagonist in Cees Nooteboom's philosophical time novel *Die folgende Geschichte* gives his personal account of the meaning of clocks and temporality. By the end of the day whenever he watches a clock, he is reminded that this is just another failed attempt to represent the inconceivable. It illustrates the riddle of time, "an unbridled, measureless phenomenon which escapes our understanding and to which we have given—having no better options—the appearance of an order" (Nooteboom, 1991, pp. 46–47).

There are many ways of attributing a preconditional status to time, in order to understand its order or, at least, the appearance of it. This includes, first, ontological options, which picture time as its own special kind of being. There are also epistemological views in which time appears constituted as a notional presupposition, biological–nativistic conceptions, cognitive science models, cultural psychology approaches, theories of sociohistory and construction of discourse. Therefore Kant's own epistemic view that time is a transcendental condition a priori of any possible experience, is only one possible perspective. But it is a view which could open up a much wider perspective if one does not feel bound to adopt it in the narrow philosophical framework of subjective idealism (and to see it as an intelligible quality of the "transcendental subject"). There is also a broader reading of Kant's a priori argument according to which the concept "transcendental" refers—on an elementary epistemological and anthropological level—to the necessary conditions of possible human experience, while dispensing with the notion of a transcendental subject.

Once freed from this notion, these conditions rather reveal themselves as included in the very process of activity in which we make our experience. Then the concrete connection between the nature of our experiences and the corresponding processes of activity becomes the focus of our attention. And this brings in the practical, social, and intellectual actions in which the cultural process of negotiating and constructing *meanings* is embedded.

If we are interested in "human temporalities"—to use J. T. Fraser's (1992; 1987, chap. 3) distinction between human and nonhuman (biological and physical) temporalities—and want to know above all how human beings experience diachronicity and how they make sense of it, then we should start by examining which of our experiences is connected with which of our time metaphors. For this investigation, the approach which I have just outlined can be a good starting point, which allows for a reading of the a priori argument in a quite different light. In this light the preconditions of our individual constructions of time become merged into the *contexts of meaning* by which our culture is constituted and by which we participate in the process of this constitution.

The concept of *culture*, to which I am referring here, offers two useful insights into the quest for the meaning of time. It presupposes, first, that the structure of social life is an irreducible complex of symbolic systems. That is to say, in this perspective of culture the significants are not only symptoms or syndromes of something else but they are themselves symbolic actions or bundles of symbolic activities (Geertz, 1973). In this way one does not view "culture" as just a social or anthropological concept but also emphasizes the character of human societies as essentially mediated through symbols and signs. It is, in short, the symbolic and semiotic synthesis of a civilization on which it is focused. Second, this concept of culture includes the idea that every individual, in order to become an active participant of the social and societal context, enters into it as a *novice* who has to undergo an ontogenetic development, which is specific for the culture. In this process the novice is supported by the more initiated and acquires both the meanings shared by his or her society and, as Lev Vygotsky (1986, p. 60) has pointed out, the "functional use of signs" to handle them.

From this point of view the entire society is a whole of many different transgenerational development processes which all serve to hand down a set of symbolic means and instruments. Each individual takes over the tool kit of the culture—a Vygotskyian idea which forms the main theme in Jerome Bruner's (1986, 1990) studies on a meaning-centered cultural psychology. As every child comes into a cultural fabric of "given" practices, conventions, discourses, and language games, he or she learns how to use and to understand meanings which keep this texture together. Learning by doing means that the child has to find out how to create and to interpret meanings, how to negotiate them, and how to share them.

I would like to put forward two arguments in favor of this perspective. First, it offers both a philosophically and psychologically well-justified way to conceive of the individual acquisition of our culturally specific time concepts; it answers the question of how we develop "time" as the most important metaphor for the diachronic dynamics of our being; and it draws our attention to time as an aspect of the general human business of meaning negotiations. Second, within this conception of a culturally constituted matrix in which we understand diachronicity, it focuses on the semiotic systems and above all on language, since it is this which embodies the most powerful symbolic tool for the acquisition of the meanings of a culture—including the meanings of time (Brockmeier, 1991a).


## Meanings and Time


The concept of meaning as a conceptual instrument for the analysis of time involves a dual figuration; it has a psychosocial double existence. That is to say, the meanings through which I am related to the world result from both a cultural and an individual development. This explains why there can be a common horizon of negotiation at all, even in the case of ambiguity, polysemy, misunderstanding, or failed compromises; human communication in one determinate cultural community may be disturbed or interrupted for many reasons but only in extreme circumstances is it semiotically impossible (due, for example, to completely incompatible symbolic codes). An underlying prelinguistic *semantic confidence* always exists among the members of one cultural community: some elementary form of understanding "other minds" always allows the participant to "read" the intentions, emotions, and thoughts of the others, and vice versa, to "decipher" the personal codes of the individual. This semantic confidence emerges as a basic pattern of interaction and communication in early childhood; in the course of development it leads to an increasingly differentiated understanding of the possibly shared meanings of an action, an object (referred to in a common activity or an act of attention sharing), or a speech act.

The symbolism we are most familiar with is, of course, our language. We have the widest knowledge of linguistic meanings and of the communicative processes by which children come to understand and handle them. It is not enough, however, to take into account only linguistic semantics. I would argue that one must leave the standpoint of psycholinguistics and developmental psychology in the narrow sense in order to grasp the construction of our time concepts. From the perspective of a psychological semantics (Brockmeier, 1988), the objective meanings of the a priori web of time, however it may be "given," must always be subjectively acquired and individually reconstructed. Otherwise time could never assume the role of a social synthesis which is, of course, in no way external to the individual's inner being. This represents indeed one of the most striking qualities of time, namely that it has simultaneously a social and a personal context, that it is outside and inside, public and private, open to everybody, and intimately close to an individual self. Even the idea of my very own time, my individual *Eigenzeit*, as Helga Nowotny (1989) has pointed out, has only come into being as a result of those sociohistorical transformations of modernity which fundamentally changed our whole time culture.

To explain this fusing power we have to consider how nearly every human achieves this synthesis function in his or her own personal life. For example, how does a person become integrated in the individual time synthesis of the past, future, and present which we construct continuously? Most significantly, this synthesis can only serve as a social order system because it operates as an inner concern for every single human being. It constitutes an intrinsic individual psychological structure. Although it may be objectively "given," it has to become a subjective affair and be painted in personal colors.

For this reason, while drawing upon the idea of the objective "givenness" of time, I would like to put the emphasis on how we as individual subjects "take" that which is given to us. The question is: How does one become familiar with the variety of meanings through which this "familiar stranger" appears and—to see the same process in a different perspective—how does one re-create these meanings as one's own? How do we grasp this a priori nature of time which "in itself," to borrow another Kantian term, is, at least psychologically, without any independent being?

## Temporal Framing of Life

To make my arguments clearer, let me point out explicitly the working hypothesis which underlies them. Against the background of the philosophical and psychological sketch that I have given so far, I suggest that we should conceptualize time as an organizing principle of human life, an instrument by means of which we structure our activities and, on the basis of these, order our thought about the world and about ourselves. I suppose one of the main psychological functions of our time concepts is to allow us to find a "theory" for our lives. That is to say, it offers a framework to create a meaningful life story in which we can settle the random succession of diachronic events in such a way that we can call them our "life": continuity, autobiography, identity. One could call this a *temporal framing of life* which makes possible intentionality and meaning. Picturing activities in such a symbolic representation of diachronicity is a kind of social and individual organization which must be regarded as an anthropological characteristic, similar to our language with which it is connected in manifold ways. If I had to encapsulate this idea in one phrase, I would say that through the order of time, the meanings of the world and of ourselves come to have structure, coherence, and sense. In helping us to draw a pattern for our experiences, the metaphorical construct of time supplies the raw material which we use to weave together the threads of our life.

Where the need for this kind of order and meaningful coherence comes from is a fascinating question. Is it, as Seneca, the Roman philosopher of time and life, thought, that we cannot bear the burden of the idea that our life could be put out at every moment? Obviously we have invented a wide range of possible answers to this quest for order and measure. Chronology, the numeric structure of diachronicity (including all kinds of calendar time and history time), embodies one of the answers. Hours, days, years—the most striking units—represent only the tip of the iceberg of a huge variety of temporal order systems (Aveni, 1990); they are the grids which human beings have created in order to match their activities and to interweave the threads of their lives.We use a multitude of stable natural or at least regular rhythms, cycles, and unities of more or less linear duration in order to orientate ourselves by social as well as individual coordinates of reference. The result is a ''metronomic society,'' as Michael Young has described it, which consists of a ''concatenation of informal and formal agreements, founded on the increasing sophistication of measurement and secured, above all, by the greater willingness of people to undergo entrainment of their time habits than entrainment of any others.'' (Young, 1988, p. 228).

We never completely lose the feeling of the artificiality of this metronomic world. All social time syntheses are themselves transient time constructions, as Norbert Elias (1984) has pointed out. They correspond to local and temporal necessities and they disappear with them. And in a sense—maybe in a repressed sense—we know about this. We never completely get rid of the feeling of fragility which goes through all the carefully erected temporal scaffolding of our lives. Everybody has had this experience: the order of time and with it the order of the self dissolves. Nooteboom describes in his novel a strange journey of this sort which starts in the midst of everyday life and which begins almost imperceptibly to become centerless:

> Days—now when I say this word aloud, I am listening to how it sounds ethereal. If you were to ask me what is the most difficult thing, I would say it's the farewell to measure. We cannot bear to be without it. Life is too empty for us, too open, we have devised everything possible in order to keep hold of it—names, sizes, anecdotes. So let me continue to do this—I don't have anything other than my conventions—go on saying day and hour, although our journey seems not to bother about its reign of terror [Nooteboom, 1991, p. 87].

I do not think that it is worthwhile arguing whether the metaphor of time offers a pattern to order our experience, or if it is just the attempt to guarantee the *appearance* of order. Once again we owe to Kant the proof that every attempt to give an ontological answer to the question whether we have here a "real" or an "apparent" order will face unsolvable epistemological difficulties. Neither of these concepts is able to grasp the meaning of a metaphor, and least of all the meanings of those metaphors by which we try to portray the fluent form of our life as well as of the diachronic dimension of being in general.

The point is rather how can we perceive the world we live in within a framework of human dimensions? How do we manage to understand our experiences of the world and our self in such a way that we can locate them in a picture of transience which shows human sizes and proportions? How do we settle or include them in a view which is under control, in lines of sound which offer "music" to our ears? We know from *Richard II* that this is a truly existential question: "how sour sweet music is / When time is broke and no proportion kept! / So is it in the music of men's lives" (Act V, scene 5, lines 42–44).

Proportion is limitation. That we can slow down or even stop the continuous flowing of the *hic et nunc*, that we can stabilize it within the grids of our systematized time metaphors is, of course, an illusion. An instance of that self-deception is created by all art: a petrification of time. As Young has put it, so are all our measures of permanence: designed to trick time and its passing. We know about the illusion of this petrification; "but the illusion, nurtured by habit and memory, is a rock on which all our personalities and all our societies have been built" (Young, 1988, p. 12).

A kind of elementary teleological projection seems to have emerged in the wake of a deeply rooted *horror vacui* in matters of sense and coherence (Brockmeier, 1992a, pp. 31–36). Fulfilling a need for the reassuring lines of a closed weltanschauung it offers a cognitive and intellectual pattern which also can embrace the meaning and sense of time. This feature of the human condition has been, oddly enough, rather unexplored so far; and so has been its anthropogenetic origin. Does it belong to the history of our particular type of civilization? Is it part of our mythical and religious heritage? Or is it a general quality of human "meaning-making," as one could think adopting the view of Bruner (1990)? This would entail conceiving time as a necessary aspect of the organization of meanings that human beings create out of their encounters with the world. So it could be understood as an intrinsic momentum of the constructive power of our "acts of meaning': it operates through the creative symbolic activities that we employ in making sense both of the world and of ourselves.

Or should it be understood—to switch this perspective only slightly—as a kind of side-effect construction of our language games; as, so to speak, an unintended but unavoidable result of the social organization of our linguistic discourse, as one might put it on a radical Wittgensteinian reading, following the suggestion of Rom Harré (1989)? According to this view, we could identify the temporal framing of life as the temporal form of discursive "positioning" ourselves: a social as well as intellectual practice which is outlined in jointly produced story lines. Thus, the metaphor of time, as with many other rule systems, would appear only as one of the "explicit formulations of the normative order which is immanent in concrete human productions, such as actual conversations between particular people on particular occasions'' (Davis and Harré, 1990, p. 44).

## Time-Self, Narrative, Autobiography

I present my paper now because I knew the deadline which the editor fixed for submitting it. This is an everyday experience: timing as an elementary form of social coordination. But this is not all. The temporal construction of the self includes more. The project of creating one's identity means nothing else than coming to terms with one's own *time-self*; that is to say, a self-concept which embraces different diachronic moments. This time-self—we might say remembering our ''Kantian'' starting point—is to be seen as a necessary condition for the possibility of personal identity. Without it no ascription of diachronic experiences to one and the same subject could be thought of, at least not when we take the epistemological ''standpoint within a given mental life'' (Cassam, 1989, p. 105). A frame of mind, though open and fluent, dialectically results from (and makes possible) many meaningful temporal constructions like personal continuity, change, development and, not least, memory. In these constructions we synthesize spatiotemporally very distinct aspects to those unities we call our lives—just as elementary events such as ''having a new experience'' or ''seeing an old friend'' involve the differentiation and recoordination of several directed times.

It seems evident that this construction of the time-self, which occurs immanently in most of our activities and experiences, demands further, or, let me say, psychologically more differentiated explanations than the manifest demands of social coordination and communication. The time-self implies not only intersubjective matching but also inner-subjective matching, even though a precise borderline between them is difficult to define.

This becomes particularly striking when we consider the narrative instruments and devices which we employ in the construction of our time-self. The linguistic structures and the whole narrative framework, in which I tell others as well as myself the ''story'' of my life, also reveals itself as a fundamental form of time-ordering. Every autobiography (or fragment of it) projects a narrative construction onto the temporalities of my life. In so doing, it outlines a particular ''tacit knowledge,'' an implicit *autobiographical time-theory* which permits us to make sense of the countless events, the ''data'' of experience, which we encounter in the course of life. Examining this time-theory more closely, we can discover that it is articulated not least through many *spatial* references, organized by the several locative systems of our language as well as by the cognitive map of the narrative scenes (Bruner, 1993). Moreover, these deictic and iconic references are often synaesthetically linked with auditive colors (from natural, linguistic, and musical contexts of experience), yes, even with smells, tastes, and temperatures (Brockmeier and Treichel, 1990).

In any case, every temporal order—whether it be a dominantly chronological one (as in our social life), whether it be a dominantly narrative one (as in the autobiographical process)—is a way to construct and to make sense of the meanings of the world we live in. So one could be inclined to say, time is a means to an end; we employ it in order to come to terms with the meaning structures of our culture. We use the various orders of time as social frameworks within which we can match our activities—the practical and material ones as well as the symbolic ones—with those of others, as well as using it as a framework within which we can outline our self.

But on the other hand (and this makes it even more complicated), time does not, of course, exist as a somehow separate means. It cannot be seen, at least not in the perspective outlined here, as an ontologically independent being. Although we are often tempted to take the clock in our hand as such a means, we should not forget that the clock is itself just an artificial representation, a metaphor for something else. It is only the mechanical or electronic shorthand of a social symbolism, which is psychogenetically, anthropogenetically and historically originated in different natural and cosmological rhythms. The history of time measurement and the technology of timing systems offers rich material for the study of this development (see e.g., Macey, 1989; Wilcox, 1987; Aveni, 1990). So we must say (and that reveals the limits of this comparison) time is a means which is a part (or an aspect or a quality) of the end itself: it exists only in and through the world of human meanings. It is only through psychologically meaningful operations that we acquire our concepts of time. And we might add: without giving diachronicity a personal dimension, a human being would never be able to cast anchor in the world—even if it be only for the span between two "nows."

In order to indicate this ambivalent or even contradictory status of time—as a means to organize the meanings of the culture we live in and as an intrinsic dimension of the meanings and their creation themselves—I would like to call those "acts of meanings," which are responsible for our temporal frames of minds, *anthropomorphic operators* of time.

## Chronology, Activity, Language, and Space

Anthropomorphic operators of time are metaphors which help us to construe our social as well as individual time syntheses. I have already mentioned three of these operators: chronology, activity, language. *Space* is a fourth operator which serves as a three-dimensional metaphor of diachronicity. The space of time describes an *Anschauungsmodell*, a model of intuition; it visualizes a horizon *for*, as well as *within* the human measures of chronology, activity, and language.

The cognitive spatialization of the temporal dimension is a very important psychological device. Locating ourselves in space is normally a fairly familiar way of locating (and that means identifying) our transient selves in time: whether at a "stage" in the individual "course of life," on the "field" of history, or under the "slowly grinding mills of God." In antiquity the entrance to the Pantheon in Rome was situated 10 meters below its present level, offering a voluminous time layer which was regarded in the late Middle Ages and in the Renaissance as an obvious *etalon*, a standard for the space the course of history literally leaves behind. Throughout the whole history of culture we find numerous spatial ideas and metaphors in which we can encounter all kinds of imaginable connections between the course of our life time(s) and that (of those) of world time(s) (Blumenberg, 1986).

Semiotically and psychologically the spatialization of time is based on the transformation of diachronic signs in synchronic signs (Jakobson, 1971). The sequential flux of the now in our stream of conciousness is apparently "slowed down" or even "held up" for a moment which allows us to perceive, conceive, or project it synchronically together in order to understand and create a meaning structure. It seems that different media of cognitive representation and communication (such as oral and written linguistic texts, images, film) embody temporal succession in different ways (Brockmeier, 1991b). Furthermore, they may include several kinds of "time-release mechanisms" and forms to transform temporal sequences into spatial layouts and vice versa (Gross, 1992).

In our epoch *Chronology* has become the best known and most conventionalized metaphor for time. It is some kind of chronological order that probably most people refer to when they are confronted with the Augustinian question. In whatever form it might appear, chronology pictures diachronicity as something imaginable and intellectually controllable; representing it as an act and a result of that symbolic modeling which happens in measuring. Criterion and *etalon* for the measurement—the "measure of time"—is some kind of regular and (intersubjectively) objectifiable matrix. It is amazing how inventive people have been in thinking out numeric, calendar, and historic registration systems. It is equally striking how these orders have been established or literally imposed by power and force (Zerubavel, 1982; Rifkin, 1987). The more they have become conventionalized, the more they have constituted social, political, and economical syntheses. They range from singular aspects of the multilayered interplay between the different natural rhythms, cycles, and spirals to the mathematical arrow of Newtonian time, or, as in our days of electronic time technologies, to the global unity and diversity of times and multiple temporalities. Chronology, once merged as a plural aspect into nearly all human existence and experience, can also become unlimitedly "multivoiced," to borrow Bakhtin's suggestive term.

Although chronology metaphors dominate phenomenologically most of our social life (nonetheless in culturally and historically very specific ways), one must not regard this kind of order as the most fundamental anthropomorphic operator of time. Considering human temporalities and focusing on the construction of our self and our identity, I think we have to take *language* primarily into account. I would like to finish this paper by distinguishing three levels on which the time-framing and worldmaking power of language can be studied: the spoken, the written, the narrated.

1. *Spoken* time is evoked in oral discourses. Every oral (as well as written) genre can be analyzed as a special kind of "interpretive system" (Fleisher Feldmann, 1991, p. 61) of the world and of its times. One of the psychological and linguistic grounds of "spoken time" is the specific phonemic coherence which is

constituted by what Jakobson (1971) has described as "simultaneous synthesis"; this is the (although limitedly) synchronic *gestalt* of perception which is engendered by the auditive mode of language.

2. *Written* time is that diachronic order which is constructed through the tense system of literacy grammars. The abstract and multilayered tense systems of, for example, the "standard average European languages" presuppose a reflexive mode of cognitive operation and coding of knowledge which is closely linked with the cultural basis of the alphabetic system (Olson, 1977). Just the (linguistic and cognitive) difference between the letter and the word, which is fundamentally based on the experience of writing, opens up a time span (and a span of imagination) which could never be perceived in a pure oral culture (Brockmeier, 1992b, chap. 2).

3. *Narrated* time is the specific temporal dimension of narratives—ranging from fairy tales, (oral) myths, and storytelling to literary fiction, history and to the (already mentioned) "genres" of autobiography and narrative constructions of the self. For example, the narrated time of literature is a way not only to project time in the imaginative space of "fictive experience" which is, as Paul Ricoeur (1985) explains, at first sight a paradoxical phenomenon, "an experience certainly, but a fictive one, since the work alone projects it" (p. 101). The narrative composition also fuses several temporal (and spatial) orders in one individual mind: "Each fictive temporal experience unfolds its world, and each of these worlds is singular, incomparable, unique" (Ricoeur, 1988, p. 128). The times which are condensed in these worlds of literary narrative do not respect the borders between subjective and objective, real and possible, virtual and impossible worlds, which are so carefully erected in our logicoscientific mode of thinking (Bruner, 1986, chap. 1).

There is good reason to hope that the recent discursive and narrative turn in linguistics, psychology, anthropology, and philosophy will open up many fascinating insights in how these different time worlds merge in our minds and lives.

## References

Aveni, A. F. (1990), *Empires of Time: Calenders, Clocks, and Cultures*. London & New York: Tauris.

Blumenberg, H. (1986), *Lebenszeit und Weltzeit*. Frankfurt am Main: Suhrkamp.

Brockmeier, J. (1988), Was bedeutet dem Subjekt die Welt? Fragen einer psychologischen Semantik. In: *Hamburger Ringvorlesung Kritische Psychologie, Wissenschaftskritik. Kategorien, Anwendungsgebiete*, ed. N. Kruse & M. Ramme. Hamburg: Ergebnisse-Verlag, pp. 141—184.

———— (1991a), The construction of time, language, and self. *Quart. Newsletter Lab. of Compar. Human Cognit.*, 2:42–52.

———— (1991b), Medien des Raums und Medien der Zeit. In: *Brecht 90. Kulturtheoretische Aspekte der Brechtschen Medienprogrammatik*, ed. I. Gellert & B. Wallburg. New York: Peter Lang Europäischer Verlag der Wissenschaften, pp. 105–124.

———— (1992a), *"Reines Denken." Zur Kritik der teleologischen Denkform*. Amsterdam & Philadelphia: Grüner-John Benjamins.

———— (1992b), Zwischen Buchstabe und Wort. *Psychologie in Österreich*, 3:56–61.

———— Treichel, H.-U. (1990), Worte, Klänge, Farben. Erkundungen in "Synaesthesia." In: *Die Chiffren. Musik und Sprache. Neue Aspekte einer musikalischen Ästhetik IV*, ed. H. W. Henze. Frankfurt am Main: Fischer, pp. 71–120.

Bruner, J. (1986), *Actual Minds, Possible Worlds.* Cambridge, MA.: Harvard University Press.

—— (1990), Two modes of thinking. In: *Acts of Meaning.* Cambridge, MA: Harvard University Press.

—— (1993), The autobiographical process. In: *The Culture of Autobiography: Constructions of Self-Representations,* ed. R. Folkenflik. Stanford, CA: Stanford University Press, pp. 38–56.

Cassam, Q. (1989), Kant and reductionism. *Rev. Metaphysics,* 43:72–106.

Davis, B., & Harré, R. (1990), Positioning: The discursive production of selves. *J. Theory Soc. Behav.,* 1:43–63.

Elias, N. (1984), *Time: An Essay.* Oxford: Blackwell, 1992.

Fleisher Feldman, C. (1991), Oral metalanguage. In: *Literacy and Orality,* ed. D. R. Olson & N. Torrance. Cambridge, MA: Harvard University Press, pp. 47–65.

Fraser, J. T. (1987), *Time, the Familiar Stranger.* Amherst: University of Massachusetts Press.

—— (1992), Human temporality in a nowless universe. *Time & Soc.,* 2:159–173.

Geertz, C. (1973), *The Interpretation of Cultures.* New York: Basic Books.

Gross, S. (1992), Reading time—Text, image, film. *Time & Soc.,* 2:207–222.

Harré, R. (1989), Language games and the texts of identity. In: *Texts of Identity,* ed. J. Shotter & K. J. Gergen. London: Sage, pp. 19–35.

Hernadi, P. (1992), Objective, subjective, intersubjective times. *Time & Soc.,* 2:147–158.

Jakobson, R. (1971), Visual and auditory signs. On the relation of visual and auditory signs. In: *Selected Writings,* Vol. 2. The Hague: Mouton, pp. 334–344.

Macey, S. L. (1987), *Patriarchs of Time: Dualism in Saturn-Cronos, the Father Time, the Watchmaker God, and Father Christmas.* Athens: University of Georgia Press.

—— (1989), *The Dynamics of Progress: Time, Method, and Measure.* Athens: University of Georgia Press.

Nooteboom, C. (1991), *Die folgende Geschichte.* Frankfurt am Main: Suhrkamp.

Nowotny, H. (1989), *Eigenzeit, Entstehung und Strukturierung eines Zeitgefühls.* Frankfurt am Main: Suhrkamp.

Olson, D. R. (1977), From utterance to text: The bias of language in speech and writing. *Harvard Ed. Rev.,* 47:257–281.

Ricoeur, P. (1985), *Time and Narrative,* Vol 2. Chicago: University of Chicago Press.

—— (1988). *Time and Narrative,* Vol 3. Chicago: University of Chicago Press.

Rifkin, J. (1987), *Time Wars: The Primary Conflict in Human History.* New York: Simon & Schuster.

Rilke, R. M. (1963), *Duino Elegies,* tr. J. B. Leishman & S. Spender. New York: W. W. Norton.

Shakespeare, W. (1597), King Richard II. *The Arden Shakespeare.* New York: Routledge.

Vygotsky, L. S. (1986), *Thought and Language,* ed. A. Kozulin. Cambridge, MA: MIT Press.

Wilcox, D. J. (1987), *The Measure of Times Past: Pre-Newtonian Chronologies and the Rhetoric of Relative Time.* Chicago: University of Chicago Press.

Young, M. (1988), *The Metronomic Society: Natural Rhythms and Human Time-Tables.* London: Thames & Hudson.

Zerubavel, E. (1982), The standardization of time: A sociohistorical perspective. *Amer. J. Sociol.,* 1:1–23.

20

# Malinowski and the Birth of Functionalism
# or,
# Zarathustra in the London School of Economics

*Robert Thornton*

*Abstract*  Bronislaw Malinowski is known as the creator of the theories of functionalism and myth as social charter. Though Polish born, he became one of the founders of 20th-century (British) social anthropology. His classic ethnographic monographs on economy, language, kinship, and sexuality among the Trobriand Islanders of Melanesia, published between 1922 and 1942, are the locus classicii of these theories. Based on my book, *The Early Writings of Bronislaw Malinowski, 1904–1914* (1993), this paper shows that Malinowski's social theories, especially his "presentism," his apparent hostility to history, and his rejection of speculations about origins, is closely related to Friedrich Nietzsche's philosophy of the meaning and value of time, origins, history, and genealogy. Malinowski also drew on Ernst Mach's positivism (the forerunner of the Viennese "logical positivism") and his relativism (which Einstein found so provocative). By tracing the use Malinowski made of these Central European ideas in formulating a theory of functionalism in anthropology, we are able to see how all of these ideas are related to a common problem: How is time experienced by life-forms and how may we account for the coherence and internal regulation of those entities that resist time by means of "life"—that which is vital, healthy, whole, and thus stands against time while being in time? The attitude toward time, life, and society was expressed differently in the context of the Austro-Hungarian empire than it was in Britain. The fundamental concepts of perspective, relativity, equilibrium, and recurrence-recursion distinguished this European "eastern" tradition from that of the British "western" tradition (individualism, epistemological absolutism, progress, evolution), and this is what made Malinowski's intellectual intervention so powerful.

Bronislaw Malinowski is known as the creator of the theories of functionalism (Malinowski, 1957; Gellner, 1958) and myth as social charter (Malinowski, 1922, 1926a), of a pragmatic theory of language as action (especially "magical language")[1] and as one who insisted on the ahistorical or synchronic character of

---

[1]His theory of language is implicit throughout his work, but is stated in Malinowski, (1923, pp. 1–74).

primitive society. As Gellner has remarked (1985a), it is the fusion of these into a relatively coherent theory that is the hallmark of Malinowski's contribution. Unlike the previous "functionalism" implicit in biological and evolutionary theories of human society and behavior, Malinowski's fusion of functionalism with ahistoricism is "an unusual combination . . . his a-historicism . . . the obverse of [the] evolutionary approach" while his functionalism was a "denial" of the rationalism implicit in 19th-century evolutionism. Though Polish by birth and training, Malinowski became one of the founders of 19th-century (British) social anthropology, a field founded by Sir James Frazer of Trinity College, Cambridge. Malinowski's classic ethnographic monographs on economy, language, kinship, and sexuality among the Trobriand Islanders of Melanesia, published between 1922 and 1944, are the locus classicii of these theories. His teaching at the London School of Economics over two decades before World War II produced a cohort of researchers and academicians who ensured the continuity of his ideas into the future, and provided the discipline of anthropology with a rich, although contradictory and compromised, body of theory and data.

Malinowski's treatment of time in the life of small-scale nonliterate societies was a departure from earlier historical speculations about human prehistory, especially as this was expressed in evolutionism, the dominant school of thought when Malinowski began to write. Its principal tenets were that societies inevitably moved from simple to complex, and from unfree to free, from status to contract, or from kinship, ritual, and religion to law, rationalism, and science. Malinowski criticized these ideas on a number of grounds, but especially on the grounds that they were not founded on empirical evidence and were therefore speculative. First, he insisted on giving anthropology a genuinely empirical basis, but, more importantly, he sought to develop a research method that would generate reliable empirical data from which more valid theories could be developed. (Malinowski's primary methodological statement concerning the nature of ethnographic research is contained in the first chapter of *Argonauts of the Western Pacific* [1922].) The theory he developed was as much about the nature of data as it was about the social processes of the society itself (Gellner, 1973, pp. ix–xii, 90–91). While he refused to try to situate the societies he studied in the speculative chronologies of human evolution, he introduced ways of comprehending the small-scale and repetitive time of daily living in such societies and placed his own modes of understanding them in the context of the history of Western science. (In doing this, he was following one of his early teachers, Ernst Mach [1872], whose history of the theory of thermodynamics was a path-breaking work in the history and philosophy of science.)

Malinowski himself made a tremendous impact on anthropology in this century, and through his teaching and research, he very largely shaped much of the contemporary view of primitive and nonindustrial or nonliterate society. His impact was strongest on anthropology, but since he was read and reinterpreted by philosophers from John Dewey (1925; see also Thornton and Skalnik, 1993, p. 29) to Richard Rorty, by psychologists, sociologists, historians, and classicists,[2] and even

---

[2]The classicists Edith Hamilton, E. R. Dodds, and J. Huntington Cairns were strongly influenced by Malinowski in their interpretations of ancient Greek society. Huntingdon Cairns became Malinowski's

today, by literary critics (Marganero, 1990; Torgovnik, 1991), his ideas have had great impact beyond the field of anthropology itself. Within anthropology, however, his influence has been profound. He personally trained most of the first cohorts of British and European anthropologists who worked in Africa, Asia, Australasia, Australia, and Oceania during the ''classic'' period of 20th-century anthropology from the early 1920s until the early 1950s. His students included African heads of state such as Jomo Kenyatta, and prominent African educators and political activists such as Z. K. Matthews in South Africa. Gregory Bateson, for instance, was his student, but the American anthropologist, Margaret Mead, Bateson's first wife and long-time research colleague, was also influenced strongly though indirectly by Malinowski.

Malinowski's influence was exercised primarily through his post at the London School of Economics (LSE), where he went in 1920 after his return from intensive and long-term field work in the Trobriand Islands. At the LSE, he directed a seminar which was attended by virtually all anthropologists and students of anthropology at the LSE, or those who were passing through London. As he developed his theories, and wrote his books that were based on the detail of his Trobriand researches, he brought each chapter to this seminar where he presented it to his students. His theoretical reach was large, so chapters and works on primitive economics, on psychology and sexuality, on crime and social order, on kinship and the family, on morality, nationalism, culture, social change, ethnic identity, and so on brought different and overlapping intellectual circles within the field of his influence.

Malinowski championed a radically empirical approach to anthropology during a time when the study of mankind was almost exclusively philosophical, yet he declared that data were created by theory, that is, that observations only became scientific data in the presence of a theory that made sense of them; in fact, that data were never more than the theory that constructed them. Thus, Malinowski insisted that valid description of primitive and nonliterate societies must be driven by specific questions. Functionalism posed several general and several specific questions about the nature of primitive society, and thus provided both a program for research and an explanatory framework for the results. At its most general level, Malinowski's functionalism demanded an answer to the question, What makes society run, or what makes society run so smoothly (Malinowski, 1926b; Thornton, 1992). This question, though simple, was radically different from the previous philosophical anthropologies of non-European societies that asked, for instance, *Is* there primitive *society* (or just a ''state of nature'')? What is the origin of primitive society? What is the historical contribution of primitive society to European industrial society? What is the rank, place, or priority of primitive societies in the history of mankind, in the history of human morality, science, or religion? All of these questions were framed in terms of the assumptions of evolution, loosely based on Darwinian theories, but more firmly founded in older European ideas of history as progress and

---

literary executor after Malinowski's death at Yale University in 1944, and saw Malinowski's posthumously published *A Scientific Theory of Culture* (Malinowski, 1944) through press.

"the great chain of being." Instead, he sought to know in detail what makes society run, that is, *How does it work and why?* His approach prejudged the data but in a different way from previous theories. He conceived of primitive societies as cultural wholes, self-contained and self-maintaining. He held that the origins of customs were irrelevant to the anthropologist who sought to answer the question what makes society run, since, he argued, only the *functions* of those institutions in the context of the values, desires, and needs of the present could have any bearing on their continued existence. Since his primary goal was to achieve adequate description of areas and aspects of human life that had so far been neglected by science, he rejected any attempt to account for *what was* before it was known *what is.* Consequently, he adopted a radical ahistoricism, really a kind of radical "presentism," that ran entirely counter to the pervasive historicism of most of European social thought up to that time. This presentism or ahistoricism is somewhat paradoxical, however, since an account of what makes society "run" would seem to suggest that it must have somewhere to run to, that is, it must imply a period of time in which to run. Malinowski's functionalism seems to deny this by postulating that it is the functional or working relationships amongst the institutions, beliefs, and rituals of society that may be held to account for *how* it runs. How long it runs, or where or when it "runs from," that is, the origins of those institutions, he held to be irrelevant to his version of a social science.

The apparent ahistoricism of Malinowski's functionalism posed conceptual problems then, and it continues to do so today. Specifically, social institutions must have had origins, and these origins must have some bearing on why certain institutions exist in some places, cultures, or contexts, and not in others. Primitive and nonliterate societies had long been assumed to be "primordial" and to lack history, but they are not outside of time, and therefore must have had histories even if we do not or cannot know them (or cannot know their details in full).[3] These histories, moreover, must determine in some way those factors that make them run. How can a specifically ahistorical method shed light on these problems? These questions continue to be asked, and some important answers have emerged, but Malinowski's functionalism was remarkably successful in generating valid and detailed descriptions of how these societies worked, and its overall success in making sense of what had often seemed senseless about much of human life has been profound.

This is not to say that previous nontheoretically informed descriptions were worthless, but rather to note that they were informed by a different set of questions and governed by a different logic, especially a different temporal logic. Descriptions by travelers were constrained by travel itineraries; the temporality of a progress across a landscape took descriptive priority over whatever might have been the case in any single location (Thornton, 1988). For the missionary, the temporality of salvation and the implicit historicity of the bible itself (a historicity always declared by the missionaries although there were some who doubted it even so) was imposed

---

[3]Fabian (1983) gives a full account of how Western historians and philosophers have assumed the ahistoricity of societies that lack written records. While they lack the *documents* of history, or archives that would make an *account* of history possible, they do not lack *history* (meaning change over time).

on the temporality (historical or otherwise) of the native myths and religious dogmas they sought to overcome. (I develop this argument with respect to South Africa in Thornton [1980].) For the colonial official, the limited bureaucratic temporality of reporting, of the campaign, of the harvest season, and the term of office, imposed a different temporal framework on the view of the primitive (Fabian, 1983). In all cases, the unexamined postulates of European practices dictated the anthropological understanding. Malinowski, guided in part by Ernst Mach's attempt to put his own understanding of physical science in the context of the history and philosophy of physical science, sought to examine the philosophical bases of the European's own understandings of myth in comparison with the "primitive"or "savage" understandings. This, in effect, was a thoroughgoing epistemological relativism that placed European and primitive concepts at the same moral and epistemological "level" in order better to understand the latter. Of course, the result was to open the way to a critical and deeper understanding of the European metaphysics, too. The implicit comparison of myth and mythical time, historical time, the temporal correlates of social and psychological practices for both European and primitive culture, shed a great deal of light on both. From the standpoint of a theory of history and of the role of the observer, Malinowski was not only to greatly improve the significant detail of anthropological reporting—that is to say more precisely, what it was that actual nonliterate people actually did and thought about economy, kinship, sexuality, death, magic, and so on—but was able to locate these facts within a specific theory of how these kinds of societies worked. His ability to do this relied on two central European philosophers and historians of science that most people of his day, and of ours, would have thought to be not only mutually incompatible with one another, but also remote from the immediate concerns of anthropological fieldwork. The first of these was the German classicist and philosopher Friedrich Nietzsche, while the second was the Moravian physicist, psychologist, and historian of science Ernst Mach.

Indeed, Malinowski's presentism, and his rejection of speculations about origins is derived primarily from Friedrich Nietzsche's diffuse philosophy of the meaning and value of time, origins, history, and genealogy (especially in *On the Genealogy of Morals* [1887]), while his insistence on studying mankind through collection and analysis of empirical "positive" data derives from Mach. For Malinowski these two positions informed each other since to insist on an empirical method was to reject the evolutionist speculation about origins, while to reject the search for origins per se was to assert the relevance of Nietzsche's perspectivism and presentism, and with both Mach and Nietzsche, to reject the English utilitarianism and realism in favor of a radical constructivism. Malinowski introduced these ideas to his English students at the London School of Economics, the very seat of English economic philosophy, and thus became both a prophet and a revolutionary.

In the *On the Genealogy of Morals*, in particular, Nietzsche reacted to what seemed to him the failures of the approach to law, crime, and punishment taken by the English Utilitarians. Believing that "they are no philosophical race, these English," Nietzsche argues that the "purpose of law," that is its "utility"—either its (good) usefulness or its (bad) abuse—could not be accounted for by looking, as the

English Utilitarians had done, for its origins in previous laws or in previous legal systems of earlier times. The origins of a law tell us nothing, he argued, about how and what it does, *or how it works* in the *present*. The historical approach, naturally, was the ordinary practice of English jurisprudence which, lacking a constitution, necessarily referred to the past for legitimation. The tradition of English jurisprudence was intensely historical and sought to justify the laws of the land on the basis of their origins and their histories rather than in terms of their logical relation to a standard as defined in a constitution. This was both functional for a society that lacked a constitution, as well as consistent with the overwhelming historicism of European thought in general. English jurisprudence, like that of the continent, was confirmed, moreover, by the speculative histories that explained contemporary legal practices in terms of Roman or classical precedent. Nietzsche argues:

> [O]n the contrary . . . the cause or the origin of a thing and its eventual utility, its actual employment and place in a system or purposes, lie worlds apart. Whatever exists, having somehow come into being, is again and again reinterpreted to new ends, taken over, transformed, and redirected by some power superior to it; all events in the organic world are a subduing, a *becoming master* . . . However well one has understood the *utility* of any physiological organ (or of a legal institution, a social custom, a political usage, a form in art or in a religious cult) this means nothing regarding its origins: however uncomfortable and disagreeable this may sound to older ears—for one had always believed that to understand the demonstrable purpose, the utility of a thing, a form or an institution, was also to understand the reason why it originated [1887, II:12, p. 513].

Michel Foucault, the French historian of punishment in the European tradition, has made this passage famous much more recently since it is the inspiration for his own historical anthropology of European practices (1975). Indeed, Foucault's *Discipline and Punish* (the French title, *Surveiller et Punir,* conveys a slightly different sense), as well as other works, can be seen as working out of the implications and ideas of Nietzsche's *Genealogy,* especially concerning the problem of the broader function of punishment as a form of "theatre of hell" (Foucault, 1975, pp. 32–69) approached by a "historian of the present" (Merquior, 1985, p. 1). Malinowski seems to have been similarly inspired, although his inspiration occurred 50 years earlier, and led him to investigate the distant culture of the Trobriand Islands rather than the exotic that lay within Europe's own cultures.

Michel Foucault, however, like Nietzsche, remained within the textual tradition of European scholarship and of European documents. Malinowski escaped this through a somewhat paradoxical commitment to empiricism. It was primarily Malinowski's understanding and parallel commitment to the positivism and relativism of Ernst Mach that made this move from text to the context of empirical social research possible. Significantly, this led to the distinctive differentiation of an empirically based fieldwork anthropology from a text-based "speculative" form of universal history into which anthropology was in imminent danger of collapsing. While this had the organizational consequence of establishing anthropology as a discipline among others in European and American universities, it also distinguished two different approaches to the temporality of primitive and small-scale society. The discipline of history, which is founded on the method of investigation of textual resources, continued to ignore nonliterate societies, while the discipline of anthropology, which uses methods that are not unlike history's but investigates the present rather than the past, introduced a new intellectual project. History and anthropology share their fundamental problem, however. They are both concerned with how time is experienced by life-forms that resist time by means of "life"—that which is vital, healthy, whole, and thus stands against time while being in time—and how we may account for the coherence and internal regulation of what we call social life. History and anthropology diverge in their approaches to this problem. Malinowski's theoretical intervention at the time when anthropology was indeed becoming more like history was partly responsible for creating this divergence.

The anthropologist, philosopher, and historian Ernest Gellner has observed that Malinowski's ahistoricism or synchronism is complex and multilayered (1958, 1985a,b, 1987a,b). Malinowski's impact on the social sciences has led a number of people to investigate the nature of his so-called ahistoricism, and to speculate about its intellectual origins. But until Malinowski's earliest writings were brought to the attention of a larger public, and translated into English (Thornton and Skalnik, 1993), these efforts relied on the internal evidence of Malinowski's own writings in English, and what was known about his background and the formative influences that must have been present during his youth and studies in Cracow, Poland. Even this was often fragmentary and contradictory, since Malinowski himself seems to have deliberately concealed much of his own intellectual past from his students and colleagues in England (Gellner, 1985b, Geertz, 1988). To date, the best discussion of the nature of Malinowski's understanding of time and its role in Malinowski's functionalism, and of Malinowski's general attitude to history, is Ernest Gellner's "Zeno of Cracow" (1987b). In this article, Gellner dissects Malinowski's functionalism, in the light of much recent knowledge about his education in Cracow and about the cultural and political climate in this Polish city in the middle of a dismembered Poland that was part of the Austro-Hungarian Empire. The Zeno of Gellner's title is the 5th-century B.C. Greek philosopher Zeno of Elea.[4] Zeno proposed a number of paradoxes that were meant to demonstrate the validity of his monistic philosophy and his denial of the reality of time and motion (Ariotti, 1975, pp. 69–80) by proving the unity and stability of Being. If Being is unitary and all that there is, he reasoned, then it must occupy only one space, namely all space, and be infinite in temporal and spatial extent. But if this is the case, then there are several logical problems that emerge, some of them apparently lacking solution within the frame of monism. One of these paradoxes shows that an arrow cannot move. Zeno

---

[4]Zeno was a pupil of Parmenides who advanced the earliest form of Monism, that is, that all Being is One, and that all thought, language, action and reality were the same, or, more precisely, that one mode of understanding and explanation should be sufficient for all sorts of being and becoming. Zeno is best known as a partner in dialogue with Socrates in several of Plato's dialogues (Sophist, Parmenides; see H. Cairns and E. Hamilton, eds., 1961).

assumed that any object occupies a space that is identical in extent to the object itself. If this is the case then the arrow cannot move since if an arrow occupies a space equal to the space it in fact occupies, it cannot at the same time occupy an extension of that space. If this is true, then it cannot be in motion since to be in motion it would have to occupy a space larger than itself, which it cannot do. Therefore, motion is impossible, and an arrow in flight is necessarily also at rest. This is a paradox, known as Zeno's paradox (one of four attributed to him), that was not solved, or even considered to be solvable until the invention of calculus by Isaac Newton and Gotfried Leibniz around 1675.

Gellner offers a plausible explanation of Malinowski's ahistoricism based on such first principles. He argues that Malinowski, reasoning in a way similar to Zeno's reasoning about Being, was constrained by his own monism to argue that primitive society can never be different than what it is since it would lose its ability to maintain itself, that is, it would cease to run altogether, if the customs that sustained it ceased to function. Gellner calls Malinowski the Zeno of Cracow because he argued that, in a sense, primitive society was like Zeno's arrow: it could not move (i.e., change historically) since to do so it would have to be different than what it was, and his principles of holism and functional integration of the social whole did not permit this.

Gellner also offers a secondary explanation of the ahistoricism of Malinowski's functionalism as being partly the result of the historical thwarting of Malinowski's strong sense of a Polish cultural nationalism. Malinowski's father had been the founder of a new department of folklore studies in the ancient Jagiellonian University in Cracow, one of Europe's first universities. The studies of Slavic folkfore, similar to and inspired by the studies of German folklore such as those by the Grimm brothers, was taken up by nationalists as both justification and impetus for their cause. But a real Polish nationalism in Malinowski's day was politically hopeless. The power of the Austrian-Hungarian Empire, and of the Russian Tsar made the idea of an independent Poland nothing more than a myth. But it was a sustaining myth, and one that had all the more value since it could not be perverted by a real state. For Malinowski, it remained a pure cultural nationalism. As such, Gellner argues, it remained outside of history, or, at least, it remained outside of a Hegelian conception of history as the story of the realization of the Idea in the evolution of the State. The Hegelian theory of the State—what Gellner has elsewhere called a "cozy, homely metaphysics . . . An Absolute in braces''—had provided other states of Europe with a justification for their existence as a moral and inevitable historical right, however corrupt or violent or evil the real State might in fact be. The fully realized State was not to be for Poland. Malinowski's ahistoricism, then, seems to have been doubly determined by the logical impossibility of change in the closed, functional society of the primitive, and the moral superiority of a cultural nationalism that remained unperverted by the presence of a (real) corrupt State. Both ideas have some merit. Nevertheless, Gellner's attempts to understand Malinowski's approach to time were uninformed by the powerful and real intellectual influences that Nietzsche and Mach exerted on Malinowski's thought. Rather than a Zeno from Cracow, a philosopher of paradox, and a cultural nationalist, Malinowski was instead a Zarathustra from central Europe, a prophet and a wanderer from Poland who brought Nietzsche's ideas into the seminars of the London School of Economics, the very heart of English Utilitarianism, of classical economics, and of bourgeois rationality.

Nietzsche's *The Birth of Tragedy* (1872) was one of the texts that influenced Malinowski most strongly (1904–1905 ca.). Nietzsche was struggling with many problems when he wrote *The Birth of Tragedy*. One of these was his effort to integrate the notion of aesthetic judgment with the idea of scientific understanding. In *The Birth of Tragedy*, he posed this problem as a conflict between the reason of Socrates and the passionate unreason of the authors of the great Greek tragedies, especially Sophocles. He saw in the underlying myths on which the Greek tragedies were constructed a template for the structure of consciousness and culture. For Nietzsche (as for Freud, who probably also drew on Nietzsche for this inspiration as well as on the Greek myths themselves) Socratic philosophy, mathematics, and science were the methods of reason, while art, dreams, and drama were the methods of unreason. Nietzsche argued that each method—reason or unreason, science or art—produced a different kind of result since each was driven by different "forces" (we cannot yet call them "motives" since Nietzsche was apparently unsure of how these things were driven except that they were the result of will, the famous "will to power").[5] More importantly, however, each was intrinsically opposed to the other as a different kind or different order of mental process. In *The Birth of Tragedy*, as it was first published in 1872, Nietzsche struggled with the relationship between rationalistic philosophy and tragedy as the relationship between two sorts of consciousness. Each sort of consciousness, the Socratic and the Sophoclean, were themselves founded on a struggle. Tragedy grew out of the struggle between an illusion or dream of order, which Nietzsche placed under the sign of Apollo, and the passion for disordering and the submersion of the individual consciousness in the ecstasy of the universal will, which Nietzsche summarized under the figure of Dionysus. The Apollonian–Dionesian struggle was a psychological struggle that became manifest in the ritual display that was the performance of the tragedy on the ancient Greek stage. Nietzsche's ideas, although bound to the text (which the near-sighted and sickly Nietzsche could never escape) nevertheless suggested to Malinowski both an ethnographic context and a process of psychological change. In his early essay on Nietzsche, Malinowski mused on the possibility of inquiring into the real social context in which Greek tragedy took place. Indeed, several classicists such as E. R. Dodds and Huntington Cairns, who were influenced by and friendly with Malinowski, set about doing precisely this. Malinowski, however, thought that the same project might be fruitfully pursued among the primitive societies of Australia and New Guinea. Nietzsche's problems, like those posed by

---

[5]Nietzsche's doctrine of Will comes directly from Arthur Schopenhauer, especially from his magnum opus, *The World as Will and Representation* (1818). Nietzsche makes his debt to Schopenhauer explicit in his work, and Malinowski comments on this in his essay on Nietzsche just cited (Malinowski, 1904–1905 ca.). Thus it is clear that Malinowski was aware of this intellectual thread, and of his place in it.

other speculative historians and philosophers of the primitive or of the Ancient, "Serve to pose the problem," Malinowski wrote, but we "cannot consider them . . . to be something positive, if only because such psychological, introspective assertions can neither be confirmed nor refuted by any proof" (1904–1905 ca.). What he proposed in his essay on Nietzsche in 1904–1905, while he was still a student at the Jagiellonian University in Cracow, was that Nietzsche's "questions" be put to a positive test. As epigrammatic assertions—the form in which Nietzsche stated them—these propositions were "so general and subjective," Malinowski wrote, "that the only things that we can do with them directly is either to agree with them or to reject them" (1904–1905 ca.). The principles of a Mach-inspired observational and descriptive science, however, could be applied to some of the problems that Nietzsche posed. This was a radical idea: Nietzsche and Mach for most of Malinowski's contemporaries must have seemed at opposite ends of the intellectual universe. Malinowski saw, however, that both were concerned with the problem of perspective or relativity. For Nietzsche this was a problem of the perspective of "power" or "mastery"; for Mach it was a problem of relative positions of observers in observing the behavior of physical systems. Though apparently concerned with radically different subject matter, Malinowski reasoned that Mach's methods and philosophy of observation could be applied to a "positive," scientific-descriptive investigation of problems that Nietzsche raised with respect to myth and its transformation, to concepts of fate, cultural creativity, and human suffering. While this could not be done among the dead Ancients, it could be done among living "primitive" people who possessed myths of similar complexity and intensity. The "cognitive value" of these myths, Malinowski argued, could be:

> [D]emonstrated by the fact that by observing, classifying and analyzing various rather objective facts from the standpoint of their assertions we arrive at the formulation of some general principles. We see these facts arranged in clear, far-reaching perspective. Above all, such facts as these would be the various forms of myth and the influence of mythical thinking on creativity; . . . then we would analyze dramatic art, whose direct artistic elements are the word on one hand and mimetics on the other and, in general, the image of the human body and groups of people. Here the use of the mask as a mimetic devise is interesting, as well as . . . the complete schematicization of the scenery in Greek tragedy [1904–1905 ca., p. 87].

To carry out this positivistic investigation of Nietzsche's questions, however, involves a special attitude to time, that of the physicist who conducts his experiments without reference to a larger historical time and only with reference to the limited time, *t*, of the experiment itself. Mach's positive methods, therefore, are one source of Malinowski's ahistoricism.

In *The Birth of Tragedy,* the work that seemed to Malinowski most susceptible to positive investigation, Nietzsche concentrated on the problem of how Socratic reason was destined to destroy the grounds of the special kind of consciousness on which tragedy was constructed, and with it, to destroy a unique and never-to-be-repeated stage in the growth of European civilization. As is well known, this idea was taken up and repeated with elaboration by many scholars after Nietzsche, including Sigmund Freud, Oswald Spengler, Ernst Cassirer, and Michel Foucault (Thiele, 1990). Each of these, in keeping with the historical and historicizing mode of rigorous scientific discourse of this century, expanded on the periodization and evolution of consciousness. Nietzsche appears to have accepted the large historical framework, else the emergence of a new consciousness such as the one he claimed to possess, or the appearance of the "Superman" (*Übermensch*) would make no sense (Dauer, 1975). Such a "historical" drive toward logical coherence and the narrative sequence of historical writing was less compelling than the logic that drove Nietzsche toward the concept of the "eternal return" or recurrence in cyclical time. Nietzsche had put the value of history itself in question, so it was of little importance whether the history ran from a pre-Socratic, nonrational age of myth, music, and tragedy, to an age, after Socratic philosophy, of reason and rationality. Implicit in Nietzsche's account, of course, is the actual historical sequence of before and after Socrates: of an age before, when tragedies *could be written,* and an age afterwards, when they not only were not writable but would make no compelling sense to an ancient audience if they had been. After Socrates, that is, after reason "set its stamp" on consciousness, the causal narrative of *explanation* privileged both moral and logical reason. This eclipsed the tragic narrative of *fate*, which satisfied the consciousness by virtue of the completion of ordained cycles that revealed little moral purpose, and less reason. In other words, while Socrates favored logical conclusions drawn by force of formal reason, Sophocles favored the narrative closure of events driven by force of human passion, itself the consequence of sexuality and the rules of kinship. But kinship, sexuality, and myth were already the substance of an empirical anthropology by the time Malinowski read Nietzsche.

Nietzsche himself glimpsed the fact that if he could only have been present in Ancient Greece at the time of Sophocles, that it would be possible to determine—to experience and to test—whether his ideas were true. Malinowski was to recognize that some of these ideas could in fact be tested empirically in contemporary societies that still relied, as the Ancient Greeks had done, on public ritual and narrative as their primary means of integration and resolution of political and psychological conflict. Nietzsche's work, of course, ultimately ended in his own madness and silence, and in the inherent contradiction within his own philosophy between the historical time in which "supermen" emerge and the eternal return of the same (Dauer, 1975). Malinowski does not solve these problems, but masters them by an application of a pragmatic presentism.

For Nietzsche, however, it was not until his own later appraisal of this early work published together with a second edition of *The Birth of Tragedy*, that he realized what he had been struggling with. The problem that is declared in the new "self-critique" of the work published in 1878, six years after its original publication, is not the relation between reason and unreason, or between the birth of tragedy in the spirit of amoral Greek myth cycles, but is instead the nature of scientific judgment versus the nature of artistic or aesthetic judgment. In other words, science sought the *cause* of an event while artistic judgment sought the *value* of the pattern of events. He asks whether the world can be judged scientifically, that is, whether there are really any rational grounds at all for assessing the *value of life*. Notoriously, Nietzsche decided that there were none. The value and pattern of life could not be assessed, or judged, on scientific grounds, but only evaluated and analyzed. Life as a whole and as value, Nietzsche believed, was forever closed to the scientific gaze. It could only be judged *aesthetically*. Thus, Nietzsche declared, six years after having originally written *The Birth of Tragedy,* that the real problem that he had sought to address in the work was whether life could be judged scientifically or aesthetically.[6] The "gravest question of all," Nietzsche wrote in the second preface to *The Birth of Tragedy,* was "What, seen in the perspective of *life*, is the significance of morality?" (1878). For the most part, those who have read Nietzsche in the 20th century have taken this dictum to mean that human history has no reason, no overall causal pattern or teleological impulse. To judge life aesthetically, to say that life can *only* be judged aesthetically, has been taken to mean that it cannot be explained but can only be understood as more or less "good" or "evil," and then *only from a particular perspective.*[7] Moreover, the categories of "good" and "evil" were not moral absolutes but were determined by whoever held mastery or power: all moral values were relative! This view has been taken to be the very origin of the blight of moral relativism, of cultural relativism, and moral paralysis in the face of objective evil, and, in sum, the cause of the moral decline of the West. Above all, it introduced the seeming impossibility of an atemporal, ahistorical approach to history itself. In *Thus Spoke Zarathustra* (1883–1891) Nietzsche introduced the idea of eternal return, the cyclical recursion of time back on (and into) itself. This concept provided a suprahistorical concept which was nevertheless temporal. Following his teacher Schopenhauer's forays into Indian philosophy, Nietzsche partially introduced the Indian concepts of rebirth and eternal return of the spirit to this world, determined by *karma*, into the Western historical narrative governed by the metaphysics of progress and "history."

There is, I think, another interpretation of this that Nietzsche himself was perhaps only partly aware of, but which Malinowski made manifest when he applied Nietzsche's insights to the problems of anthropology in the second decade of the 20th century. Malinowski reasoned that there is indeed a fundamental difference between the explanations that a causal narrative offers and the appreciation that an aesthetic assessment offers. That difference, however, is fundamentally a difference between the role that time is made to play in each. The causal account demands origins and causal sequences over a relatively long period of time in order to be entirely convincing (especially after Darwin and Charles Lyell had extended the time frame of biological history). The aesthetic account relies on a truncated "moment" of time in which meaning happens, communication occurs, a gift is exchanged, and during which consciousness grasps a concept, or social groups round

---

[6]Kaufmann (1978) notes that Nietzsche, in *The Birth of Tragedy,* considered aesthetic values to be "supra-historical" and independent of history, while history itself could best be understood "as a work of art" (p. 148).

[7]Nietzsche's philosophy of power culminates in this perspectivism: "Every force center and not only man—construes the whole rest of the world from its own point of view . . . " *(Wille zur Mach,* section 636, quoted in Kaufmann (1978, p. 264).

out their perception of an event. If history is causal, then functionalism is aesthetic in the sense that it is, in Nietzsche's terms, suprahistorical. This, then, is the charter for Malinowski's approach to the "primitive."

According to the methods of historical writing, we must construct a history in order to account for a cultural trait, an institution, or a social form. A history is precisely a narrative of social states and events that stand in a sequential relation to each other, and that therefore cause the latter by virtue of the presence of the former. What is distinctively historical, and what Malinowski rejected in the case of primitive societies without texts, is the secondary construction of a causal relationship between events and social states that do stand in a such sequential relationships. Malinowski stands with David Hume and Nietzsche in seeing these causal linkages as themselves a product of cultural convention. In this respect, too, he accepts Ernst Mach's radical empiricism which saw causality as a construct that could not be demonstrated empirically. Depending on the school of history one belongs to, and depending on the kind of force or impulse that the historian imagines to be the fundamental cause of events and of subsequent states of the social order, the historical narrative will select certain kinds of events and certain kinds of facts as relevant. The assignment of cause is thus conventional and itself part of historical process. These procedures of historical explanation also depends on there being some genuinely historical facts, that is, some genuinely observable traces of the past that have survived into the present but which remain uncontaminated by the present. For the historian, these facts are provided by the archives, and the guarantee of noncontamination is provided by the fact that texts themselves are independent of the cultures and consciousness in terms of which they are subsequently read. Neither the data itself, nor the guarantee of noncontamination by the present exist for the nonliterate, nontextmaking society. Therefore history, as such, is impossible for these societies. The attempt to construct "histories" of marriage, of totemism, of patriarchy, religion, and other institutions for societies that lacked texts was therefore fatally flawed, according to Malinowski. Scholars like Frazer could validly compile fact after fact of primitive beliefs and customs, but the attempt to place these in historical, temporal, or evolutionary sequence was merely "speculative," and therefore to be despised.

Before Malinowski's theoretical intervention, American and European ethnologists constructed a set of hypothetical social states, and sought to describe societies-that-might-have-been, and attributed causal links between hypothesized states of society. Such hypothetical histories maintained the rules of reason in history, and thus seemed to possess the compelling rhetorical claim to being adequate explanations. But the empirical status of the data itself was neglected. Malinowski recognized that the data of these "histories" of mankind were not data at all since it was not genuinely and directly from the past, but rather of the present, and bearing the full stamp of the present in the dogmas and ideologies which informed it. This was true of Sir James Frazer, Adolf Bachofen, Louis Henry Morgan, Henry Sumner Maine, and all of the other conjectural prehistorians of mankind in the 19th and early 20th centuries.

Malinowski, in effect, discovered another way that did not rely on the seemingly compelling rhetoric and methods of the historian. He sought to evaluate illiterate society aesthetically as Nietzsche had done for the Greeks. Unlike Nietzsche, however, Malinowski stepped well beyond the texts that imprisoned Nietzsche, and, guided by the positivism of Ernst Mach, created his own empirical data in the moment of his direct experience of Trobriand society. That was the basis for an ahistorical method, but he needed, as well, to show why the historical accounts that had been offered before were inadequate. He needed to do this in order to clear the space for his own theoretical intervention. He also needed to do it for the same rhetorical reasons that had led Nietzsche to pose the scientific against the artistic view of life. To accomplish the destruction of the former historical accounts, all he needed to do was to expose the facts on which they relied as bogus. He did this easily, following Ernst Mach's methods, by showing that they did not stand up to positive testing. They could not be replicated or demonstrated in the field. To fulfill the later requirement—the same that lies at the root of Nietzsche's construction of the argument of *The Birth of Tragedy*—he needed to develop a different understanding of the nature of understanding itself. The causal account of history could not serve in a rhetoric of description that deliberately neglects the flow of historical time and offers instead an account of the networks of social relations and chains of significance that function in the observable present. For this, an idea of time that is more like a picture than an arrow is required; in other words, a genealogy rather than a history, and an ethnographic description rather than an explanatory story. This "picture" of primitive society was provided, for Malinowski, by the concepts of reciprocities, mutual rights, and obligation, and above all by "function." It made sense of a chaotic but repetitive reality of village life but it said nothing about its value or its place in history. For Nietzsche, the sense of aesthetic coherence was provided by the concept of fate and the eternal return. Both concepts permit, even require, an aesthetic judgment in place of the causal account. Thus, Malinowski's functionalism and his ahistoricism, is a metaphysics of the now, and fieldwork is the scientific method for the investigation and description of the human present. It implies a time that is relative to human purposes, values, relationships, and will, but which rejected the hypothetical world-historical time.

Malinowski grew up and received his early and most important intellectual training in the intellectual context of Central Europe at a time when the arts and sciences were flourishing there. The attitude toward time, life, and society was expressed differently in the context of the universities and coffeehouses of the Austro-Hungarian and Russian empires and in eastern Germany than it was in Britain, to which Malinowski went. The fundamental concepts of perspective, relativity, equilibration, and recurrence-recursion distinguished this part of a European "eastern" tradition from that of the British tradition with its notions of individualism, epistemological absolutism, progress, and evolution. This contrast is what made Malinowski's intellectual intervention so powerful. By combining the methods of Mach's positivism and relativism with the insights of Nietzsche's aestheticism and perspectives, Malinowski became the Zarathustra of the London School of Economics.

# References

Ariotti, P. E. (1975), The concept of time in western antiquity. In: *The Study of Time*, Vol. 2, ed. J. T. Fraser. Berlin: Springer, pp. 69–80.

Cairns, H., & Hamilton, E., Eds. (1961), *The Collected Dialogues of Plato*. Princeton, NJ: Princeton University Press.

Dauer, D. W. (1975), Nietzsche and the concept of time. In: *The Study of Time*, Vol. 2, ed. J. T. Fraser. Berlin: Springer, pp. 81–97.

Dewey, J. (1925), *Experience and Nature*. The Paul Carus Lecture. Chicago: Open Court.

Fabian, J. (1983), *Time and the Other: How Anthropology Makes Its Object*. New York: Columbia University Press.

Firth, R. (1957), *Man and Culture: An Evaluation of the Work of Malinowski*. London: Routledge & Kegan Paul.

Foucault, M. (1975), *Discipline and Punish: The Birth of the Prison*, tr. A. Sheridan. London: Penguin.

Geertz, C. (1988), I-witnessing: Malinowski's children. In: *Works and Lives: The Anthropologist as Author*. Stanford, CA: Stanford University Press.

Gellner, E. (1958), On Malinowski. *Universities Quart.*, 13:86–92. In: *The Concept of Kinship*, ed. E. Gellner. Oxford: Basil Blackwell, 1973.

———— (1973), *The Concept of Kinship*. Oxford: Basil Blackwell.

———— (1985a), Malinowski and the dialectic of past and present. *Times Lit. Suppl.*, June 7:288.

———— (1985b), Malinowski goes home. *Anthropol. Today*, 1:5–7.

———— (1987a), Original sin: The legacy of Bronislaw Malinowski and the future of anthropology. *Times Ed. Suppl.*, 727:13.

———— (1987b), Zeno of Cracow, or revolution at Nemi, or the Polish revenge. In: *Culture, Identity and Politics*. Cambridge, U.K.: Cambridge University Press.

Kaufmann, W. (1978), *Nietzsche: Philosopher, Psychologist, Anti-christ*, 4th ed. Princeton, NJ: Princeton University Press.

Mach, E. (1872), *History and Root of the Principle of the Conservation of Energy*, tr. T. J. McCormack. Chicago: Open Court, 1911.

Malinowski, B. (1904–1905 ca.), Observations on Friedrich Nietzsche's *Birth of Tragedy*. In: *The Early Writings of Bronislaw Malinowski*, ed. R. Thornton & P. Skalnik. Cambridge, U.K.: Cambridge University Press, 1993.

———— (1922), *Argonauts of the Western Pacific: An Account of Native Enterprise and Adventure in the Archipelagoes of Melanesian New Guinea*. London: Routledge.

———— (1923), The problem of meaning in primitive languages. In: *The Meaning of Meaning*, ed. C. K. Ogden & I. A. Richards. London: Kegan Paul.

———— (1926a), *Myth in Primitive Psychology*. London: Kegan Paul.

———— (1926b), *Crime and Custom in Savage Society*. Paterson, NJ: Littlefield, Adams & Co.

———— (1935), An ethnographic theory of language. In: *Coral Gardens and their Magic*, Vol. 2. London: Allen & Unwin, pp. 4–74.

———— (1944), *A Scientific Theory of Culture and Other Essays*. Chapel Hill: University of North Carolina Press.

———— (1957), *Man and Culture*. London: Routledge & Kegan Paul.

Marganero, M. (1990), *Modernist Anthropologies*. Princeton, NJ: Princeton University Press.

Merquior, J. G. (1985), *Foucault*. London: Fontana.

Nietzsche, F. (1872), The Birth of Tragedy. In: *Basic Writings of Nietzsche*, ed. & tr. W. Kaufmann. New York: Random House, 1968, pp. 3–146.

———— (1878), Attempt at self-criticism. In: *Basic Writings of Nietzsche*, ed. & tr. W. Kaufmann. New York: Random House, 1968.

———— (1883–1891), Thus Spoke Zarathustra. In: *The Portable Nietzsche*, ed. & tr. W. Kaufmann. New York: Viking Press, 1954.

———— (1887), On the Genealogy of Morals. In: *Basic Writings of Nietzsche*, ed. & tr. W. Kaufmann. New York: Random House, 1968.

Schopenhauer, A. (1818), *The World as Will and Representation,* tr. E. F. J. Payne. New York: Dover, 1969.

Thiele, L. P. (1990), The agony of politics: The Nietzschean roots of Foucault's thought. *Amer. Pol. Sci. Rev.,* 84:907–925.

Thornton, R. (1980), Evolution, salvation and history: The early ethnography of southern Africa, 1860–1920. *Soc. Dynam.,* 6:14–23.

────── (1988), The rhetoric of ethnographic holism. *Cult. Anthropol.,* 3:288, 297–298.

────── (1992), Chains of reciprocity: The impact of Nietzsche's *Genealogy of Morals* on Malinowski's *Crime and Custom in Savage Society. Polish Sociolog. Bull.,* 1:19–33.

────── Skalnik, P., Eds. (1993), *The Early Writings of Bronislaw Malinowski: 1904–1914.* Cambridge, U.K.: Cambridge University Press.

Torgovnik, M. (1991), *Gone Primitive.* Durham, NC: Duke University Press.

# 21

# Pragmatism, Life, and the Politics of Time

*John J. Stuhr*

*Abstract*   In this paper, I develop (1) the outlines of a pragmatic account of time and (2) the political implications of this pragmatic view. Any pragmatic account of time must be set forth in the context of a comprehensive pragmatic metaphysics. I sketch this experiential metaphysics, drawing on the work of philosopher John Dewey, and show how pragmatism undercuts traditional philosophical problems and categories in a manner that allows us to take time seriously. Here I argue that pragmatism provides not only a temporal theory of experience but also an experiential theory of time, and I develop the outlines of this theory through a discussion of the irreducible temporality of experience. On this view, as I explain, experience is eventful, continuous, historical, and temporally qualitative (such that there is no time like the present). In this light, I turn to the political implications of a pragmatic account of time, showing that pragmatism makes possible a new temporal understanding of the nature and value of individuality. At the same time, this theory links individuality to community where community is understood temporally as a fully democratic way of life. I conclude by arguing that just as an experiential account of time leads to understanding individuality and community as temporal, so too this pragmatic social theory leads to commitment to democratic social action (for which, again, there is no time like the present).

## Time in Theory and Time in Life

This essay begins with a commonplace: *There is no time like the present.* Of course, to raise issues concerning the nature of the present time with readers of this volume of *The Study of Time,* members of the International Society for the Study of Time, or other researchers and students of time, is to preach to the already converted. This is not to suggest, obviously, that there are not deep and live disagreements among scholars of time. Instead, it is merely to note that such scholars share certain overlapping interests in, and inquiries into, the nature of time. Put simply, they take the study or theory of time seriously.

Of course, theories of time—time in studies such as this—and practices of time—time in life—frequently seem, and actually may be, quite different, separate, and even unrelated. Moreover, these gaps between the theory and practice of time exist in theory in different ways than they do in practice. Awareness of these differences often leads to general skepticism about the pragmatic value of studies of time. What practical, real life difference do they make? What, if anything, is their "cash-value," to use the phrase of American philosopher William James?

Let me briefly provide a concrete illustration of this skepticism. I mentioned to my neighbor, a physician, that I was planning to travel to a conference in France on time. A bit perplexed, she asked me if I usually planned to arrive at international conferences late. I explained that the conference was on the *topic* of time. At this, she became a bit more perplexed and suggested that I might as well arrive late: "They can't tell you anything that will make a difference. If you don't already know about time, no conference is going to help. If you already do know, there is no need for a conference. I bet the papers will vibrate between the abstract and the trivial. And when the learned voices quiet, people will still hope and remember, prosper and suffer, change and die—already aware of time in ways no conference can produce." She concluded, "And I thought you were a pragmatist!"

I have a lot of sympathy for this sort of skepticism about theory and theorists—at least in the humanities and human sciences. Still, I do not think that it is a contradiction in terms to take time pragmatically. To do so must involve two moments or tasks. First, the skeptic about the pragmatic value of the study of time does get one thing right: The temporal quality of experience is ontologically pervasive and primary; moreover, it is ineffable. (This last point has been understood much better by novelists and poets than by most philosophers.) In this light, to take time pragmatically in theory requires the critical articulation of a genuinely temporalist theory of experience—a temporalist metaphysics.

Second, the skeptic about the pragmatic value of the study of time also gets one thing wrong: An honest, unflinching awareness of time as a generic, pervasive trait of experience has far-reaching practical consequences for individual, community, and social action. (This point has been understood much better by pragmatic and phenomenological philosophers than by social scientists.) In this context, to take time pragmatically is to set forth a politics of time that outlines intelligent ends and the means for their ongoing realization in a manner consonant with the irreducible temporality of experience. In this essay, I will address briefly these two issues.

## A Pragmatic Metaphysics of Life and Time

At present, there is a need not so much to invent from scratch but to recover and reappropriate a pragmatic theory of time. The main features of such a theory are set forth richly in the classical American philosophical tradition: in the works of Charles Peirce, William James, Josiah Royce, and George Santayana; in George Herbert Mead's *The Philosophy of the Present* (1932), *The Philosophy of the Act*

(1938), and other writings; and, above all, in the philosophy of John Dewey. Dewey called problems of time the most fundamental in philosophy, and throughout his long life addressed these problems in detail in his many books and articles, including: *Experience and Nature* (1925); *Reconstruction in Philosophy* (1920); *The Quest for Certainty* (1929); "Time and Individuality," "Events and the Future"; and "Events and Meanings."

The centerpiece of this work is Dewey's mature theory of experience, and it is this theory that contains his clearest and most comprehensive account of time. At the outset, it is crucial to understand that Dewey does not use this word "experience" in any of its traditional scholarly or ordinary popular senses. For Dewey, experience is not to be understood as subjective—as the inner or private states of an experiencing subject. Nor is it to be understood as objective—as the outer or external condition of an object experienced. Moreover, it is not even to be understood as interactive—as a combination of subject and object that exists as such independently of their interaction with one another.

For Dewey, the subject–object dualisms so dear to the hearts of materialists and idealists, realists and antirealists, and empiricists and rationalists, are merely metaphysical fictions. Dewey replaces these metaphysics of fictional subjective and/ or objective substances with an organic, holistic metaphysics of temporal activity or process. From this standpoint, the ontological dualisms of traditional philosophies are merely intellectual distinctions—distinctions made by reflection for assorted practical purposes. These distinctions may have a functional or useful status (and they may not!), but never an ontological status.

If we fail to understand this, we merely make up problems. This creates a gap between, on the one hand, philosophy and the pseudoproblems of philosophers, and on the other hand, life and the real problems of men and women. Thus, in *Knowing and the Known*, Dewey (1989a) warns:

> What has been completely divided in philosophical discourse into man and world, inner and outer, self and not-self, subject and object, individual and social, private and public, etc., are in actuality parties in life trans-actions. The philosophical problem of trying to get them together is artificial. On the basis of fact, it needs to be replaced by consideration of the conditions under which they occur as distinctions, and of the special use served by these distinctions [p. 248].[1]

Instead, taking his lead from Darwin, Heisenberg, Maxwell, Einstein, Peirce, and William James, Dewey views experience in a radically empirical way. This approach is oriented to the evolutionary, transient, uncertain, and open, rather than the supposedly eternal, permanent, secure, and closed. "Experience" is:

> "double-barreled" in that it recognizes in its primary integrity no division between act and material, subject and object, but contains them both in an unanalyzed

---

[1]For a more recent analysis of the ways in which these traditional philosophical dualisms plague social science as well as the humanities and natural sciences, see Adam (1990). Although framed without explicit awareness of Dewey and pragmatism, Adam's treatment of these dualisms is remarkably Deweyan.

totality. "Thing" and "thought," as James says in the same connection, are single-barreled; they refer to products discriminated by reflection out of primary experience [1925, pp. 18–19].

For Dewey, then, experience is a process in which subject and object are temporally unified and constituted as partial features and relations within this ongoing, unanalyzed totality. Experience is an ontologically primary temporal unity: "An organism does not live *in* an environment; it lives by means of an environment.... The processes of living are enacted by the environment as truly as by the organism; for they *are* an integration" (1986a, p. 32).

Dewey constantly struggled to arrive at adequate language to express this point about experience. He thus abandoned the term "interaction" in favor of "transaction," trying to stress that the parties in a transaction are defined by their interrelation and have no independent existence. Late in his life, near 90, he even considered abandoning the term "experience" in favor of the word "culture" to designate the "inclusive subject-matter" that modern philosophy breaks up into dualism.

Of course, this account is not meant as a finished portrait of Dewey's view of experience, but I trust that it is an adequate working sketch (for a fuller treatment, see Stuhr [1979, 1991, 1992]). It is meant to function as a point of entry for the development and extension of a Deweyan, pragmatic account of time. Dewey understood metaphysics as the description of the generic and pervasive features of experience. Given Dewey's general view of experience, then, what are the generic and pervasive features of experience?

For Dewey, *experience is intrinsically and irreducibly temporal.* Equally important, *time is intrinsically and irreducibly experiential.* From the standpoint of pragmatism, to deny either of these two claims is to fail to take time seriously. That is, any notion of time as supposedly *wholly* independent of experience—whether this notion of time is intuitive, mystical, scientistic, or other—is nonsensical and literally meaningless. For all pragmatists about life and time, this means that there are no timeless realities or existents.

In the context of this pragmatic *temporal account of experience* and reciprocal *experiential account of time,* I want to develop five key points. First, *experience is intrinsically and irreducibly social.* Moreover, since time is intrinsically and irreducibly experiential, all *time is social.* Dewey thus argues that the social is *the* inclusive philosophic idea, and that there is no existence ontologically independent of the sociotemporal. There is, for example, no existence which is physical, organic, or mental and not also social as well. This is so because the physical, organic, and mental are not levels of being ontologically separate from experience, but rather features of experience distinguished for various purposes:

> Timeless laws, *taken by themselves,* like all universals, express dialectic intent, not any matter of fact existence.... They are out of time in the sense that a particular temporal quality is irrelevant to them. If anybody feels relieved by calling them eternal, let them be called eternal. But let not "eternal" be then conceived as a

kind of absolute perduring existence or Being. It denotes just what it denotes: irrelevance to existence in its temporal quality. These non-temporal, mathematical or logical qualities are capable of abstraction, and of conversion into relations, into temporal, numerical and spatial order. As such they are dialectical, non-existential [Dewey, 1925, pp. 148–149].

For pragmatists, then, the universe is a sociotemporal universe through and through, and all temporality is qualitatively sociotemporality. Sociotemporality, that is, is not merely one among many levels of time. It is not a container that encloses other levels of time. Experience does not cover or surround nature; rather, it penetrates and permeates nature:

> [E]xperience, if scientific inquiry is justified, is no infinitesimally thin layer or foreground of nature.... [It] reaches down into nature; it has depth. It also has breadth to an indefinitely elastic extent. It stretches. That stretch constitutes inference.... The very existence of science is evidence that experience is such an occurrence that it penetrates into nature and expands without limit through it [Dewey, 1925, p. 4].

At this point, it is worth noting that this pragmatic emphasis on the ontological primacy of temporality as qualitatively social temporality and the accompanying rejection of ontological hierarchies seem to stand in sharp contrast to J. T. Fraser's account of "nowless universes" and his well-known and influential theory of time as a nested hierarchy of unresolvable, creative conflicts (Grünbaum, 1968; Fraser, 1978, 1981, 1987, chap. 3, 4, 1992). For pragmatists, a "nowless universe" is either a useless ontological abstraction or a reflective distinction (with useful, but not ontological, status) made from within our now-filled universe. Either way, a "nowless universe" is no "universe" at all.

Dewey thus argues that the social is *the* inclusive philosophic idea, and that there is no existence ontologically independent of the sociotemporal (though, of course, there was in the past existence before the beginning of social life and there may be in the future existence after the end of social life):

> And, of course, when events-without-meaning are referred to, that very fact brings them within the field of thought and discourse and in so far confers meaning upon them, if only the meaning of being without meaning. One could go further: to refer to anything as an *event* is in so far to ascribe character or nature and hence a meaning to it [1984a, p. 88; see also Dewey, 1984b, pp. 41–54].

> It is a situation of which, by scientific warrant, it always is to be said that it is on its way to the present situation, that is, to "experience," and that this way is its own way. The conditions which antecede experience are, in other words, already in transition towards the state of affairs in which they are experienced. Suppose one keep in mind the fact of *qualitative-transformation-towards* and keep in mind that this fact has the same objective warrant as any other assigned trait (mechanical and chemical characteristics and relations, etc.) [Dewey, 1977, p. 101].

Thus, when Fraser (1987) notes that our experience of time's passage "is a notion that we must bring to physical science as living beings" since "it cannot be extracted from what is known about time in the physical world" (p. 222), Dewey would reply that if the physical world is a world without time's passage, then it is a world without time and, thus, is an abstraction and not an existential "world" at all.

This line of thought has been voiced repeatedly by other pragmatists. Dewey's colleague, George Herbert Mead, for example, in his brilliant but neglected extended analysis of "the social nature of the present," writes that a scientific object is "an abstraction of that within experience which is subject to exact measurement" (1932, p. 141). In a sweeping essay that ranges from Galileo, Kant, and Rousseau to Einstein, Lorentz, and Minkowski, Mead argues that emergent social life marks not an additional level or kind of temporality but a penetration or permeation or *transformation* of temporality: "That is, we recognize that emergent life changes the character of the world just as emergent velocities change the character of masses." Life "extends its influence to the environment about it" (1932, pp. 234–235).

By contrast, the abstract conception of a timeless universe eliminates human life, its influence, and, paradoxically, the phenomenon of relativity now often thought to support such an abstract conception of time: "If at this point there were no time, there could be no temporal perspectives, and, if reality could be located at such a temporal zero point, the experiences of relativity would be just what it was in an instant of no temporal spread" (Mead, 1938, p. 236). This sort of point has been made by more recent philosophers who write in a broadly pragmatic vein. For example, in his "Introduction" to a collection of essays on postmodernism, pragmatism, and process philosophy, David Ray Griffin critically examines Fraser's hierarchical theory of time from the perspective of philosophers such as Peirce, James, and Whitehead, and asserts that we need not and must not speak of a "genesis" of time. Griffin (1993) argues that Fraser's view, as Fraser himself seems to Griffin in some respects to admit, is self-contradictory and commits Whitehead's "fallacy of misplaced concreteness"—misidentifying the products of thought as existents prior to and independent of that thought (pp. 12–14). However, as Mead concludes, the genuinely experimental scientist, apart from unwarranted philosophic bias, "is not a positivist" and has no inclination to build up a universe out of abstract scientific data: "The reference of his data is always to the solution of problems in the world that is there about him, the world that tests the validity of his hypothetical reconstructions" (Mead, 1938, p. 62).

This need for ongoing reconstruction points to a second aspect of a pragmatic theory of time: Experience is *eventful, always changing, active, precarious, and hazardous*. Even that which is stable and fixed is merely stable relative to specifiable changes, and is not absolutely stable or permanent. As G. J. Whitrow (1981) has remarked, "The essence of time is its transitional nature, and no theory of time can be complete that does not account for the fact that everything does not happen at once" (p. 581). Traditional philosophers (and many contemporary thinkers) often deny this in their explicit and implicit quests for certainty—their attempts through theory alone to make life, meaning, and values timeless and safe, eternally prevailing over the instability of actual life. This denial takes the form of converting precarious ideals that may be realized (if realized at all) only through temporal action and without advance guarantee into fixed actualities realized only in metaphysical theories of existence sure and fixed prior to action. In longing for a perfect or safe world, that is, philosophers have claimed ultimate reality for their values, and have turned the goals of experience into the antecedently existing causal conditions of that experience itself. Despite this, experience—the unity of experiencing subject and experienced object—is undeniably precarious, changing, unsettled. The stablest thing we can speak of is not free from the many conditions set to it by other things. It is subject to continual tests imposed upon it by surroundings which are only in part compatible and reinforcing. As Dewey (1925) says succinctly, "Every existence is an event" and objects are complexes of events: "Nothing but unfamiliarity stands in the way of thinking of both mind and matter as different characters of natural events in which matter expresses their sequential order and mind the order of meanings in their logical connections and dependencies" (p. 66).

Third, *experience is continuous.* Experience is not only eventful; it undergoes change itself. That is, experience is not simply a succession or seamless sequence of events. Rather, these events themselves have temporal connections and relations—connection directly experienced and immediately real. Experience is connected, conjoining and disjoining, resisting and yielding, modifying and being influenced, organized and confused, planned and surprising. Connections, changes, continuities, relations, changing-in-the-direction-of, and being-in-transition-toward are pervasive experienced features of experience. Events interact and transform one another. Each event itself not only passes, but undergoes change and altering. Whatever influences the changes of other things or features of experience is itself changed. Variation or continuous change is a feature of each event. For example, Dewey notes that an indispensable character of anything which may be termed an event is "a qualitative variation of parts with respect to the whole which requires duration to display itself." Every event is a passing into and out of other things in such a way that a later occurrence is an integral part of the character or nature of present experience—of the so-called "specious present" (Shereover, 1981, pp. 166–146; Dewey, 1986b, pp. 62, 66). Every present is a presence of a past no longer present and a future not yet present. This is what Charles Shereover (1975) aptly has called the "spread" of time that "takes my perspective of future and selected recall of the relevant past into constituting what I take to be the present situation" (p. 225). Temporal continuity, like precariousness, then, is an experienced fact. It also constitutes a postulate or methodological starting point from the perspective of which experienced facts are facts. In an important sense, that is, the temporal continuity of experience cannot be denied. All efforts at denial must always link that which is being denied (continuity) with the rest of experience, with the experience of naming and understanding in the denying itself. Each denial of continuity, then, only serves to establish additional connections and continuities. Thus, any theory that sets up a breach of continuity, a complete rupture, is not so much false as nonsensical, absurd, and meaningless (Dewey, 1988a, pp. 141–154).

Experience, however, is not merely changing and continuous. In the fourth place, *experience is historical.* Experience is not, that is, an undifferentiated flow or mere happening of events. Rather, experience itself consists of experiences, affairs with beginnings and endings, initiations, and consummations. Life is an affair of affairs, a "thing of histories, each with its own plot, its own inception and movement toward close, each having its own particular rhythmic movement" (Dewey, 1987a, pp. 42–43). These existential ends of experiences, of course, are not final or moral ends *for* experience (as traditional teleological philosophies and religions hold), but only ends-in-view (and means to them) within experience. Nor do they amount to fate (as mechanists hold), but only orders within historical events—orders that we potentially can control so as to achieve particular ends and deepen the significance of our experience. Temporal order or succession is one such example. Importantly, this view of experience as overlapping temporal histories permits both humanists and scientists together to "apprehend causal mechanisms and temporal finalities as phases of the same natural processes, instead of as competitors where the gain of one is the loss of the other" (Dewey, 1925, p. 83).

Fifth, this natural process, *experience, is qualitatively temporal.* Within and shot through each experience there is a self-sufficient, wholly immediate, individual quality that renders an experience that particular experience—that meal, that snowstorm, that philosophy lecture, and so on. This qualitative aspect of experience is not an object of knowledge. It is not *known* but *had.* It is the brute "is-ness" of experience. Indeed, much to the dismay of traditional philosophers who conflate knowledge about life with life itself, Dewey says that this qualitative dimension of experience cannot be known or communicated since the objects of knowledge are mediate rather than immediate; they concern the relations among experiences and not the immediate quality of experience. If existence in its immediacy could speak, Dewey (1925) says, "it would proclaim 'I may have relatives but I am not related' " (p. 75). This qualitative character of experience is not something subjective, for experience is not something subjective: Qualities belong as much to the thing experienced as to the experiencing subject. So, a particular experience as it occurs simply is; that particular experience, as it is connected by later reflection to things beyond itself, becomes a sign and refers. Dewey thus distinguishes carefully temporal quality from temporal order, and rightly holds temporal quality to be ontologically more fundamental—an immediate, ineffable feature of all experience. Temporal order, on the other hand, is not immediate. It is the result of reflection, a matter of science, a consequence of inquiry into the conditions upon which occurrence of particular temporal qualities depend. But temporal order does not explain or explain away temporal quality. The temporal quality and presentness of experience cannot be reduced to temporal order. This qualitative character of experience, as all neighborhood skeptics know, whether or not they've ever studied pragmatism or any other philosophy, cannot be captured fully in later reflection on experience (no matter how "phenomenological" the reflection):

> Things in their immediacy are unknown and unknowable, not because they are remote or behind some impenetrable veil of sensation or ideas, but because knowledge is a memorandum of conditions of their appearance, concerned that is with

sequences, coexistences, relations. Immediate things may be pointed to by words but not described or defined [Dewey, 1925, pp. 74–75].[2]

Thus, as this outline of an experiential view of time and a temporal view of experience indicates, in practice as in pragmatic theory, there really is no time like the present. This, Dewey notes in "Time and Individuality" (1988b), an essay originally presented as part of a public lecture series on "Time and Its Mysteries," is "genuine time" (pp. 98–114).

## A Pragmatic Politics of Time

So what? Does any of this matter? What are the practical bearings of a pragmatic account of time?

I want to consider two sorts of practical consequences of a temporal account of experience and an experiential account of time. These consequences concern (1) individuality and (2) community (and their interrelationship in and through action).

First, to take time pragmatically is to grasp the intrinsically and irreducibly temporal character of individuality. As Dewey (1988b) writes: "The unescapable conclusion is that . . . human individuality can be understood only in terms of time as fundamental reality" (p. 107). Any adequate account of individuality must be historicist. As Charles Sherover (1989) puts it:

> As social beings, any individual thinking thus necessarily incorporates particular-ized time predicates, as particular differentiating descriptions of the historically developed social milieu out of which whatever individuality we each manifest has arisen. Because each individuality is idiosyncratically reflective of the nourishing culture, it is historical, thereby temporal, from the outset [p. 284].[3]

This means that individuality is precarious and never static or secure. Individuality is fragile and delicate. When fortunate conditions and appropriate actions support its emergence and operation, individuality arises as an achievement—and not as something "ready-made," given, provided once and for all at birth, secured by mere physical separation and difference from all others.

Like most political philosophers and theorists, we often take individuality for granted and fail to see that not only do human beings "go out of existence" (Fraser, 1981, p. 589), but that while they exist they can and in fact often do fail to be genuine individuals. This is not because individuals somehow at times fail to *be*

---

[2]For a brief discussion of Dewey's distinction between temporal quality and temporal order in the context of several major American philosophers' views of time, see Helm (1985). For a fuller discussion of temporal quality and experiential time in this same context of American philosophers, see Sherover (1975).

[3]Dewey, unlike Sherover, differentiates between the self and the individual as a self who achieves individuality.

themselves, as though individuality were a matter of permanently embodying an original personal essence. Instead, it is because individuals often fail to *become* themselves, because individuality is a matter of continuity, and the ongoing development and realization of possibilities and capacities that are not self-identical or unchanging over time.

The identity of the individual, long sought after by philosophers in theory, exists in life only in time. Put this way, it should be clear that an individual does not *have* a history; rather, an individual *is* a history. As Dewey points out, time does not merely surround an individual externally:

> The human individual is himself a history, a career, and for this reason his biography can be related only as a temporal event. That which comes later explains the earlier quite as truly as the earlier explains the later. . . . Temporal seriality is the very essence, then, of the human individual [Dewey, 1988b, p. 102].

The continuing development of an individual's possibilities and capacities, then, can take place only if there exist in time the social conditions necessary to support this development.

When this development is thought to be the natural unfolding of some innate, natural end that controls development toward it, then the sociopolitical agenda is largely defined and constituted by a negative task. This task is that of removing restrictions on, and getting out of the way of, the individual. It is the task of making sure that society, and especially big government, does not impede the natural development of the individual.

This once liberal but now reactionary position rightly prizes the individual, but lacks historical and temporal sense. As a result, this view wrongly identifies the necessary conditions of the individual's development. When it is recognized that the capacities and potentials of individuals are themselves intrinsically temporal matters, then a different, less traditional and more radical political agenda arises. This agenda is more positive than negative in character; it is not a matter of getting out of the way of the individual's supposedly natural development or teleology, but of creating and recreating the conditions needed for the development and realization of individual potentials. Moreover, this agenda is more fully social than merely political in scope; it is not simply a matter of sustaining a democratic form of politics or government, but of fostering a thoroughly democratic way of community or social life.

Dewey correctly draws "the only reasonable conclusion":

> [P]otentialities are not fixed and intrinsic, but are a matter of an indefinite range of interactions in which an individual may engage. . . . There are at a given time unactualized potentialities in an individual because and in so far as there are in existence other things with which it has not as yet interacted [Dewey, 1988b, p. 109].

These "other things" are the social environment of every individual, and this social environment (for better or worse) is the indispensable condition of the formation and realization of the individual. The pragmatic point, then, is not simply that the individual inhabits a social environment (and is related to that environment as a part to a whole). Rather, the point is that the individual, internally and intrinsically, is social (and is related to the environment as reciprocally and mutually constituting).

This means that individuality, understood temporally, requires freedom—where freedom is understood not simply as the absence of restriction but as the presence of those resources necessary for the temporal development of the self. It also means that individuality requires intelligence in the direction of this freedom—where intelligence is understood as the temporal (and thus fallible) application of methods of critical, experimental inquiry to human life. But above all, it means that individuality requires social arrangements such that the intelligent free action of one person contributes to the development of the individuality of others. This is the requirement of community. In short, a pragmatic theory of time makes possible a temporal account of individuality that in turn requires a social theory committed to the free and intelligent development of genuine communities.

This pragmatic social theory that defines individuality in terms of community stands in a reciprocal relation to a pragmatic theory of time that defines temporality as sociotemporality. That is, a pragmatic social theory and a pragmatic theory of time mutually imply one another. From the standpoint of this social theory, then, and against all reductionist, abstractionist, and dualistic views of time, it is clear that *all* temporality is sociotemporality, qualitative differences among experiences notwithstanding. Put more simply, to the extent that human societies fail to be genuine communities, human beings fail to be genuine individuals. The realization in life of individuality and the realization of community are one. On this issue, as well as others, the differences between pragmatists (such as Dewey) and the influential 20th-century existential phenomenologists (such as Martin Heidegger) could not be greater or more obvious. Heidegger fails to grasp the intrinsic connection between individuality and community, and so identifies authentic individuality with the very social conditions that Dewey thinks constitute the loss of individuality and the absence of a democratic way of life. As Hubert Dreyfus (1975) has aptly put it, individuality emerges for Heidegger only when one member of society "experiences an anxiety attack" (p. 158).

Now, this notion of community lies at the heart of the pragmatic link between individuality and democracy—where democracy is understood primarily and most fully as a way of life rather than only a form of government. As Dewey argued, democratic government is only a part of a democratic way of life (I develop this pragmatic point in Stuhr [1993]). It is a means for realizing democratic ends in individual lives and social relationships. And, while it is the best and most effective means yet devised to achieve these ends, still it is only a means. Thus, Dewey warned that we must not treat temporal means, that is, institutions and practices of democratic government such as majority rule, broad suffrage, and periodic elections, as ends, final or complete in themselves. These practices, the political dimension of democracy, are "external and very largely mechanical symbols and expressions of a fully democratic way of life" (Dewey, 1987b, p. 218; 1988c, p. 295).

Because democratic structures and institutions of government have a value that is temporal rather than eternal, they must be dynamic rather than static if they are to continue to nurture a democratic way of life under changing conditions and through changing times. This means that democracy itself must be understood temporally: It is not something fixed, "a kind of lump sum that we could live off and upon," something finished and final that simply can be "handed on from one person or generation to another," or something so natural that it simply forever maintains itself once established (Dewey, 1988c, pp. 298–299; 1989b, p. 259). Accordingly, we can't afford passively to idolize practice and institutions that proved instrumental in the past. We constantly must appraise and be ready to revise them when necessary in terms of their present and projected future contributions to a democratic way of life.

The basis of this democratic way of life is the conviction that no person or group of persons is sufficiently wise and good to govern others without their consent, "without some expression on their own part of their own needs, their own desires and their own conception of how social affairs should go on and social problems should be handled" (Dewey, 1988d, p. 295). This conviction, in turn, requires both equality (such that social decisions are cooperative expressions) and opportunity (such that all people have both a right and a duty to participate in this decision-making process). This means that all persons involved in and affected by social practices should participate in their formation and direction. Dewey (1987b) viewed this as the generic meaning of democracy and termed it the "key-note of democracy as a way of life, . . . necessary from the standpoint of both the general social welfare and the full development of human beings as individuals" (pp. 217–218). As Dewey said succinctly and compellingly (1984c):

> From the standpoint of the individual, it [the idea of democracy] consists in having a responsible share according to capacity in forming and directing the activities of the groups to which one belongs and in participating according to need in the values which the groups sustain. From the standpoint of the groups, it demands a liberation of the potentialities of members of a group in harmony with the interests and goods which are common. . . . Regarded as an idea, democracy is not an alternative to other principles of associated life. It is the idea of community life itself. . . . Wherever there is conjoint activity whose consequences are appreciated as good by all singular persons who take part in it, and where the realization of the good is such as to effect an energetic desire and effort to sustain it in being just because it is a good shared by all, there is in so far a community. The clear consciousness of a communal life, in all its implications, constitutes the idea of democracy [pp. 327–328].

Both this idea of democracy as temporal community and the consequences of attempting to implement this idea are revolutionary. Thus Dewey wrote (1987c):

> *The fundamental principle of democracy is that the ends of freedom and individuality for all can be attained only by means that accord with those ends. . . . The end of democracy is a radical end. For it is an end that has not been adequately*

*realized in any country at any time.* It is radical because it requires a great change in existing social institutions, economic, legal and cultural. A democratic liberalism that does not recognize these things in thought and action is not awake to its own meaning and to what that meaning demands [pp. 298–299].[4]

In this context, then, to the extent to which a given person does not participate consistently and fully in the consideration, formation, and implementation of social values, decisions, and policies, democracy as an individual's self-determining and self-realizing way of life simply does not exist—in actual life. And, to the extent that given social practices, groups, and institutions do not foster shared interests, harmonious differences, and individual growth, democracy as a free community's way of life does not really exist—again, in life.

Just as an experiential account of time and a temporal account of experience—a pragmatic metaphysics—leads directly to an understanding of individuality and community as irreducibly temporal—a pragmatic social theory—so too this pragmatic social theory leads directly to a commitment to social action. To take time seriously—in life and not just in theory—is to be committed to social action that strives to secure the changing conditions necessary for the realization of genuine individuality and community.

This social *action*, in turn, requires or presupposes a democratic *faith*. It requires that we make democracy as a way of life an *ideal*. Now, to say that a democratic way of life is an ideal is not to say that it is something unreal. As real ideals, individuality and community would be deep commitments, grasped by our imagination, that unify our lives, make meaningful our efforts, and direct our actions. As ideals, they are generated through imagination, but are not "made out of imaginary stuff." Instead, anything but "unreal" or "imaginary," ideals are "made out of the hard stuff of the world of physical and social experience—the material and energies and capacities that are the conditions for its existence" (Dewey, 1986c, pp. 33–34; see also Dewey, 1988d, p. 174).

To describe individuality and community as ideals, however, is not so much to state a present fact as it is to recommend a future course of action—an admittedly radical course of action. To quietly favor, idly wish and hope for, or routinely assent to such a democratic course of action, however, is not thereby to make it an ideal. Today, we still must make individuality and community real ideals. We have not done so yet. Thus, idealizing individuality and community is the first step in the task of realizing individuality and community. When, and if, our idealizing imagination does seize upon individuality and community, personal life will express this ideal in action. In Dewey's terms, individuality, freedom, intelligence, communication, cooperation, and community will become loyalties or values-in-action, instead

---

[4]It is this radical insistence on understanding democracy in terms of temporal, *participatory community* that separates Dewey's social philosophy from more conservative, "republican" political theories (usually grounded not in pragmatic accounts of experience but in idealist and phenomenological accounts of the subject). Charles Sherover's (1989, 1993) excellent work articulates this "republican" position, rejected (rightly, I think) by Dewey.

of mere values-in-name only.[5] As ideals, individuality and community require this committed expression and action from each of us.

It is neither possible nor desirable to detail in advance or in the abstract just what form this action should take in the future. As Ralph Waldo Emerson noted, each age must write its own books; as Dewey echoed this sentiment, each age knows its own ills. Still, it is possible to indicate the general outline and spirit of this action and, at the same time, to indicate briefly the special social value at this time of a pragmatic politics of time. First, pragmatism constitutes a middle path between, and an alternative to, both the pervasive pessimism and the rampant optimism of our day. In the face of complex, interdependent, and massive contemporary social problems—injustice, poverty, disease, violence, conflict, destruction, and death—many people readily feel overwhelmed and powerless. Our sense of impotence makes for pessimism—things are bad and they will continue to be bad because we can't do anything about them. This, in turn, often leads to self-confirming apathy—''since we can't do anything about it, why bother?''—, self-defeating withdrawal from the community—''the best we can do is look out for ourselves and keep the barbarians at the gate''—, and self-destructive nihilism—''it doesn't matter what we do, nothing is better than anything else, at least we won't be fooled again.'' By contrast, many other people embrace absolutism in their response to these same large social problems. They are convinced that they have the Truth, that they have seen Reality, that their cause is Just, that God has chosen them. Supremely confident, they are optimists, convinced that time is on their side.

Pragmatism provides a political alternative to both pessimism and optimism. The alternative is meliorism. Meliorism is the spirit of the politics of time: It rejects the belief that things will be bad (pessimism) and the belief that things will be good (optimism). Instead, it asserts that however bad or good things now may be, intelligent action can make them better. As a political theory, pragmatism offers no guarantees about the future. It provides only the demand to act in the present—on behalf of the future and in light of the past. The need for this intelligent social inquiry and action has never been greater.

Second, pragmatism also constitutes a middle path between, and an alternative to, both short-term thinking and the view from eternity. For many people today, the only way of thinking is short-term thinking, and the only future worth thinking about is the immediate future. We seem to have concluded that we can avoid unwanted long-term consequences simply by refusing to think about them in the present. Unfortunately, our social policies—from the environment to the economy, from education to health care, and from urban planning to foreign affairs—demonstrate that this is not the case. Short-term thinking now fails to ameliorate our present ills. Moreover, it creates complex new, additional difficulties. These long-term difficulties are precisely the sort that cannot be resolved by the short-term thinking that created them. Nor can these long-term problems be resolved by our

---

[5]I take this pragmatic point about individuality and ethics to parallel Nathaniel Lawrence's remark about time and metaphysics: ''Time is as time does'' (1979, p. 11).

public leaders who recognize the paradox that in order to be long-term office holders, they must be short-term thinkers. By contrast, many other people, concerned to reject short-term thinking and its future-free present, rush to embrace the view from eternity and its present-free future. For persons who adopt the view from eternity, present problems, satisfactions, and desires don't matter much. Like life itself, they will pass. In the meantime, we must be patient—"that's just the way it is"—, anesthetized—"its only pain"—, or other-worldly (for surely eternity would be another world)—"deliver us from evil." The prophets of the view from eternity are familiar: Middle Eastern leaders who call on young men to fight and die in a "holy war"; charismatic cult leaders who counsel mass suicide; battered women who feel they simply should bear their burden; and, ordinary citizens who daily recognize and ignore suffering—embodying an easy stoicism in the face of other people's pain.

Pragmatism provides a political alternative to short-term thinking that sacrifices the future in the name of the present, and the view from eternity that justifies looking past the present in the name of the future. Pragmatist politics is not a politics of the present; nor is it a politics of the future. From the standpoint of pragmatic ethics, neither present nor future is a value or end-in-itself. Instead, pragmatism provides a politics of time, a politics of temporal continuity. Present and future both are aspects of a temporally integrated means–end continuum of value. For pragmatic politics, that is, present satisfaction is not simply an end-in-itself. No rational valuation of the present can be made if the present is considered in isolation from the future to which it is a means. In the same way, future satisfaction or gratification is not simply an end-in-itself. No rational valuation of the future can be made if the future is considered in isolation from the present means to it. Pragmatist politics is decidedly this-worldly, but it insists that this world is a temporal world. Any community in the present must be at once a community of memory and a community of hope—in the present.

Of course, this philosophy, as pragmatists stress and skeptical neighbors feel, *cannot be justified* (or, for that matter, falsified) *by anything written in any essay in any book or by anything said in any paper at any conference*. Its justification, if any, is a function of the results of the actions it may call forth. Today, this philosophy remains largely untested. *But today, like any other day, there is no time like the present.*

## References

Adam, B. (1990), *Time and Social Theory*. Oxford: Polity Press.

Dewey, J. (1920), Reconstruction in Philosophy. In: *John Dewey: The Middle Works, 1899–1924*, Vol. 10. Carbondale, IL: Southern Illinois University Press, 1980.

——— (1925), Experience and Nature. In: *John Dewey: The Later Works, 1925–1953*, Vol. 1. Carbondale, IL: Southern Illinois University Press, 1981.

——— (1977), Reality as experience. In: *John Dewey: The Middle Works, 1899–1924*, Vol. 3. Carbondale, IL: Southern Illinois University Press, 1977.

———— (1984a), Meaning and existence. In: *John Dewey: The Later Works, 1925–1953,* Vol. 3. Carbondale, IL: Southern Illinois University Press, 1984.

———— (1984b), The inclusive philosophic idea. In: *John Dewey: The Later Works, 1925–1953,* Vol. 3. Carbondale, IL: Southern Illinois University Press, 1984.

———— (1984c), The public and its problems. In: *John Dewey: The Later Works, 1925–1953,* Vol. 2. Carbondale, IL: Southern Illinois University Press, 1984.

———— (1986a), Logic: The Theory of Inquiry. In: *John Dewey: The Later Works, 1925–1953,* Vol. 12. Carbondale, IL: Southern Illinois University Press, 1986.

———— (1986b), Events and the future. In: *John Dewey: The Later Works, 1925–1953,* Vol. 12. Carbondale, IL: Southern Illinois University Press, 1986.

———— (1986c), A common faith. In: *John Dewey: The Later Works, 1925–1953,* Vol. 9. Carbondale, IL: Southern Illinois University Press, 1986.

———— (1987a), Art as experience. In: *John Dewey: The Later Works, 1925–1953,* Vol. 10. Carbondale, IL: Southern Illinois University Press, 1987.

———— (1987b), Democracy and educational administration. In: *John Dewey: The Later Works, 1925–1953,* Vol. 11. Carbondale, IL: Southern Illinois University Press, 1987.

———— (1987c), Democracy is radical. In: *John Dewey: The Later Works, 1925–1953,* Vol. 11. Carbondale, IL: Southern Illinois University Press, 1987.

———— (1988a) Nature in experience. In: *John Dewey: The Later Works, 1925–1953,* Vol. 14. Carbondale, IL: Southern Illinois University Press, 1988.

———— (1988b), Time and individuality. In: *John Dewey: The Later Works, 1925–1953,* Vol. 14. Carbondale, IL: Southern Illinois University Press, 1988.

———— (1988c), Democracy and education in the world of today. In: *John Dewey: The Later Works, 1925–1953,* Vol. 15. Carbondale, IL: Southern Illinois University Press, 1988.

———— (1988d), Democracy and America. In: *John Dewey: The Later Works, 1925–1953,* Vol. 13. Carbondale, IL: Southern Illinois University Press, 1988.

———— (1989a), Knowing and the Known. In: *John Dewey: The Later Works, 1925–1953,* Vol. 16. Carbondale, IL: Southern Illinois University Press, 1989.

———— (1989b), The democratic faith and education. In: *John Dewey: The Later Works, 1925–1953,* Vol. 15. Carbondale, IL: Southern Illinois University Press, 1989.

Dreyfus, H. (1975), Human temporality. In: *The Study of Time,* Vol. 2, ed. J. T. Fraser & N. Lawrence. New York: Springer-Verlag, p. 15.

Fraser, J. T. (1978), The individual and society. In: *The Study of Time,* Vol. 3, ed. J. T. Fraser, N. Lawrence, & D. Park. New York: Springer-Verlag, pp. 419–442.

———— (1981a), Toward an integrated study of time. In: *The Voices of Time,* 2nd ed., ed. J. T. Fraser. Amherst: University of Massachusetts Press, pp. xiv–xlviii.

———— (1981b), The study of time. In: *The Voices of Time,* ed. J. T. Fraser. Amherst: University of Massachusetts Press, pp. xlv–xlviii.

———— (1987), *Time: The Familiar Stranger.* Amherst: University of Massachusetts Press.

———— (1992), Human temporality in a nowless universe. *Time & Soc.,* 1:159–173.

Griffin, D. R. (1993), *Founders of Constructive Postmodern Philosophy.* Albany, NY: State University of New York Press.

Grünbaum, A. (1968), The status of temporal becoming. In: *The Philosophy of Time,* ed. R. Gale. London: Macmillan, pp. 322–354.

Helm, B. P. (1985), *Time and Reality in American Philosophy.* Amherst: University of Massachusetts Press.

Lawrence, N. (1979), My time is your time. In: *The Study of Time,* Vol. 4, ed. J. T. Fraser, N. Lawrence, & D. Pack. New York: Springer-Verlag, p. 11.

Mead, G. H. (1932), *The Philosophy of the Present.* LaSalle, IL: Open Court.

———— (1938), *The Philosophy of the Act.* Chicago: University of Chicago Press, 1972.

Sherover, C. (1975a), Time and ethics: How is morality possible? In: *The Study of Time,* Vol. 2, ed. J. T. Fraser & N. Lawrence. New York: Springer-Verlag.

———— (1975b), *The Human Experience of Time: The Development of Its Philosophic Meaning.* New York: New York University Press.

———— (1981), Perspectivity and the principle of continuity. In:  *The Study of Time,* Vol. 4, ed. J. T. Fraser, N. Lawrence, & D. Park. New York: Springer-Verlag.

———— (1989a), Res cogitans: The time of mind. In:  *The Study of Time,* Vol. 6, ed. J. T. Fraser. Madison, CT: International Universities Press.

———— (1989b), *Time, Freedom, and the Common Good: An Essay in Public Philosophy.* Albany: State University of New York Press.

———— (1993), The process of polity. In:  *The Study of Time,* Vol. 7, ed. J. T. Fraser. Madison, CT: International Universities Press, pp. 243–261.

Stuhr, J. J. (1979), Dewey's notion of qualitative experience. *Trans. Charles S. Peirce Soc.,* 15:68–83.

———— (1991), *John Dewey.* Nashville, TN: Carmichael & Carmichael.

———— (1992), Dewey's reconstruction of metaphysics. *Trans. Charles S. Peirce Soc.,* 18:160–176.

———— (1993), Democracy as a way of life. In:  *Philosophy and the Reconstruction of Culture: Pragmatic Essays After Dewey,* ed. J. J. Stuhr. Albany: State University of New York Press.

Whitrow, G. J. (1981), Time and the universe. In: *The Voices of Time,* ed. J. T. Fraser. Amherst: University of Massachusetts Press, pp. 581.

# Name Index

# Subject Index